AF858959

Low Energy Collision Physics

Applications to cold atoms

Online at: https://doi.org/10.1088/978-0-7503-3959-9

Low Energy Collision Physics

Applications to cold atoms

Bimalendu Deb
School of Physical Sciences, Indian Association for the Cultivation of Science, Kolkata, India

IOP Publishing, Bristol, UK

© IOP Publishing Ltd 2025. All rights, including for text and data mining (TDM), artificial intelligence (AI) training, and similar technologies, are reserved.

This book is available under the terms of the IOP-Standard Books License

No part of this publication may be reproduced, stored in a retrieval system, subjected to any form of TDM or used for the training of any AI systems or similar technologies, or transmitted in any form or by any means, electronic, mechanical, photocopying, recording or otherwise, without the prior permission of the publisher, or as expressly permitted by law or under terms agreed with the appropriate rights organization. Certain types of copying may be permitted in accordance with the terms of licences issued by the Copyright Licensing Agency, the Copyright Clearance Centre and other reproduction rights organizations.

Permission to make use of IOP Publishing content other than as set out above may be sought at permissions@ioppublishing.org.

Bimalendu Deb has asserted his right to be identified as the author of this work in accordance with sections 77 and 78 of the Copyright, Designs and Patents Act 1988.

ISBN 978-0-7503-3959-9 (ebook)
ISBN 978-0-7503-3957-5 (print)
ISBN 978-0-7503-3960-5 (myPrint)
ISBN 978-0-7503-3958-2 (mobi)

DOI 10.1088/978-0-7503-3959-9

Version: 20260101

IOP ebooks

British Library Cataloguing-in-Publication Data: A catalogue record for this book is available from the British Library.

Published by IOP Publishing, wholly owned by The Institute of Physics, London

IOP Publishing, No.2 The Distillery, Glassfields, Avon Street, Bristol, BS2 0GR, UK

US Office: IOP Publishing, Inc., 190 North Independence Mall West, Suite 601, Philadelphia, PA 19106, USA

This book is dedicated to the memory of my mother Sushanta Bala Deb and to the memory of my wife Sushmita Deb.

Contents

Preface xi

Acknowledgments xiii

Author biography xiv

1 Introduction 1-1

1.1 Classical concept of collisions 1-2

1.2 Useful terminologies and mathematical preliminaries 1-4

1.3 Some basics about quantum collisions 1-6

1.3.1 Lippmann–Schwinger equation 1-17

Exercise 1-19

References 1-20

2 Single-channel scattering 2-1

2.1 Potential scattering in 3D 2-1

2.2 Partial-wave expansion 2-6

2.3 Partial-wave Green's function 2-15

2.4 Wigner threshold laws 2-18

2.5 Scattering by a short-range potential 2-22

2.6 Classical theory of collisions 2-29

2.7 Coulomb scattering 2-33

2.7.1 Classical expression for $\sigma(\theta)$ 2-33

2.7.2 Quantum mechanical expression for $\sigma(\theta)$ 2-34

2.8 High vs low-energy and classical vs quantum collisions 2-36

Exercise 2-38

References 2-39

3 Multichannel scattering 3-1

3.1 Two-particle scattering with spin 3-2

3.2 Formal theory of multichannel scattering 3-3

3.2.1 Case (i): all channels are open 3-5

3.2.2 Case (iii): some channels are open, others are closed 3-7

3.2.3 Multichannel scattering cross section 3-7

Exercise 3-10

References 3-10

4 Resonances and bound states **4-1**
4.1 Energy dependence of phase shift near a resonance 4-2
4.2 Resonances, bound states, and analyticity of S-matrix elements 4-3
4.3 Resonances and bound states for a rectangular potential well 4-4
4.4 Shape resonances 4-6
4.5 Feshbach resonances 4-7
4.5.1 Two-channel model 4-7
4.6 Fano resonances 4-11
4.6.1 Fano state 4-11
4.6.2 Doubly dressed Fano state: quantum interference 4-14
4.6.3 Two coupled bound states, only one of which is coupled to a continuum 4-18
4.7 Bound state in a continuum 4-19
4.8 Resonant mean-field interaction 4-19
Exercise 4-21
References 4-21

5 Cold atomic collisions **5-1**
5.1 Atomic hyperfine interactions 5-1
5.2 Atom–atom potentials 5-3
5.2.1 Born–Oppenheimer approximation 5-3
5.2.2 Long-range molecular potentials 5-7
5.2.3 Fundamentals of symmetry concepts in diatomic molecules 5-8
5.3 Collisions between cold atoms 5-11
5.4 Collisions in a magnetic field 5-14
5.4.1 Magnetic Feshsbach resonance 5-14
5.4.2 Outline of the advancement in MFR 5-19
5.5 Collisions in a laser field 5-21
5.5.1 Optical Feshbach resonances 5-22
5.5.2 Theory of OFR 5-24
5.5.3 PA Hamiltonian 5-28
5.5.4 T-matrix element 5-28
5.6 Collisions in both magnetic and laser fields 5-33
5.6.1 Magneto-optical Fano–Feshbach resonance 5-33
5.7 Higher partial-wave Feshbach resonances 5-34
5.8 Collisions in a static electric field 5-36
5.8.1 Scattering between a pair of homonuclear atoms 5-37

	5.8.2 Scattering between a pair of heteronuclear atoms	5-40
5.9	Ion–atom collisions at low energy	5-42
5.10	Bound state in a continuum of cold atoms	5-45
	Exercise	5-48
	References	5-49
6	**Association spectroscopy**	**6-1**
6.1	Photoassociation	6-2
	6.1.1 Single-photon photoassociation	6-3
	6.1.2 Raman photoassociation: the creation of ground-state molecules	6-8
	6.1.3 A brief overview of photoassociation spectroscopy and its applications	6-11
6.2	Magneto- and magneto-optical association: Feshbach molecules	6-19
6.3	Efimov trimers	6-21
	Exercise	6-27
	References and further reading	6-27
7	**Collisions in low dimensions**	**7-1**
7.1	Scattering in two dimensions	7-2
	7.1.1 Theory	7-2
	7.1.2 Two-dimensional zero-range potential	7-6
	7.1.3 Coulomb scattering in 2D	7-7
7.2	Scattering in one dimension	7-9
	7.2.1 One-dimensional zero-range potential	7-11
7.3	Cold atoms in low dimensions	7-12
	Exercise	7-16
	References	7-16
8	**Numerical methods**	**8-1**
8.1	Numerov–Cooley algorithm	8-2
8.2	Scattering-state calculations	8-6
8.3	Green's function calculation	8-8
8.4	Bound state calculation by the Numerov method	8-8
8.5	Discrete variable representation method	8-10
8.6	Quantum defect theoretic method	8-10
	Exercise	8-11
	References	8-11

Appendix A: Dirac delta functions and their properties **A-1**

Appendix B: Green's function of a free particle **B-1**

Appendix C: Expansion of the full 3D Green's function in terms of partial-wave Green's functions **C-1**

Preface

This book, consisting of eight chapters, presents basics of low energy collision physics and their applications to cold atoms. The purpose is to bridge the gap between undergraduate or graduate teaching in scattering theory and research frontiers in cold atomic and molecular sciences.

The first chapter begins with an introduction to collision phenomena in general, highlighting the distinction between low versus high energy, classical versus quantum and Coulomb versus non-Coulomb scattering. The second and third chapters present the formal theoretical methods in single- and multi-channel scattering, respectively. Resonances, bound states and their inter-dependence are described in the fourth chapter. While the first four chapters lay the foundation of low energy scattering theory, the remaining four chapters provide description on applications to cold atoms.

The prerequisites to read this book are undergraduate level classical mechanics, introductory quantum mechanics and some elementary knowledge about atomic structure. In writing this book, I have made a conscious effort to make it self-contained on the fundamental aspects of the field. The physical laws that govern low energy collision phenomena are mostly derived from first principles. To understand or appreciate the applied aspects of the field, one requires some graduate level knowledge about atomic and diatomic molecular structure and spectroscopy. I am aware that many college or university physics curricula do not cover these topics adequately. That is why fifth chapter begins with an extended discussion on atomic and molecular structure that are relevant to cold atomic collisions and related phenomena such as association spectroscopy which is discussed in the sixth chapter. The seventh chapter presents theories of scattering in low dimensions, and a brief description of low dimensional collision phenomena in trapped atoms. The eighth chapter provides some important numerical methods to calculate scattering amplitudes and bound states.

With the advancement in cooling and trapping of atoms since 1980s, atomic collisions at near-zero energy and near-zero temperature have become important in atomic and molecular science today. Although, basic theoretical framework for low energy collisions was laid down in 1940s and 1950s in the context of nuclear scattering, the advent of cold atoms has brought a completely new perspective into the field. What is new is that the scattering properties of cold atoms are precisely controllable with an external magnetic or electric or electromagnetic field.

There exist several excellent textbooks on scattering theories, but most of them are written in pre-cold-atom times, that is, before the 1990s. The last three decades have witnessed tremendous advancement both in the theoretical as well as in the experimental fronts of cold atoms. There are several review articles published in journals, periodicals and conference proceedings; and advanced reference texts on the developments of the field. This book gives a bare minimum of these developments, and instead focuses mostly on the pedagogic aspects of the field for beginners who wish to initiate a research career in this or any related field. Whether I have been able to fulfill this stated purpose rests on the opinion of the readers.

I am thankful to the IOP book publishing team for being patient and supportive throughout my endeavor to see this book in the final shape. Especially, I would like to thank Ms Phoebe Hooper of IOP publishing. I know, no matter how much one checks and rechecks the errors and correct them, there may still remain errors—I would appreciate the pointing out of any error by readers.

Acknowledgments

My teaching and research experience over the last two decades at IACS have helped me to gain deeper insight into the topics discussed in this book. Academic discussions and debate with my research group members have shaped my pedagogic understanding of the field. I would like to acknowledge my past group members Jisha, Debashree, Arpita Sr., Partha, Arpita Jr., Subrata, Somnath, Subhanka, Dibuendu whose thesis works have inspired me to write this book. I would like to thank my present group members Anushree and Subhra for helping me in organizing the bibliography files.

Author biography

Bimalendu Deb

Bimalendu Deb, the author of this book, is a Professor of Physics, at the School of Physical Sciences, Indian Association for the Cultivation of Science (IACS), Jadavpur, Kolkata, India. He has more than thirty years of experience in research and teaching in atomic, molecular and optical sciences. His research interests include theoretical quantum optics, quantum computing, cold atomic collisions, and related fields. Currently, at IACS, he leads a group of researchers working in theoretical quantum optics, quantum computing and quantum simulation. Over the last two decades, he guided twelve researchers towards obtaining their PhD degrees in the areas of cold atomic collisions. He is involved with National Quantum Mission of the Department of Science & Technology, Government of India, as a Member Principal Investigator of a Technical Group on the field of quantum sensing with trapped atoms.

Chapter 1

Introduction

The microscopic basis of a physical theory originates from the analysis of collision phenomena. Particles or waves scattered by a target carry information about the structure and physical properties of the target. Historically, Rutherford's celebrated alpha-scattering experiment [1] first unraveled the structure of the subatomic world, specifically the atomic nucleus. Since the early days of quantum theory, laboratory experiments on collisions between subatomic [2] as well as atomic particles [3, 4] at various energy scales have played a key role in advancing nuclear, atomic, and molecular physics in the twentieth century. Progress in particle physics draws heavily from the outcomes of collision experiments at accelerators and colliders. Collision theory is therefore important in almost all branches of physics.

The major part of this book will describe non-Coulomb two-body collisions at low energy. At the outset, we make a distinction between Coulomb and non-Coulomb collisions. Collisions between two charged bodies or particles due to pure Coulomb forces are called Coulomb collisions, while non-Coulomb collisions occur due to forces other than Coulomb ones. In a two-particle non-Coulomb collision in free space, the two particles move freely before and after the collision event. Before the collision, when the two particles are far apart or asymptotically at a large separation, their relative kinetic energy is known as collision energy. Depending on the scale of the collision energy, the collisions can be broadly classified into two categories: high-energy and low-energy collisions. For non-Coulomb collisions at low energy, the classical theory of collisions breaks down, as we will show in chapter 2. This breakdown occurs because the de Broglie wavelength, $\lambda_{\rm dB} = h/p$ (where h is Planck's constant and p denotes the magnitude of a particle's momentum), is not negligible in comparison to other length scales of the system, such as the range of particle–particle interactions or the average interparticle distance within the ensemble. Therefore, at low energy, one describes the collision phenomena of massive particles in terms of the scattering of Schrödinger waves or matter waves. The angular momentum of relative motion is $\mathbf{L} = \mathbf{r} \times \mathbf{p}$, where $\mathbf{r}$ is the distance from the colliding particle

doi:10.1088/978-0-7503-3959-9ch1 © IOP Publishing Ltd 2025. All rights, including for text and data mining (TDM), artificial intelligence (AI) training, and similar technologies, are reserved.

(projectile) to the scattering center or the target's center, and $\mathbf{p}$ is the collision momentum. For collisions of spinless particles, this angular momentum is conserved. Classically, the magnitude of $\mathbf{L}$ is $L = pb$, where b is the impact parameter of the collision. Quantum mechanically, $L = \sqrt{\ell(\ell+1)}\,\hbar$, where $\ell = 0, 1, 2, \ldots$ is an integer quantum number and $\hbar = h/(2\pi)$. One can thus draw a classical–quantum correspondence: $b = \sqrt{\ell(\ell+1)}\,\lambda_{dB}/(2\pi)$. As we show later, at low energy, the quantum states of two particles colliding due to a non-Coulomb potential are mainly influenced by a few lower ℓ. Furthermore, if both particles have spin, low-energy collision processes are largely affected by spin-dependent forces or potentials. The approximations applicable in high-energy collision theory generally break down in low-energy collision theory. The energy scale or range that defines the domain of low-energy collisions depends on the nature of the forces acting between the colliding particles. The energy scale for nucleon–nucleon collisions at low energy is different from that for low-energy collisions between neutral atoms or between an ion and a neutral atom.

With the advancements in cooling and trapping of atoms and ions in the late twentieth century, collision physics at low energy has come to the fore. The cooling and trapping of molecules is currently in development. In the field of low-energy atomic collisions, it has become customary to express the collision energy in terms of temperature, $T \sim E/K_B$, where E is the collision energy and K_B stands for the Boltzmann constant. In the context of low-temperature trapped atomic gases, depending on the temperature range, atoms are called 'cold' or 'ultracold.' Although there is no clear-cut distinction between these two terminologies and both are often used interchangeably, we will generally use 'cold atoms' to refer to atoms having temperatures in the subkelvin to millikelvin (mK) regime, while the temperature of ultracold atoms falls within the sub-mK to nanokelvin (nK) regime.

The basic principles of low-energy collision theory were established during the early developments in nuclear physics in the 1940s and 1950s. Now, cold and ultracold atoms provide plenty of opportunities for testing the theory with high precision. New experimental methods of manipulating low-energy collisions of atoms using external fields (which include static magnetic, static electric, and dynamical electromagnetic fields) have been developed over the last three decades. These advances have opened new perspectives not only in low-energy collision physics but also in several allied and emerging fields such as cold chemistry and quantum simulations of many-body physics.

In this introductory chapter, we will start with a discussion of the classical concept of collisions. We then define the general terminologies of collision theory and discuss some mathematical preliminaries that are generally useful in describing non-Coulomb quantum collision theory. To show the basic difference between Coulomb and non-Coulomb collisions, we briefly describe Coulomb scattering.

1.1 Classical concept of collisions

We will discuss the classical theory of collisions in some detail in chapter 2. Here, we give an outline of the classical concept of collisions. In classical mechanics, the

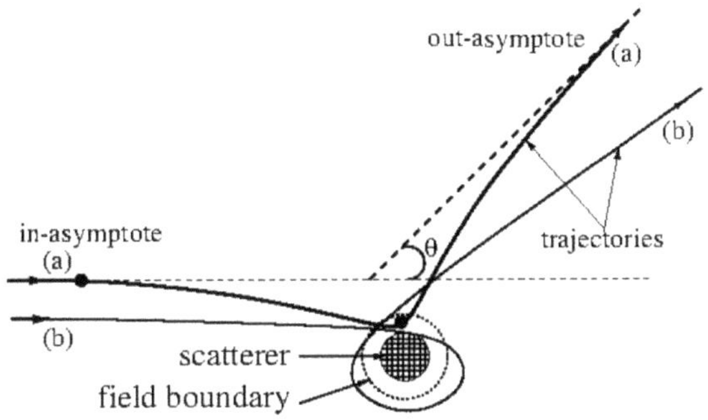

Figure 1.1. Schematic showing two possible classical trajectories (a) and (b) of a projectile due to scattering by a target. The in- and out-asymptotes refer to the asymptotic rectilinear trajectories for $t \rightarrow \pm\infty$, respectively. θ is the deflection angle, i.e. the angle between the in- and out-asymptotes for trajectory (a), for which the scattering angle is the same as the deflection angle. For trajectory (b), the projectile makes one round trip around the scatterer before going to the outbound asymptotic limit. Therefore, for this trajectory, the scattering angle exceeds the deflection angle by π.

collision phenomena are described in terms of Newton's equations of motion. The underlying assumption here is that a classical particle has a definite position and momentum at each instant of time, and given the forces that act on the particle, its time evolution is absolutely determined by Newton's laws or other variants of these laws, such as the Hamiltonian or Hamilton–Jacobi equations. For a projectile or a moving particle colliding with a fixed target in the laboratory frame of reference as schematically shown in figure 1.1, the collision event is viewed as the deflection of the classical trajectory of the particle by the forces exerted by the target on the particle. For a non-Coulomb collision, it is reasonable to assume that the forces have an effective spatial zone or range of influence over the particle; beyond this range, the forces are assumed to be negligible. Before the arrival of the particle at the boundary surface of this effective range, the particle moves freely. Within the range, the particle's motion is governed by the force field of the target. After exiting the surface of the range of influence, the particle again moves freely. If the initial trajectory of the particle, that is, the trajectory before entering the range of influence, is changed after exiting the surface of the range, then we say that the particle is scattered. In actual scattering experiments, a beam of particles rather than a single particle impacts a target. Therefore, we need a statistical description of the collision phenomena that introduces collision probability or equivalent quantities such as scattering cross sections.

The aim of classical scattering theory is to relate the incoming trajectories of particles to the outgoing ones. The trajectory a particle traces before arriving at the surface of the range of influence is known as its incoming trajectory or in-asymptote, while that after exiting the surface is called the outgoing trajectory or out-asymptote. Particles usually spend very little time (typically on the order of 0.1 nanoseconds (ns)) within the range of influence. For most of the duration of a scattering event, we have to deal with free particle motion. The microscopic details of the interaction or the forces acting between the particle and the target are *a priori* unknown.

Determining such microscopic details from the analysis of scattering data is the object of inverse scattering theory.

In the case of collisions between two moving particles, the collision process can be most conveniently described by transforming the coordinates and velocities or momenta of the particles into the center-of-mass (COM) frame, which we will discuss in the next subsection. If the interparticle force field is spherically symmetric (i.e. the field is central), then the two-particle scattering problem may be effectively reduced to that of a single particle scattering off a target located at the COM in the COM frame.

1.2 Useful terminologies and mathematical preliminaries

Here, we define some of the terminology often used in collision theory and discuss related mathematical preliminaries.

Differential and total scattering cross sections

For a single moving particle scattering off a fixed target, the laboratory fixed z-axis may be conveniently defined using the particle's incoming direction. Then, the laboratory fixed x- and y-axes are taken to be any two axes that are perpendicular to the z-axis, so that these three axes form a right-handed Cartesian coordinate system. Considering a spherical polar coordinate system with z as the polar axis, one can define a scattering solid angle $\Omega = (\theta, \phi)$ as the angle of deviation of the scattered or outgoing particle from its initial direction $\Omega_{\text{in}} = (0, 0)$. For a well-collimated monoenergetic beam, all particles have almost the same propagation direction. Therefore, the mean propagation direction is the same as an individual particle's propagation direction. Otherwise, one must consider a mean propagation direction when defining the laboratory z-axis. Let dN be the number of particles scattered off a target per unit time between the solid angles Ω and $\Omega + d\Omega$ for an incident beam of particles with flux F, which is defined as the number of particles crossing per unit area on the plane perpendicular to the beam's mean propagation direction per unit time. The differential scattering cross section $\sigma(\Omega)$ is defined by the following relation:

$$\sigma(\Omega) \equiv \frac{d\sigma}{d\Omega} = \frac{1}{F}\frac{dN}{d\Omega} \tag{1.1}$$

The total scattering cross section σ_T is obtained by integrating the differential scattering cross section over the solid angle. Explicitly,

$$\sigma_T = \int d\Omega \frac{d\sigma}{d\Omega} = \int_0^{2\pi} d\phi \int_0^{\pi} \sin\theta d\theta \frac{d\sigma}{d\Omega}. \tag{1.2}$$

For scattering due to a central force, $\frac{d\sigma}{d\Omega}$ is independent of ϕ and depends only on the polar angle θ. Both differential and total scattering cross sections have the dimensions of [length]^2 or area. The scattering rate is defined as

$$\textbf{scattering rate} = \textbf{cross section} \times \textbf{speed} \times \textbf{density},$$

where 'speed' refers to the asymptotic initial speed of the colliding particles and 'density' refers to the number density of particles in the incident beam. If the particle beam is not monoenergetic and the speed is distributed according to some distribution law, then one must integrate over that distribution of speed to obtain the average rate.

Elastic, inelastic, and reactive collisions

The collisions by which the chemical compositions of the projectile and target change are called reactive collisions. In nonreactive collisions, the chemical compositions of the colliding particles remain the same. Depending on the energy balance of the colliding particles before and after the collision, nonreactive collisions can be classified into two categories: elastic and inelastic collisions. If the kinetic energy of the projectile (in the case of a fixed target) before and after the collision event remains the same, then the collision is said to be elastic; otherwise, it is inelastic. In the case of two-particle elastic collisions, the total kinetic energy of the two particles remains the same before and after the collision event. Inelastic collisions may be characterized by a change in internal degrees of freedom (DOF), such as the spin of the colliding particles, as well as the transformation of energy from one form to another (for example, from kinetic to potential energy or vice versa). Reactive collisions may be categorized under the general category of inelastic processes. In an inelastic collision, the transformation of energy may be manifested as a change in an internal degree of freedom or in the chemical composition of the scattered particles. The latter type of inelastic collision is generally termed a reactive collision.

Two-particle collisions: the center-of-mass frame

Let us consider a collision between two moving particles, 1 and 2, of mass m_1 and m_2, respectively. In the laboratory frame of reference, let the position vectors of particles 1 and 2 at time t be denoted by $\mathbf{r}_1$ and $\mathbf{r}_2$ with corresponding velocities $\mathbf{v}_1 = \dot{\mathbf{r}}_1 = \frac{d\mathbf{r}_1}{dt}$ and $\mathbf{v}_2 = \dot{\mathbf{r}}_2$, respectively. Their total kinetic energy is given by

$$T_{12} = \frac{1}{2}m_1\mathbf{v}_1^2 + \frac{1}{2}m_2\mathbf{v}_2^2 = \frac{1}{2m_1}\mathbf{p}_1^2 + \frac{1}{2m_2}\mathbf{p}_2^2 \tag{1.3}$$

where $\mathbf{p}_i = m_i\mathbf{v}_i$ is the dynamical momentum of the ith ($i = 1, 2$) particle. Note that for a free particle, that is, a particle moving in a force-free space or in a constant potential, dynamical momentum is the same as canonical momentum. We now introduce the relative position vector $\mathbf{r}$ and the COM position vector $\mathbf{R}$ as given by

$$\mathbf{r} = \mathbf{r}_1 - \mathbf{r}_2 \tag{1.4}$$

$$\mathbf{R} = \frac{m_1\mathbf{r}_1 + m_2\mathbf{r}_2}{m_1 + m_2}. \tag{1.5}$$

One can thus define the relative velocity $\dot{\mathbf{r}} = \mathbf{v}_1 - \mathbf{v}_2$ and the COM velocity

$$\dot{\mathbf{R}} = \frac{m_1\mathbf{v}_1 + m_2\mathbf{v}_2}{m_1 + m_2}. \tag{1.6}$$

Using these expressions, Eq.(1.1) can be expressed in the following form:

$$T_{12} = \frac{1}{2}M\dot{\mathbf{R}}^2 + \frac{1}{2}\mu\dot{\mathbf{r}}^2, \tag{1.7}$$

where $M = m_1 + m_2$ is the total mass of the two particles and

$$\mu = \frac{m_1 m_2}{M} \tag{1.8}$$

is called the reduced mass of the two particles. In terms of the COM momentum $\mathbf{P} = M\dot{\mathbf{R}}$ and the relative momentum $\mathbf{p} = \mu\dot{\mathbf{r}}$, T_12 in the above equation can be written as

$$T_{12} = \frac{1}{2M}\dot{\mathbf{P}}^2 + \frac{1}{2\mu}\mu\dot{\mathbf{p}}^2. \tag{1.9}$$

It is easy to see that $\mathbf{P}$ and $\mathbf{p}$ are related to $\mathbf{p}_1$ and $\mathbf{p}_2$ by the relations

$$\mathbf{P} = \mathbf{p}_1 + \mathbf{p}_2 \tag{1.10}$$

$$\mathbf{p} = g_2\mathbf{p}_1 - g_1\mathbf{p}_2, \tag{1.11}$$

where $g_1 = m_1/M$ and $g_2 = m_2/M$. The total angular momentum $\mathbf{J} = \mathbf{J}_{\text{com}} + \mathbf{L}$, where $\mathbf{J}_{\text{com}} = \mathbf{R} \times \mathbf{P}$ and $\mathbf{L} = \mathbf{r} \times \mathbf{p}$. In the COM frame, $\mathbf{P} = 0$. The scattering states are conveniently calculated in the COM frame, and then they can be expressed in the laboratory frame through the well-known frame transformation rules [5].

Recall that in classical mechanics, the canonical or generalized momentum variables are introduced by first defining the Lagrangian $\mathscr{L}(\{q\}, \{\dot{q}\}, t)$ in terms of a collection of generalized position variables $\{q\} \equiv \{q_1, q_2, \cdots\}$ and corresponding generalized velocity variables $\{\dot{q}\} \equiv \{\dot{q}_1, \dot{q}_2, \cdots\}$; and then the canonical momentum variables p_j are defined as

$$p_j = \frac{\partial\mathscr{L}}{\partial\dot{q}_j}. \tag{1.12}$$

1.3 Some basics about quantum collisions

In the quantum theory of collisions, collision phenomena are necessarily described in terms of Schrödinger or de Broglie waves of the colliding particles. In quantum mechanics, each moving particle is associated with a traveling wave of de Broglie wavelength λ_{dB}. Thus, a two-particle collision event in quantum mechanics may be viewed as the collision between two de Broglie waves. Since the quantum state of a particle is described by the wave function governed by the Schrödinger equation, the de Broglie wave of the particle belongs to the general class of Schrödinger wave functions, which may be traveling (in the case of a free particle or a particle moving in a constant potential) or localized (in the case of a bound particle moving in a localized space under the influence of a binding potential). Since λ_{dB} is inversely proportional to p, as p decreases, λ_{dB} increases, implying that the wave nature becomes more prominent. When λ_{dB} becomes comparable to the significant spatial variation of the force or potential that scatters the particle, the scattering phenomena must be described explicitly using a quantum mechanical treatment. For a thermal ensemble of particles at temperature T, the thermal de Broglie wavelength $\lambda_T \propto 1/\sqrt{T}$. Therefore, at a very low temperature at which λ_T is appreciable, the interparticle interactions are dominated by quantum collisions. At high temperatures, collisions may also become quantum mechanical, particularly when particles collide with spin-dependent forces, as spin is a purely quantum mechanical trait of a particle. However, the approximations applicable in the methods used to model high-energy quantum collisions are usually invalid in the case of low-energy collisions. Figure 1.2 shows a schematic of a wave packet scattering off a fixed scatterer.

Since, in the case of non-Coulomb collisions, one must extract useful information about the scattering center, the target, or the scattering potential by detecting the state of motion of the freely moving scattered particles and comparing it with the known state of incident particles, it is essential to understand how the quantum states of the scattered free particles are calculated by solving the Schrödinger equation. Only within the range of the scattering force or potential does the motion of the particle deviate significantly from that of a free particle. The basic method of quantum scattering is to continuously connect the Schrödinger wave inside the scattering zone with the wave outside the zone. Thereby, one determines the scattering phase shift by comparing the scattered wave with the incident one. Figure 1.3 illustrates schematically the basic distinction between classical and quantum scattering.

Since quantum scattering is essentially the scattering of Schrödinger waves, it has many similarities to the scattering of electromagnetic waves by matter. In both Schrödinger and electromagnetic waves, the superposition principle plays a key role

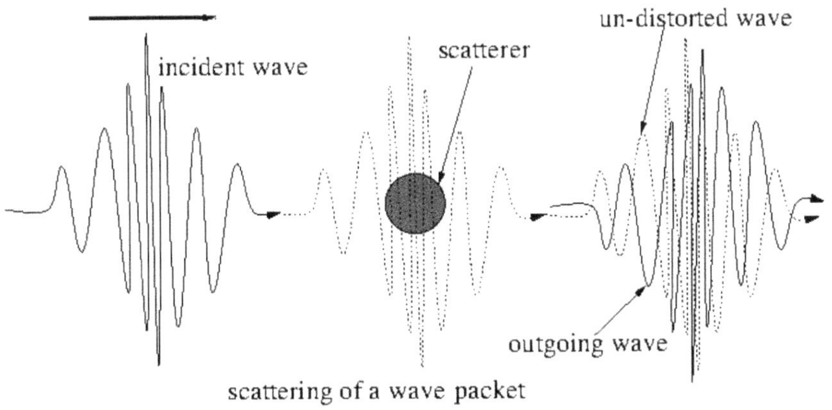

Figure 1.2. Schematic of a wave packet scattering off a fixed target scatterer. In the absence of any scatterer, the wave packet remains undistorted (dashed), while a phase shift appears in the outgoing wave due to scattering (solid curve at right).

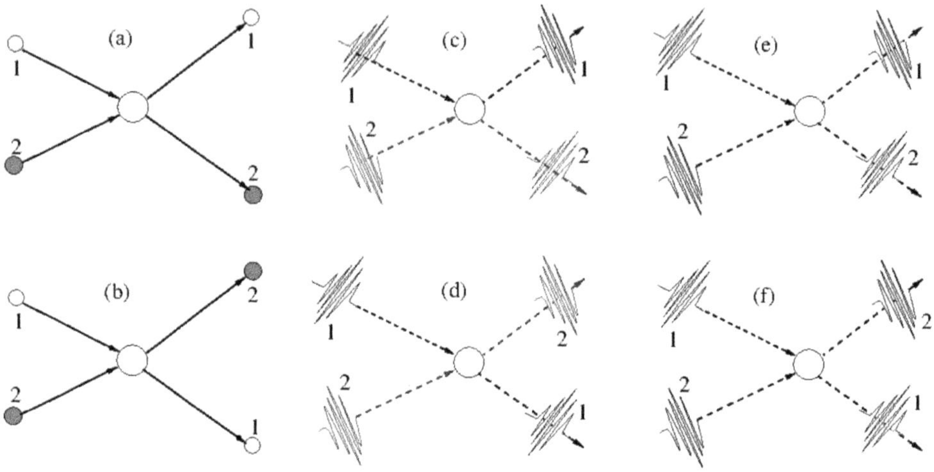

Figure 1.3. Schematics of collisions between two particles, 1 and 2. (a, b) Classical collision; (c, d) quantum collision between two distinguishable particles represented by two wave packets; and (e, f) quantum collision between two indistinguishable particles. In (e) and (f), the post-collision final two-particle states are similar and therefore must be symmetrized or anti-symmetrized, depending on the parity of the states. In other words, two transition paths that lead to the final states in (e) and (f) can interfere either constructively or destructively.

in the description of scattering phenomena. While in classical optics, the propagation of electromagnetic waves is governed by the Maxwell equations or the classical wave equations derived from the Maxwell equations, the behavior of Schrödinger waves of particles or matter waves is governed by the Schrödinger equation. For a monochromatic electromagnetic wave of angular frequency ω, wavelength λ, and wave number $k = 2\pi/\lambda$ in free space, the dispersion relation is $\omega = ck$, while that for a Schrödinger wave is $\omega = \hbar k_{\mathrm{dB}}^2/2m$ where $\hbar = h/2\pi$ and $k_{\mathrm{dB}} = 2\pi/\lambda_{\mathrm{dB}}$. This difference in dispersion relations arises from the fact that the classical wave equation is a differential equation that is second order in both time and space variables, while the Schrödinger wave equation is first order in time and second order in the space variable. The reader is referred to the book by R G Newton [6] for an excellent exposition of the comparative study of the scattering of waves and particles.

Quantum states and wave packets

The quantum states of moving particles must be described by a complex wave function because the velocity of the particle is proportional to the gradient of the phase of the wave function. Let $\psi(\mathbf{r}, t)$ be the wave function of a flux of non-interacting particles, where $\mathbf{r}$ is the position vector of a single particle in the flux measured from a fixed scattering center. Let $\phi(\mathbf{r}, t)$ be the wave function of a single particle impinging on a target. In the case of two-particle scattering, $\phi(\mathbf{r}, t)$ is the wave function of the reduced particle in the COM frame, where $\mathbf{r}$ is the relative coordinate. It is the solution of the Schrödinger equation,

$$i\hbar\frac{\partial}{\partial t}\phi(\mathbf{r}, t) = \left[-\frac{\hbar^2}{2\mu}\nabla^2 + V(\mathbf{r}, t)\right]\phi(\mathbf{r}, t), \tag{1.13}$$

where $V(\mathbf{r}, t)$ is the interaction potential between the incident particle and the scatterer or the interaction potential between two particles. Here, we have assumed that the potential does not depend on the COM coordinate. Multiplying the complex conjugate of equation (1.13), that is

$$-i\hbar\frac{\partial}{\partial t}\phi^*(\mathbf{r}, t) = \left[-\frac{\hbar^2}{2\mu}\nabla^2 + V(\mathbf{r}, t)\right]\phi^*(\mathbf{r}, t), \tag{1.14}$$

by $\phi(\mathbf{r}, t)$, and subtracting the resulting equation from the equation that results when we multiply equation (1.13) by $\phi^*(\mathbf{r}, t)$, one obtains the continuity equation

$$\frac{\partial}{\partial t}\rho_1 + \boldsymbol{\nabla}.\,\mathbf{j} = 0, \tag{1.15}$$

where $\rho_1(\mathbf{r}, t) = |\phi(\mathbf{r}, t)|^2$ is the single-particle probability density and

$$\mathbf{j} = \frac{\hbar}{2\mu}[\phi^*(\mathbf{r}, t)(-i\boldsymbol{\nabla})\phi(\mathbf{r}, t) - \phi(\mathbf{r}, t)(-i\boldsymbol{\nabla})\phi^*(\mathbf{r}, t)] \tag{1.16}$$

is the single-particle quantum current associated with the gradient of the phase of the wave function.

Suppose, in a monoenergetic beam of N particles, all particles share the same form of single-particle wave function. Then the multiparticle wave function $\psi(\mathbf{r}, t)$ may be normalized to N such that $\phi(\mathbf{r}, t) = \psi(\mathbf{r}, t)/\sqrt{N}$. Therefore, $|\psi(\mathbf{r}, t)|^2$ must be interpreted as the number density $\rho(t)$ of the particles at time t. Also, the quantum current of the particles is given by

$$\mathbf{J} = \frac{\hbar}{2\mu}[\psi^*(\mathbf{r}, t)(-i\boldsymbol{\nabla})\psi(\mathbf{r}, t) - \psi(\mathbf{r}, t)(-i\boldsymbol{\nabla})\psi^*(\mathbf{r}, t)], \tag{1.17}$$

which clearly shows that $\mathbf{J} = 0$ if $\psi(\mathbf{r}, t)$ is real. Thus, the quantum current arises from the phase of the Schrödinger wave function. It is easy to see that $\mathbf{J}$ and ρ are connected again by the continuity equation

$$\frac{\partial}{\partial t}\rho + \boldsymbol{\nabla}\cdot\mathbf{J} = 0. \tag{1.18}$$

The number of particles incident (scattered) per unit time per unit area perpendicular to the direction of the incident (scattered) beam is given by $|\mathbf{J}|$ of the incident (scattered) beam.

In a collision event, $V(\mathbf{r}, t) = 0$ for $t \to -\infty$ (before the collision) and $t \to \infty$ (after the collision). The collision starts at $t = 0$ and lasts for a small duration τ, during which $V(\mathbf{r}, t) \neq 0$. We will assume that for $0 \leqslant t \leqslant \tau, V(\mathbf{r}, t) = V(\mathbf{r})$; that is, the potential is time-independent, as is the usual case. In free space ($V(\mathbf{r}, t) = 0$), we have to deal with the quantum states of free particles. However, one needs to distinguish between the two types of free states - one before the collision, which we call the incident or ingoing state, $|\mathrm{IN}\rangle$, and the other the outgoing state, $|\mathrm{OUT}\rangle$. The two states differ by a phase shift. The kinetic energy $E_k = \hbar^2 k^2/2m$ and the wave number k of a free particle are continuous variables, and E_k equals the total energy E as long as the particle moves in potential-free space. The wave vector $\mathbf{k}$ corresponds to the de Broglie wave number $\mathbf{k}_{dB}$, but for brevity, we omit the subscript 'dB' from now on. The wave number $k = |\mathbf{k}|$, and the momentum and the velocity of a free particle are given by $\mathbf{p} = \hbar\mathbf{k}$ and $\mathbf{v} = \mathbf{p}/m$, respectively. One can define a momentum state $|\mathbf{p}\rangle$, or, equivalently, a wave vector state $|\mathbf{k}\rangle$. Their position representations are $\phi_{\mathbf{p}}(\mathbf{r}) = \langle \mathbf{r}|\mathbf{p}\rangle$ and $\phi_{\mathbf{k}}(\mathbf{r}) = \langle \mathbf{r}|\mathbf{k}\rangle$.

Using separation of variables, one can write $\phi(\mathbf{r}, t) = \exp[-iEt/\hbar]\phi(\mathbf{r})$. In free space, $\phi(\mathbf{r})$ satisfies the eigenvalue equation

$$-\frac{\hbar^2}{2\mu}\nabla^2\phi(\mathbf{r}) = E\phi(\mathbf{r}), \tag{1.19}$$

which can also be written in a convenient form

$$[\nabla^2 + k^2]\phi(\mathbf{r}) = 0. \tag{1.20}$$

In 3D, one of the solutions of this equation is the plane wave $\phi_{\mathbf{k}}(\mathbf{r}) = A\exp[i\mathbf{k}.\,\mathbf{r}]$, where A is a constant that fixes the normalization. Since it is a continuum state, the usual procedure of unit normalization of bound states does not apply here. Instead, such 3D continuum states are normalized in the following way:

$$\int d^3\mathbf{r}\phi^*_{\mathbf{k}'}(\mathbf{r})\phi_{\mathbf{k}}(\mathbf{r}) = \delta^3(\mathbf{k} - \mathbf{k}'), \tag{1.21}$$

where $\delta^3(\mathbf{k} - \mathbf{k}')$ represents the 3D Dirac delta function. Using the properties of the delta function, one obtains $A = (1/2\pi)^{3/2}$. Therefore, the normalized plane wave solution of equation (1.19) is

$$\phi_{\mathbf{k}}(\mathbf{r}) = \left(\frac{1}{2\pi}\right)^{3/2}\exp[i\mathbf{k}.\,\mathbf{r}]. \tag{1.22}$$

Alternatively, one can define momentum wave functions

$$\phi_{\mathbf{p}}(\mathbf{r}) = \left(\frac{1}{2\pi\hbar}\right)^{3/2}\exp[i\mathbf{k}.\,\mathbf{r}] = \left(\frac{1}{\hbar}\right)^{3/2}\phi_{\mathbf{k}}(\mathbf{r}), \tag{1.23}$$

which are normalized in the following way

$$\int d^3\mathbf{r}\phi^*_{\mathbf{p}'}(\mathbf{r})\phi_{\mathbf{p}}(\mathbf{r}) = \delta^3(\mathbf{p} - \mathbf{p}'). \tag{1.24}$$

Plane-wave solutions are a convenient mathematical tool for dealing with wave phenomena. However, they do not fulfill the Heisenberg uncertainty relation. Therefore, physically more acceptable solutions are wave packets that obey the uncertainty principle. Such wave packets can be constructed by superposing the plane waves, as given by

$$\Phi(\mathbf{r}, t) = \int d^3\mathbf{k}A_{\mathbf{k}}e^{-iE_k t/\hbar}\phi_{\mathbf{k}}(\mathbf{r}), \tag{1.25}$$

where A_k are the superposition coefficients. It should also be borne in mind that standing-wave solutions are mathematically valid solutions of equation (1.19), but they do not yield any current, as required for moving particles in a beam. However, since a plane wave solution can be formed by superposing standing-wave solutions, such solutions are also useful in the mathematical descriptions of scattering processes. One of the underlying features of propagating quantum wave packets is that the width of the wave packet spreads in time as it propagates. For instance, let us consider a one-dimensional wave packet moving in the x-direction. The uncertainty principle for the quantum motion of such a particle states that $\Delta x \Delta p \geqslant \hbar/2$, where $(\Delta x)^2 = \langle(\hat{x} - \langle\hat{x}\rangle)^2\rangle$ and $(\Delta p)^2 = \langle(\hat{p} - \langle\hat{p}\rangle)^2\rangle$, with $\langle A\rangle$ being the expectation value of the operator $\hat{A}$ given by

$$\langle A\rangle = \int_{-\infty}^{\infty} dx \psi^*(x, t)\hat{A}\psi(x, t). \tag{1.26}$$

A minimum-uncertainty wave packet satisfies the condition $\Delta x \Delta p = \hbar/2$. Let the wave packet at time $t = 0$ be the minimum wave packet given by [5]:

$$\psi(x, t = 0) = [2\pi(\Delta x)^2]^{-1/4} \exp\left[\frac{i\langle\hat{p}\rangle x}{\hbar} - \frac{(x - \langle\hat{x}\rangle)^2}{4(\Delta x)^2}\right]. \tag{1.27}$$

Suppose, at $t = 0$, $\langle\hat{x}\rangle = 0$ and $\langle\hat{p}\rangle = 0$. The time-dependent wave packet is then given by

$$\psi(x, t) = \frac{1}{\sqrt{2\pi}} \int A_k e^{-iE_k t} e^{ikx}, \tag{1.28}$$

where

$$A_k = \frac{1}{\sqrt{2\pi}} \int dx e^{-ikx} \psi(x, t = 0). \tag{1.29}$$

Thus, one can explicitly evaluate the time evolution of the wave packet [5], which reveals that, during time t, the variance $(\Delta\, x)$ increases by $\hbar^2 t^2/[4m^2(\Delta x)^2]$. If one assumes that this increase is the minimum, then this quantity must be equal to $(\Delta x)^2$. This means that Δx increases by $\sqrt{\hbar t/m}$. So, a quantum wave packet spreads by an amount that is at least $\sqrt{\hbar t/m}$. This quantity provides a benchmark for the validity of classical theory—if this spread is very small compared to other relevant length scales of the system, then one can safely apply classical theory. If this quantity is large, then quantum fluctuations dictate the dynamics of the system, thus invalidating the classical concepts of dynamics. We will discuss this point further in the next chapter, where we present the classical theory of collisions.

Scattering channels

A channel refers to a set of DOF of both a projectile and a scatterer, or two colliding particles that are far apart. This set of DOF characterizes an asymptotic joint state of both the projectile and scatterer, or the two particles. For example, for collisions between two spin-half particles in the absence of a magnetic field, there are four degenerate channels; three of them belong to the spin triplet state, and one is associated with a spin singlet. In the presence of a magnetic field, these channels become non-degenerate. The DOF may be classified into internal and external DOF. The term 'internal' refers to all those DOF excluding the COM motional DOF of the colliding particles. For example, in the case of scattering between two ground-state atoms, the electronic spin or hyperfine spin of the two atoms constitutes the internal DOF, while the angular momenta of their relative rotational motion about the interatomic axis constitute the external DOF. In the case of reactive scattering, different reaction pathways or different rearrangements of reacting and product particles are referred to as rearrangement channels.

In this book, we mostly consider channels of two-particle scattering in the absence as well as the presence of external fields. The asymptotic internal energy of two particles in a channel is known as the channel threshold. If a scattering phenomenon involves only one channel, then the scattering is called single-channel scattering; otherwise, it is called multichannel scattering. If the collision energy, i.e. the energy of two colliding particles at large separation, is above the threshold of a particular channel, then the corresponding channel is called an open channel; otherwise, it is called a closed channel. The observable scattering phenomena appear only in open channels; however, a closed channel can significantly influence the scattering phenomena. A single-channel scattering event is essentially elastic. The methods of multichannel scattering can account for elastic, inelastic, and reactive collision phenomena. Multichannel scattering phenomena are schematically shown in figure 1.4.

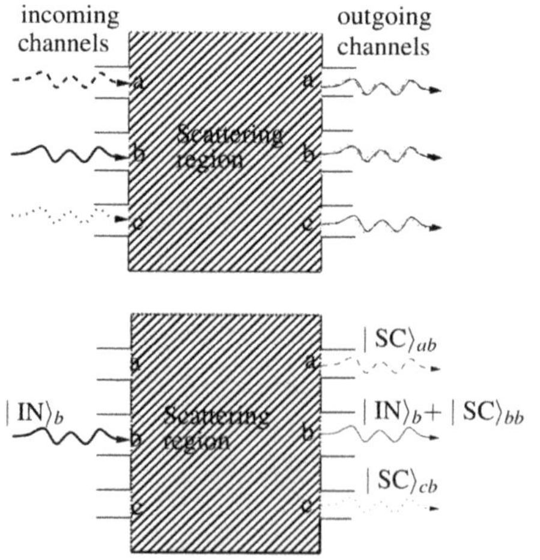

Figure 1.4. Schematic showing multichannel scattering phenomena. A three-channel case is considered as an example, with channels being indexed as a (dashed), b (solid), and c (dashed). Each of the incident channel states has a certain probability of exiting through each channel. Therefore, an outgoing state in a particular channel is a superposition of the contributions from all incident channels. In the lower panel, the wave incident in channel b is exiting through all three channels. The outgoing state of channel b is the superposition of the incident input state ($|\mathrm{IN}\rangle_b$) and the scattered state ($|\mathrm{SC}\rangle_{bb}$). The outgoing states of the other two channels, a and c, due to the incident channel b, are the scattered parts of the incident channel state.

Scattering matrices and matrix elements

To describe scattering processes and measure the associated observables, several scattering operators or matrices are introduced. The most important matrix is the $\hat{\mathbf{S}}$ matrix, which transforms an incident stationary state $|\mathrm{IN}\rangle$ into an outgoing stationary state $|\mathrm{OUT}\rangle$ through an intermediate scattering zone, as schematically shown in figure 1.5. Thus, the operator $\hat{\mathbf{S}}$ is defined by the relation

$$|\mathrm{OUT}\rangle = \hat{\mathbf{S}}|\mathrm{IN}\rangle. \tag{1.30}$$

A stationary state is an eigenstate of the Hamiltonian. For $t \to \pm\infty$, the Hamiltonian $H = H_0$ describes the joint states of a scatterer and a target when they are non-interacting. Obviously, $|\mathrm{IN}\rangle$ and $|\mathrm{OUT}\rangle$ are the eigenstates of H_0. For $0 \leqslant t \leqslant \tau$, $H = H_0 + V$, where V is the part of the Hamiltonian that describes the interaction between the scatterer and the target. Let $|\psi\rangle$ denote a stationary state of H at $t = 0$.

To obtain an explicit mathematical expression for $\hat{\mathbf{S}}$, one introduces two wave or Møller operators, Ω_+ and Ω_-. These are defined in terms of the evolution operators $U_0(t) = e^{-iH_0t/\hbar}$ and $U(t) = e^{-iHt/\hbar}$. Obviously, we have $\lim_{t\to\pm\infty} U(t) \to U_0(t)$. A scattering process leads to the relations

$$\lim_{t\to-\infty} U(t)|\psi\rangle \to U_0(t)|\mathrm{IN}\rangle \tag{1.31}$$

$$\lim_{t\to+\infty} U(t)|\psi\rangle \to U_0(t)|\mathrm{OUT}\rangle. \tag{1.32}$$

We can then write

$$\lim_{t\to-\infty} U(t)^{\dagger} U_0(t)|\mathrm{IN}\rangle = |\psi\rangle \tag{1.33}$$

$$\lim_{t\to+\infty} U(t)^{\dagger}(t) U_0(t)|\mathrm{OUT}\rangle = |\psi\rangle. \tag{1.34}$$

The Møller operators are defined as

$$\Omega_+ = \lim_{t\to-\infty} U(t)^{\dagger} U_0(t) \tag{1.35}$$

and

$$\Omega_- = \lim_{t\to+\infty} U(t)^{\dagger} U_0(t). \tag{1.36}$$

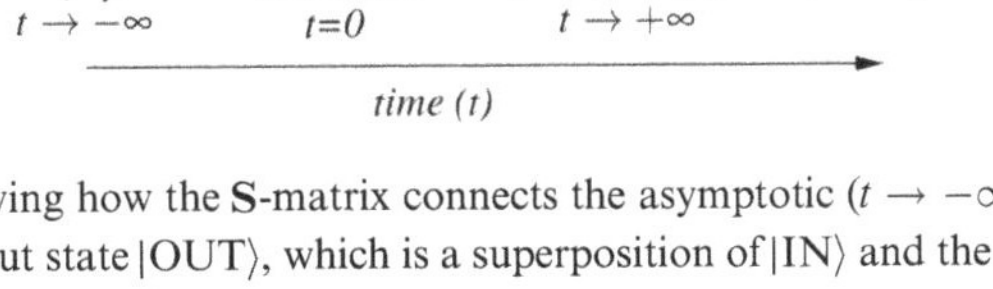

Figure 1.5. Schematic showing how the **S**-matrix connects the asymptotic ($t \to -\infty$) input state $|\mathrm{IN}\rangle$ with the asymptotic ($t \to +\infty$) output state $|\mathrm{OUT}\rangle$, which is a superposition of $|\mathrm{IN}\rangle$ and the state $|\mathrm{SCAT}\rangle$ that contains information about the scattering phenomenon.

Equations (1.33) and (1.34) imply that Ω_+ and Ω_- are the operators that transform the states $|\mathrm{IN}\rangle$ and $|\mathrm{OUT}\rangle$, respectively, to the state $|\psi\rangle$. Using equations (1.33) and (1.34), one can express $|\mathrm{OUT}\rangle$ in terms of $|\mathrm{IN}\rangle$ as given by

$$|\mathrm{OUT}\rangle = \Omega_-^\dagger|\psi\rangle = \Omega_-^\dagger\Omega_+|\mathrm{IN}\rangle. \tag{1.37}$$

Comparing this equation with equation (1.30), we obtain

$$\boxed{\hat{\mathbf{S}} = \Omega_-^\dagger\Omega_+}. \tag{1.38}$$

Both $|\mathbf{IN}\rangle$ and $|\mathbf{OUT}\rangle$ are the asymptotic states or simply asymptotes of a scattering system. In general, these states may be represented by a momentum state $|\mathbf{p}\rangle$, where $\mathbf{p}$ is the relative momentum of the incident or outgoing wave. Since $|\mathbf{p}\rangle$ is an eigenstate of H_0, we have $H_0|\mathbf{p}\rangle = E_p|\mathbf{p}\rangle$. Here, we assume that the force field or potential field responsible for scattering is spherically symmetric and therefore does not affect the COM motion of the system. Hence we do not consider COM motion explicitly. We denote the stationary states generated by the action of the operators Ω_+ and Ω_- on $|\mathbf{p}\rangle$ by $|\mathbf{p}+\rangle$ and $|\mathbf{p}-\rangle$. In order to obtain an expression for $|\mathbf{p}+\rangle$, we first rewrite Ω_+ in integral form:

$$\begin{aligned}\Omega_+ &= \lim_{t\to-\infty} U(t)^\dagger U_0(t) = \lim_{t\to\infty} U(t)U_0^\dagger(t)\\ &= 1 + \int_0^\infty \frac{d}{dt}\left[U(t)U_0^\dagger(t)\right].\end{aligned} \tag{1.39}$$

The time derivative can be calculated as follows:

$$\frac{d}{dt}\left[U(t)U_0^\dagger(t)\right] = -\frac{i}{\hbar}U(t)VU_0^\dagger(t). \tag{1.40}$$

Thus, we have

$$|\mathbf{p}+\rangle = |\mathbf{p}\rangle - \frac{i}{\hbar}\int_0^\infty e^{-i(H-E_p)/\hbar}V|\mathbf{p}\rangle. \tag{1.41}$$

To carry out the integration appearing on the rhs of the above equation, we need to multiply the integrand by a damping factor $e^{-\varepsilon t}$ and then take the limit $\varepsilon \to 0$ after carrying out the integration. One can thus obtain

$$|\mathbf{p}+\rangle = |\mathbf{p}\rangle + G^+(E_p)V|\mathbf{p}\rangle, \tag{1.42}$$

where $G^+(E_p)$ is the Green's operator defined by

$$G^+(E_p) = \frac{1}{E_p - H + i\varepsilon}, \tag{1.43}$$

where the limit $\lim_{\varepsilon\to 0}$ is assumed. Similarly, one can obtain

$$|\mathbf{p}-\rangle = |\mathbf{p}\rangle + G^-(E_p)V|\mathbf{p}\rangle, \tag{1.44}$$

where

$$G^{-}(E_p) = \frac{1}{E_p - H - i\varepsilon}. \tag{1.45}$$

Equations (1.42) and (1.44) are the Lippmann–Schwinger equation [7] in particular forms involving the Green's operators $G^{\pm}(E_p)$. Later, we will see that the Lippmann–Schwinger equation can also be expressed in other forms. The states $|\mathbf{p}+\rangle$ and $|\mathbf{p}-\rangle$ are called outgoing and ingoing states, respectively. In general, if $|\phi\rangle$ is an incident state and $|\psi\rangle$ is a scattering state, then the two states are related by

$$|\psi\rangle = (\mathbf{1} + GV)|\phi\rangle, \tag{1.46}$$

where

$$G(E) = \frac{1}{E - H} \tag{1.47}$$

is the Green's operator defined based on the assumption that the eigenstates of H have a small imaginary part. If we define the Green's operator $G_0(E) = 1/(E - H_0)$ corresponding to H_0, we can easily show that

$$G = G_0 + G_0 V G. \tag{1.48}$$

Since the rhs of equation (1.46) contains the product operator GV, using the equation (1.48), we can write

$$GV = G_0(V + VGV) = G_0 T \tag{1.49}$$

where

$$T = V + VGV \tag{1.50}$$

is called the transition operator. Again, making use of (1.48), one can make an expansion of T as given by

$$T = V + VG_0V + VG_0VG_0V + \cdots \tag{1.51}$$

This series converges if the matrix element of the operator $G_0 V$ between the eigenstates of H_0 is smaller than unity. In other words, it is convergent if the scattering potential is small compared to the collision energy. This expansion forms the basis for the Born approximation, which is valid for relatively high-energy collisions. In the low-energy limit, this approximation breaks down.

Apart from the $\hat{\mathbf{S}}$ and $\hat{\mathbf{T}}$ matrices, there exists another important matrix known as the $\hat{\mathbf{K}}$ matrix, which is particularly useful in numerical computation of the scattering amplitude. This matrix is related to the $\hat{\mathbf{S}}$ matrix by the relation

$$\hat{\mathbf{K}} = -i(\hat{\mathbf{S}} + \mathbf{1})^{-1}(\hat{\mathbf{S}} - \mathbf{1}). \tag{1.52}$$

To relate the operator $\hat{\mathbf{S}}$ to T, we first express $\hat{\mathbf{S}}$ in the following form:

$$\begin{aligned}\hat{\mathbf{S}} &= \left(\lim_{t\to+\infty} U_0^\dagger(t)U(t)\right)\left(\lim_{t'\to-\infty} U^\dagger(t')U_0(t')\right)\\ &= \lim_{t\to\infty}\left(U_0^\dagger(t)U(t)\right)\left(U(t)U_0^\dagger(t)\right)\\ &= 1 + \int_0^\infty \frac{d}{dt}\left[U_0^\dagger(t)U^2(t)U_0^\dagger(t)\right].\end{aligned} \tag{1.53}$$

Let us first consider the matrix element $S_{\mathbf{p}',\mathbf{p}} = \langle \mathbf{p}'|\hat{\mathbf{S}}|\mathbf{p}\rangle$ of single-channel scattering, where $\mathbf{p}'$ and $\mathbf{p}$ are the scattered and incident momenta, respectively. The derivative in equation (1.53) can be calculated using

$$\frac{d}{dt}\left[U_0^\dagger(t)U^2(t)U_0^\dagger(t)\right] = -\frac{i}{\hbar}\left[U_0^\dagger(t)VU^2(t)U_0^\dagger(t) + U_0^\dagger(t)U^2(t)VU_0^\dagger(t)\right]. \tag{1.54}$$

We multiply the equation (1.54) by the damping factor $e^{-\varepsilon t}$ and carry out the integration on the rhs of equation (1.53). Explicitly,

$$S_{\mathbf{p}',\mathbf{p}} = \delta^3(\mathbf{p} - \mathbf{p}') + \frac{1}{2}\langle \mathbf{p}'|[V\hat{G}^+((E_p + E_{p'})/2) + \hat{G}^+((E_p + E_{p'})/2)V]|\mathbf{p}\rangle \tag{1.55}$$

where we have used $H_0|\mathbf{p}(\mathbf{p}')\rangle = E_{p(p')}|\mathbf{p}(\mathbf{p}')\rangle$. If we now substitute the relations $\hat{G}^+\hat{V} = \hat{G}_0^+\hat{T}$ and $\hat{V}\hat{G}^+ = \hat{T}\hat{G}_0^+$ into the above equation, we obtain

$$S_{\mathbf{p}',\mathbf{p}} = \delta^3(\mathbf{p} - \mathbf{p}') + \langle \mathbf{p}'|\left[\hat{T}\frac{1}{E_{p'} - E_p + i\varepsilon'} - \frac{1}{E_{p'} - E_p - i\varepsilon'}\hat{T}\right]|\mathbf{p}\rangle, \tag{1.56}$$

where $\varepsilon' = 2\varepsilon$. Now, taking the limit $\varepsilon' \to 0$ and using the relation

$$\lim_{\varepsilon\to 0}\frac{1}{x \pm i\varepsilon} = \mathbf{P}(1/x) \mp i\pi\delta(x), \tag{1.57}$$

we obtain

$$\boxed{S_{\mathbf{p}',\mathbf{p}} = \delta^3(\mathbf{p} - \mathbf{p}') - 2i\pi\delta\left(E_p - E_{p'}\right)T_{\mathbf{p}',\mathbf{p}}}, \tag{1.58}$$

where $T_{\mathbf{p}',\mathbf{p}} = \langle \mathbf{p}|\hat{T}|\mathbf{p}'\rangle$. Since a single-channel collision is an elastic collision, we have $E_p = E_{p'}$. The appearance of the delta function $\delta(E - E_{p'})$ in the second term on the rhs of the above equation implies that the second term is nonzero only when $E_p = E_{p'} = E$. Therefore, the relevant $\hat{\mathbf{T}}$-matrix element has to be calculated on the energy shell defined by $E_p = E_{p'} = E = \hbar^2k^2/2\mu$. One then defines the on-the-energy-shell transition matrix element denoted by the relation $t_{\mathbf{p}',\mathbf{p}} = \pi T_{\mathbf{p}',\mathbf{p}}|_{E_p = E_{p'}}$. The matrix $\hat{\mathbf{t}}$ is related to the matrix $\hat{\mathbf{T}}$ defined by equation (1.50) by the relation

$$\hat{\mathbf{t}} = \pi\hat{\mathbf{T}}(E + i0^+) = \pi[\hat{V} + \hat{V}\hat{G}(E + i0^+)\hat{V}], \tag{1.59}$$

where $\hat{G}(E + i0^+)$ implies that the Green's function has to be evaluated at a complex energy with infinitesimally small positive imaginary part. Some authors write the above equation in different forms, with a prefactor of π or without the factor of two; accordingly, the relation between the $\hat{\mathbf{t}}$ matrix element and the $\hat{\mathbf{T}}$ matrix element changes. Note that $\hat{\mathbf{t}}$ is not a proper matrix, since its element is defined only on the energy shell, while the $\hat{\mathbf{T}}$ matrix is defined for both on- and off-the-energy shells. As we will derive in the next chapter, the relation between the $\hat{\mathbf{t}}$-matrix element and the scattering amplitude $f_{\mathbf{p},\mathbf{p}'}$ is given by

$$(2\pi)^3 t_{\mathbf{p}',\,\mathbf{p}} = -\frac{4\pi^2\hbar^2}{2\mu} f_{\mathbf{p}',\,\mathbf{p}}. \tag{1.60}$$

1.3.1 Lippmann–Schwinger equation

Equation (1.46) is one form of the Lippmann–Schwinger equation. Using equations (1.49) and (1.50), one can express equation (1.46) in the following form:

$$|\psi\rangle = |\phi\rangle + G_0 V|\psi\rangle, \tag{1.61}$$

which is another form of the Lippmann–Schwinger equation that is quite useful for perturbative calculations.

A scattering event may be viewed as a physical process by which an incoming wave $|\phi\rangle$ arrives and is then coherently (in the case of elastic scattering) or incoherently (in the case of inelastic scattering) scattered or emitted by the interaction zone into an outgoing wave. The outgoing state is a superposition of the incoming (non-scattered part) and scattered states: $|\mathrm{OUT}\rangle = |\mathrm{IN}\rangle + |\mathrm{SC}\rangle$, where $|\mathrm{SC}\rangle \equiv |\psi_{\mathrm{sc}}\rangle$. Obviously, $|\psi_{\mathrm{sc}}\rangle$ carries information about the forces or interactions that influenced the wave, which are described by the operator $\hat{\mathbf{T}}$. The state $|\psi\rangle$ satisfies the eigenvalue equation

$$[\hat{H}_0 + \hat{V}]|\psi\rangle = E|\psi\rangle. \tag{1.62}$$

Here, $|\psi\rangle$ must reduce to $|\phi\rangle$ for $\hat{V} = 0$. If $|\phi\rangle \equiv |\mathbf{p}\rangle$, then equation (1.61) can be written in the form

$$|\mathbf{p}\pm\rangle = |\mathbf{p}\rangle + G_0^{\pm} V|\mathbf{p}\pm\rangle. \tag{1.63}$$

This is the first form of the Lippmann–Schwinger equation. One can obtain the second form of the Lippmann–Schwinger equation by substituting equation (1.49) into equation (1.46) and replacing $|\phi\rangle \to |\mathbf{p}\rangle$. Thus, the second form of the Lippmann–Schwinger equation is

$$|\mathbf{p}\pm\rangle = |\mathbf{p}\rangle + \hat{G}_0^{\pm}\hat{\mathbf{T}}|\mathbf{p}\rangle. \tag{1.64}$$

In the next chapter, we will show that the first form of the Lippmann–Schwinger equation leads to an asymptotic scattering boundary condition in terms of the scattering amplitude, while the second form of the equation leads to an asymptotic boundary condition in terms of the $\hat{\mathbf{T}}$-matrix element. Thus, the use of both forms provides a relationship between the scattering amplitude and the $\hat{\mathbf{T}}$-matrix element. Since the $\hat{\mathbf{T}}$-matrix element is related to the $\hat{\mathbf{S}}$-matrix element by equation (1.58), one can eventually relate the scattering amplitude to the $\hat{\mathbf{S}}$-matrix element.

Optical theorem

For a single-channel problem of scattering by a real potential, the $\hat{\mathbf{S}}$ matrix is always unitary. As a consequence, the scattering phase shift is real and the total scattering cross section has only an elastic part with no inelastic component. In the case of scattering by a complex potential, the scattering phase shift is complex, and the total scattering cross section consists of both elastic and inelastic parts. Here, we prove the optical theorem in the former case. This theorem relates the total scattering cross section to the imaginary part of the scattering amplitude.

Using the identity operator $\int d^3\mathbf{p}|\mathbf{p}\rangle\langle\mathbf{p}| = \mathbf{1}$ and equation (1.59), we obtain

$$\begin{aligned}\int d^3\mathbf{p}''\hat{\mathbf{S}}^\dagger_{\mathbf{p}',\mathbf{p}''}\hat{\mathbf{S}}_{\mathbf{p}'',\mathbf{p}} = \int d^3\mathbf{p}''[\delta^3(\mathbf{p}'-\mathbf{p}'') + 2i\delta(E-E_{p'})\hat{\mathbf{t}}_{\mathbf{p}''\leftarrow\mathbf{p}'}] \\ \times[\delta^3(\mathbf{p}''-\mathbf{p}) - 2i\delta(E-E_{p''})\hat{\mathbf{t}}_{\mathbf{p}''\leftarrow\mathbf{p}}],\end{aligned} \tag{1.65}$$

where we have used $\hat{\mathbf{t}}^*_{\mathbf{p}'\leftarrow\mathbf{p}} = \hat{\mathbf{t}}_{\mathbf{p}\leftarrow\mathbf{p}'}$. Applying the unitarity condition $\hat{\mathbf{S}}^\dagger\hat{\mathbf{S}} = \hat{\mathbf{S}}\hat{\mathbf{S}}^\dagger = 1$, the lhs of the above equation is $\delta^3(\mathbf{p}-\mathbf{p}')$. The integration over $\mathbf{p}''$ of the product $\delta^3(\mathbf{p}'-\mathbf{p}'')\delta^3(\mathbf{p}''-\mathbf{p})$ of the two delta functions on the rhs of the equation yields $\delta^3(\mathbf{p}-\mathbf{p}')$. Thus, canceling this delta function that appears on both the sides of the equation, we obtain

$$\begin{aligned}4\int d^3\mathbf{p}''\delta(E-E_{p'})\hat{\mathbf{t}}_{\mathbf{p}''\leftarrow\mathbf{p}'}\delta(E-E_{p''})\hat{\mathbf{t}}_{\mathbf{p}''\leftarrow\mathbf{p}} = 2i\int d^3\mathbf{p}'' \\ \times[\delta^3(\mathbf{p}'-\mathbf{p}'')\delta(E-E_{p''})\hat{\mathbf{t}}_{\mathbf{p}''\leftarrow\mathbf{p}} - \delta^3(\mathbf{p}''-\mathbf{p})\delta(E-E_{p'})\hat{\mathbf{t}}_{\mathbf{p}''\leftarrow\mathbf{p}'}],\end{aligned} \tag{1.66}$$

which reduces to

$$4\int d^3\mathbf{p}''\hat{\mathbf{t}}^*_{\mathbf{p}'\leftarrow\mathbf{p}''}\delta(E-E_{p''})\hat{\mathbf{t}}_{\mathbf{p}''\leftarrow\mathbf{p}} = 2i[\hat{\mathbf{t}}_{\mathbf{p}'\leftarrow\mathbf{p}} - \hat{\mathbf{t}}_{\mathbf{p}\leftarrow\mathbf{p}'}]. \tag{1.67}$$

The rhs of the above equation is equal to $-4\mathrm{Im}[\hat{\mathbf{t}}_{\mathbf{p}'\leftarrow\mathbf{p}}]$. The properties of delta functions, as discussed in appendix A, provide

$$\delta(E-E_{p'}) = \frac{m}{p}\delta(p-p'), \tag{1.68}$$

where $E = p^2/2m$. Using spherical polar coordinates $\mathbf{p}'' = (p'', \theta'', \phi'')$ and integrating over $\mathbf{p}''$, we obtain

$$mp\int\int \sin\theta'' d\theta'' d\phi'' \hat{\mathbf{t}}^*_{\mathbf{p}'\leftarrow\mathbf{p}''}\hat{\mathbf{t}}_{\mathbf{p}''\leftarrow\mathbf{p}} = -\mathrm{Im}[\hat{\mathbf{t}}_{\mathbf{p}'\leftarrow\mathbf{p}}]. \tag{1.69}$$

Substituting equation (1.60) into the above equation, we get

$$\iint \sin\theta'' d\theta'' d\phi'' f^*_{\mathbf{p}'\leftarrow\mathbf{p}''} f_{\mathbf{p}''\leftarrow\mathbf{p}} = \frac{4\pi}{\hbar^2 p} \mathrm{Im}\,[f_{\mathbf{p}'\leftarrow\mathbf{p}}]. \tag{1.70}$$

If we assume that $\mathbf{p} = \mathbf{p}'$, then the above equation reduces to

$$\iint \sin\theta'' d\theta'' d\phi'' |f_{\mathbf{p}''\leftarrow\mathbf{p}}|^2 = \frac{4\pi}{\hbar^2 p} \mathrm{Im}\left[f_{\mathbf{p}\leftarrow\mathbf{p}}\right] \tag{1.71}$$

In terms of the normalized state vector $|\mathbf{k}\rangle$ of equation (1.22), we can rewrite the above equation in the form

$$\iint \sin\theta'' d\theta'' d\phi'' |f_{\mathbf{k}''\leftarrow\mathbf{k}}|^2 = \frac{4\pi}{k} \mathrm{Im}\left[f_{\mathbf{k}\leftarrow\mathbf{k}}\right] \tag{1.72}$$

In scattering geometry, the direction of the incident momentum $\mathbf{p}$ is usually considered to be the z-axis of the laboratory frame of reference. The scattering angle Ω is the angle between the incident momentum and the scattered momentum. Here, the angle $(\theta'', \phi'') = \Omega''$ appearing on the lhs of equation (1.72) is the angle of scattering, $|f_{\mathbf{k}''\leftarrow\mathbf{k}}|^2 = |f_k(\Omega'')|^2$ is the differential scattering cross section, and $f_k(\Omega'')$ is the differential elastic scattering amplitude, where $k = |\mathbf{k}| = |\mathbf{k}''|$. Therefore, the lhs $\int d\Omega'' |f_k(\Omega'')|^2$ is nothing but the total elastic scattering cross section. Thus, equation (1.72) states that the total scattering cross section is proportional to the imaginary part of the forward scattering amplitude $f_k(0) = f_{\mathbf{k}\leftarrow\mathbf{k}}$. Physically, this means that the particles that are removed or lost (hence the imaginary part of $f_k(0)$) from the forward direction appear as scattered particles. In deriving equations (1.71) and (1.72), we have made use of the unitarity of the $\hat{\mathbf{S}}$ matrix, so the total scattering cross section here is purely elastic. As we will see in the next chapter, the optical theorem can be generalized to include the inelastic part in the total scattering cross section by considering an effective complex potential. The total scattering cross section in the generalized optical theorem is again proportional to the imaginary part of the forward scattering amplitude.

Chapter 2 will begin with a description of 3D potential scattering and the generalized optical theorem. We will then present a detailed analysis of 3D, 2D, and 1D single-channel potential scattering phenomena at low energy, with some specific applications to cold atoms.

Exercise

- By drawing a schematic diagram of classical scattering of a particle by a target, introduce the concept of impact parameter b. From classical-quantum correspondence, show that $b = \sqrt{\ell(\ell+1)}\,\lambda_{dB}/(2\pi)$, where ℓ is an integer and λ_{dB} is the de Broglie wavelength.
- Define differential and total scattering cross sections. What is scattering rate?
- Show that for two particles interacting via a central potential, the Hamiltonian is separable in relative and center-of-mass coordinates.

- Writing down the Schroedinger equation of a particle moving in a potential $V(\mathrm{r}, t)$, derive the equation of continuity.
- Discuss the normalization procedure of plane waves in terms of Dirac delta function.
- What is a minimum uncertainty wave packet? Discuss how to construct a wave packet using plane waves as basis functions.
- What are the open and closed channels of scattering?
- Define **S**, **K** and **T** matrices of scattering. Write down Lippmann-Schwinger equation in terms of free-particle Green operator.
- Show that for scattering by a conservative force, the total elastic scattering cross section is proportional to the imaginary part of the forward scattering amplitude.

References

[1] Rutherford E 1911 The scattering of alpha and beta particles by matter and the structure of the atom *Phil. Mag. Ser.* **6** 669–88

[2] Blatt J M and Weisskopf V F 1952 *Theoretical Nuclear Physics* (New York: Springer)

[3] Mott N F and Massey H S W (ed) 1965 The theory of atomic collisions ed N F Mott and H S W Massey *The International Series of Monographs on Physics* 3rd edn (Oxford: Clarendon Press)

[4] Child M S 1996 *Molecular Collision Theory* (Mineola, NY: Dover)

[5] Schiff L I 1968 *Quantum Mechanics* (New York: McGraw-Hill)

[6] Newton R G 1982 *Scattering Theory of Waves and Particles* (New York: Springer)

[7] Lippmann B A 1950 Variational principles for scattering processes. II. Scattering of slow neutrons by para-hydrogen *Phys. Rev.* **79** 481–6

Chapter 2

Single-channel scattering

In the previous chapter, we defined the scattering channel. In this chapter, we will focus only on two-particle single-channel scattering. Single-channel scattering implies that the scattering event involves only one asymptotic joint state; that is, only a single set of degrees of freedom or quantum numbers is sufficient to characterize the scattering state completely. The validity of single-channel scattering rests on the fact that the forces or potentials responsible for scattering the particles cannot couple the single channel of interest to any other channel; alternatively, the other channels are energetically sufficiently far from the threshold of the single channel that the collision energy is too low to excite any other channel.

2.1 Potential scattering in 3D

By potential scattering, we mean scattering due to a conservative force field derivable from a time-independent scalar potential. We assume that the interaction potential is a function of the relative coordinate **r** only and does not depend on the center-of-mass (COM) coordinate **R**. This is the usual case for two-particle collisions in free space. For two particles interacting with such a potential in an isotropic confining potential, such as a 3D isotropic harmonic oscillator, the Hamiltonian is separable in relative and COM coordinates, implying that the wave function is also separable in these two coordinates. Thus, in such isotropic confining potentials, one can discuss the scattering in terms of **r** only. However, for an anisotropic confining potential, such a separation is not possible. Here, we are only studying the separable problems in free space and pay attention only to the relative motion of the two particles, while the COM wave function contributes only an overall phase to the total wave function of the two particles. Under these conditions, the problem of two-particle scattering in the COM frame reduces to that of a single effective particle of reduced mass μ moving in the interaction potential, known as the reduced particle.

doi:10.1088/978-0-7503-3959-9ch2

© IOP Publishing Ltd 2025. All rights, including for text and data mining (TDM), artificial intelligence (AI) training, and similar technologies, are reserved.

To describe potential scattering, we start with the Lippmann–Schwinger equation (1.61) or (1.64) introduced in the previous chapter, where we discussed the equation in momentum space. However, potential scattering can be described conveniently in coordinate representation. Let $|\mathbf{r}\rangle$ represent the continuum basis set in relative coordinates. This is an eigenstate of the operator $\hat{\mathbf{r}}$, meaning $\hat{\mathbf{r}}|\mathbf{r}\rangle = \mathbf{r}|\mathbf{r}\rangle$. The normalization of $|\mathbf{r}\rangle$ is expressed as $\langle\mathbf{r}'|\mathbf{r}\rangle = \delta^3(\mathbf{r} - \mathbf{r}')$. The completeness relation for the coordinate representation is given by

$$\int d^3\mathbf{r}|\mathbf{r}\rangle\langle\mathbf{r}| = \mathbf{1}, \tag{2.1}$$

where $\mathbf{1}$ is an identity operator. The coordinate representation of a wave vector $|\psi\rangle$ is $\psi(\mathbf{r}) = \langle\mathbf{r}|\psi\rangle$. Thus, in coordinate representation, equation (1.61) can be expressed in the form

$$\psi(\mathbf{r}) = \phi(\mathbf{r}) + \int d^3\mathbf{r}'\langle\mathbf{r}|\hat{G}_0|\mathbf{r}'\rangle V(\mathbf{r}')\psi(\mathbf{r}'). \tag{2.2}$$

Similarly, the coordinate representation of equation (1.63) is given by

$$\psi_{\mathbf{k}}^{\pm}(\mathbf{r}) = \frac{1}{(2\pi)^{3/2}}e^{\mathbf{k}\cdot\mathbf{r}} + \int d^3\mathbf{r}'\langle\mathbf{r}|\hat{G}_0^{\pm}(E_p)|\mathbf{r}'\rangle V(\mathbf{r}')\psi_{\mathbf{k}}^{\pm}(\mathbf{r}'). \tag{2.3}$$

In this chapter, we discuss time-independent potential scattering due to a central potential, which is a function of the separation r between two particles. Basically, the problem here is to solve the time-independent Schrödinger equation (1.62) in coordinate representation under appropriate scattering boundary conditions. We consider the outgoing boundary condition, implying that we are interested in evaluating the wave function $\psi_{\mathbf{k}}^{+}(\mathbf{r})$. For this, we need to solve the equation

$$\left[-\frac{\hbar^2}{2\mu}\nabla^2 + V(r)\right]\psi_{\mathbf{k}}(\mathbf{r}) = E\psi_{\mathbf{k}}(\mathbf{r}), \tag{2.4}$$

where $V(r)$ is the potential and $E = \hbar^2k^2/2\mu$ denotes the collision energy. Here, we have omitted the superscript '+' for simplicity. If $V(r)$ vanishes faster than $1/r^2$ as $r \to \infty$, then it is possible to define asymptotic scattering boundary conditions at large r in terms of free-particle basis states. If the potential supports N_b bound states $\psi_n(r)$, where $n = 1, 2, 3 \cdots N_b$, then this set of bound states, along with the scattering states, forms a complete basis. Symbolically, this statement can be expressed as

$$\sum_n |n\rangle\langle n| + \int d^3\mathbf{k}|\psi_{\mathbf{k}}\rangle\langle\psi_{\mathbf{k}}| = \mathbf{1}. \tag{2.5}$$

As we will see later, in the case of Coulomb scattering, that is, for a $1/r$ potential, one can define differential scattering but not the total scattering cross section (equation (2.16)). For a $1/r^2$ potential, the total scattering cross section exists, but one cannot define the differential cross section (equation (2.11)).

Asymptotic boundary conditions

In the asymptotically large separation ($r \to \infty$), the nature of the outgoing wave emanating from a scattering center depends on whether the potential is of the Coulomb or non-Coulomb type. By 'non-Coulomb potential,' we mean any potential that is short-ranged, falls off exponentially, or behaves as $1/r^n$, where $n > 1$ as $r \to \infty$. As we will prove later, for non-Coulomb potentials, the scattering wave function is sinusoidal at $r \to \infty$ but has a phase shift relative to the incident wave. For Coulomb potentials, the wave function does not attain a sinusoidal form at any separation, including the asymptotic regime. Here, we focus only on non-Coulomb potentials, and later we will discuss Coulomb scattering.

For simplicity, it is usually assumed that the incident wave $\phi(\mathbf{r})$ is a plane wave $\phi_{bfk}(\mathbf{r})$ with a well-defined wave vector $\mathbf{k}$. It is given by

$$\phi_{\mathbf{k}}(\mathbf{r}) = \frac{1}{(2\pi)^{3/2}} \exp(i\mathbf{k} \cdot \mathbf{r}). \tag{2.6}$$

The outgoing wave is a superposition of the incident plane wave with wave vector $\mathbf{k}$ and a spherical wave scattered along the wave vector $\mathbf{k}' = k\hat{r}$. An asymptotic form of the outgoing wave function consistent with these boundary conditions can be derived from equation (2.3). The position representation of the Green's operator $G_0^+(\mathbf{r}, \mathbf{r}') = \langle \mathbf{r} | \hat{G}_0^+(E_p) | \mathbf{r}' \rangle$, also known as the Green's function of a free particle, is given by

$$G_0^+(\mathbf{r}, \mathbf{r}') = -\frac{2\mu}{4\pi\hbar^2} \frac{\exp(ik|\mathbf{r} - \mathbf{r}'|)}{|\mathbf{r} - \mathbf{r}'|}. \tag{2.7}$$

An explicit derivation of $G_0^+(\mathbf{r}, \mathbf{r}')$ is given in appendix B. Upon substituting equation (2.7) into equation (1.13) with outgoing boundary conditions (superscript '+'), we obtain

$$\psi_{\mathbf{k}}(\mathbf{r}) = \phi_{\mathbf{k}}(\mathbf{r}) - \frac{2\mu}{4\pi\hbar^2} \int d^3\mathbf{r}' \frac{\exp(ik|\mathbf{r} - \mathbf{r}'|)}{|\mathbf{r} - \mathbf{r}'|} V(\mathbf{r}') \psi_{\mathbf{k}}(\mathbf{r}'). \tag{2.8}$$

Here, we have omitted the superscript '+' for notational simplicity. To find an expression for $\psi_{\mathbf{k}}(\mathbf{r})$ in the limit $r \to \infty$ (i.e. in the asymptotically large separation where the potential $V(r)$ vanishes or becomes negligibly small compared to the collision energy $E = \hbar^2 k^2 / 2\mu$), we first approximate $\exp(ik|\mathbf{r} - \mathbf{r}'|)$ by noting that

$$\begin{aligned} |\mathbf{r} - \mathbf{r}'| &= [(\mathbf{r} - \mathbf{r}')^2]^{1/2} \\ &\simeq r - \frac{\mathbf{r}' \cdot \mathbf{r}}{r} = r - \mathbf{r}' \cdot \hat{r}, \quad r \gg r'. \end{aligned} \tag{2.9}$$

Thus, we can write

$$\exp(ik|\mathbf{r} - \mathbf{r}'|) \simeq \exp(ikr)\exp(-i\mathbf{k}' \cdot \mathbf{r}'). \tag{2.10}$$

Now, writing $\exp(-i\mathbf{k}'\cdot\mathbf{r}') = (2\pi)^{3/2}\phi^*_{\mathbf{k}'}(\mathbf{r}')$, one obtains the asymptotic relation

$$\psi_{\mathbf{k}}(\mathbf{r}) \xrightarrow{r\to\infty} \frac{1}{(2\pi)^{3/2}}\left[\exp(i\mathbf{k}\cdot\mathbf{r}) + f_{\mathbf{k}'\mathbf{k}}\frac{e^{ikr}}{r}\right], \tag{2.11}$$

where $f_{\mathbf{k}',\mathbf{k}}$ is the scattering amplitude given by

$$f_{\mathbf{k}',\mathbf{k}} = -(2\pi)^3\frac{2\mu}{4\pi\hbar^2}\int d^3\mathbf{r}'\phi^*_{\mathbf{k}'}(\mathbf{r}')V(r')\psi_{\mathbf{k}}(\mathbf{r}'). \tag{2.12}$$

Now, in order to relate $f_{\mathbf{k}',\mathbf{k}}$ to a $\hat{\mathbf{T}}$-matrix element, we express equation (1.44) in the position space in the following way:

$$\psi_{\mathbf{k}}(\mathbf{r}) = \phi_{\mathbf{k}}(\mathbf{r}) + \int d^3\mathbf{k}''\int d^3\mathbf{r}'\langle r|\hat{G}_0^+|\mathbf{r}'\rangle\langle\mathbf{r}'|\mathbf{k}''\rangle\langle\mathbf{k}''|\hat{\mathbf{T}}|\mathbf{k}\rangle. \tag{2.13}$$

As before, upon substituting the asymptotic form of the Green's function

$$\langle r|\hat{G}_0^+|\mathbf{r}'\rangle \simeq -\frac{2\mu}{4\pi\hbar^2}e^{-i\mathbf{k}'\cdot\mathbf{r}\prime}\frac{e^{ikr}}{r} \tag{2.14}$$

and also $\langle\mathbf{r}'|\mathbf{k}''\rangle = (2\pi)^{-3/2}e^{i\mathbf{k}''\cdot\mathbf{r}\prime}$ into equation (2.13), we obtain the asymptotic boundary condition

$$\psi_{\mathbf{k}}(\mathbf{r}) \xrightarrow{r\to\infty} \frac{1}{(2\pi)^{3/2}}\left[\exp(i\mathbf{k}\cdot\mathbf{r}) - \frac{2\mu}{4\pi\hbar^2}\int d^3\mathbf{k}''\int d^3\mathbf{r}'e^{-i(\mathbf{k}'-\mathbf{k}'')\cdot\mathbf{r}\prime}\langle\mathbf{k}''|\hat{\mathbf{T}}|\mathbf{k}\rangle\frac{e^{ikr}}{r}\right]. \tag{2.15}$$

Now, substituting $\int d^3\mathbf{r}'e^{-i(\mathbf{k}'-\mathbf{k}'')\cdot\mathbf{r}\prime} = (2\pi)^3\delta^3(\mathbf{k}'-\mathbf{k}'')$, we get

$$\psi_{\mathbf{k}}(\mathbf{r}) \xrightarrow{r\to\infty} \frac{1}{(2\pi)^{3/2}}\left[\exp(i\mathbf{k}\cdot\mathbf{r}) - (2\pi)^3\frac{2\mu}{4\pi\hbar^2}T_{\mathbf{k}',\mathbf{k}}\frac{e^{ikr}}{r}\right]. \tag{2.16}$$

Comparing equation (2.16) with (2.11), we obtain

$$(2\pi)^3 T_{\mathbf{k}',\mathbf{k}} = -\frac{4\pi\hbar^2}{2\mu}f_{\mathbf{k}',\mathbf{k}}. \tag{2.17}$$

Substituting equation (2.39) into the above equation yields

$$T_{\mathbf{k}',\mathbf{k}} = \int d^3\mathbf{r}'\phi^*_{\mathbf{k}'}(\mathbf{r}')V(r')\psi_{\mathbf{k}}(\mathbf{r}'). \tag{2.18}$$

Using $t_{\mathbf{k}',\mathbf{k}} = \pi T_{\mathbf{k}',\mathbf{k}} \equiv t_{\mathbf{p}',\mathbf{p}}$ and $f_{\mathbf{k}',\mathbf{k}} \equiv f_{\mathbf{p}',\mathbf{p}}$, we can rewrite the above equation in the form

$$(2\pi)^3 t_{\mathbf{p}',\mathbf{p}} = -\frac{4\pi^2\hbar^2}{2\mu}f_{\mathbf{p}',\mathbf{p}}. \tag{2.19}$$

For another incident wave vector $\mathbf{k}''$, we can write the boundary condition

$$\psi_{\mathbf{k}''}(\mathbf{r}) \xrightarrow{r\to\infty} \exp(i\mathbf{k}\cdot\mathbf{r}) + f_{\mathbf{k}'\mathbf{k}''}\frac{e^{ikr}}{r}. \tag{2.20}$$

If both $\psi_{\mathbf{k}}$ and $\psi_{\mathbf{k}''}$ satisfy the time-independent Schrödinger equation (2.4) for the same collision energy E and potential $V(r)$ with the boundary conditions of equations (2.16) and (2.20), respectively, one can easily obtain the relation

$$\psi_{\mathbf{k}''}\,\nabla^2\,\psi_{\mathbf{k}} - \psi_{\mathbf{k}}\,\nabla^2\,\psi_{\mathbf{k}''} = 0. \tag{2.21}$$

This equation is true whether the potential V is real or complex.

Born approximation

If the collision energy is much larger than the strength of the potential, then the collision energy rather than the potential has a more prominent role in determining the scattering wave function. Under this condition, one can develop an approximate or perturbative method to solve the Lippmann–Schwinger equation. This perturbative approximation is known as the Born approximation. The essence of this approximation is to consider the plane wave (or free-particle wave function) as the zeroth-order wave function, allowing one to calculate the higher-order corrections to the wave function using the Lippmann–Schwinger integral equation. Accordingly, one can calculate term by term higher-order **T**-matrix elements or scattering amplitudes.

Generalized optical theorem

Any central potential of physical interest is real, so the single-channel scattering is, by default, of the elastic type. However, the single-channel Schrödinger equation can be solved for a model complex potential, allowing us to discuss the small inelastic part of the scattering in a phenomenological way. Such a model potential is generally called the optical model potential, in analogy with the complex refractive index used in optics to describe the absorption of an optical wave propagating through a medium. An elegant description of the use of complex potentials in scattering theory is given in Schiff's Quantum Mechanics [1]. The optical theorem, as discussed in the previous chapter for elastic scattering, can only be generalized for scattering involving both elastic and inelastic processes using an optical potential.

The continuity equation for particles scattered by a real potential was derived in equation (1.8). Let us now see how this equation is modified due to a complex potential given by

$$V = V_R - iV_I. \tag{2.22}$$

Following the same method as that used to derive equation (1.18), we obtain

$$\frac{\partial}{\partial t}\rho + \boldsymbol{\nabla}.\,\mathbf{J} = -\frac{2V_I}{\hbar}\rho. \tag{2.23}$$

The rhs of the above equation acts as a source (or sink) of the density $\rho(\mathbf{r}, t)$ if V_I is negative (or positive), respectively. For inelastic scattering, it is usually assumed that the complex potentials act as a sink, implying that the incident particle flux is attenuated due to inelastic processes. Suppose, in a region of space, that $\nabla \cdot \mathbf{j} = 0$. Then, in that region, for $V_I > 0$, we have $\rho_1(\mathbf{r}, t \to \infty) = 0$. However, for a scattering problem, $\rho_1(\mathbf{r}, t \to \infty) \neq 0$ in the asymptotic region. This means that there exists no region in space where $V_I \neq 0$ and $\nabla \cdot \mathbf{j} = 0$ in the context of scattering problems [1].

Integrating equation (2.23) over a large volume $\mathcal{V}$ bounded by the closed surface area A and making use of Gauss's theorem, one obtains

$$\frac{\partial}{\partial t}\int_{\mathcal{V}} \rho(\mathbf{r}, t)d^3\mathbf{r} = -\oint_A \mathbf{J}(\mathbf{r}, t) \cdot d\mathbf{S} - \int_{\mathcal{V}} \frac{2V_I}{\hbar}\rho(\mathbf{r}, t)d^3\mathbf{r}. \tag{2.24}$$

If $\mathcal{V}$ encompasses the entire space, then the quantity $\int_{\mathcal{V}} \rho(\mathbf{r}, t)d^3\mathbf{r}$ is the total probability $P(t)$ that must be conserved. Hence, the lhs of the above equation vanishes. The surface integral $-\oint_A \mathbf{J}(\mathbf{r}, t) \cdot d\mathbf{S}$ on the rhs is proportional to the flux of particles that are elastically scattered plus the particles that are removed from the incident beam due to inelastic processes (for instance, absorption of particles by the potential). If $V_I = 0$, then this surface integral vanishes, implying that the particles scattered outward are supplied by the incident wave. Therefore, in the stationary scattering problems, this integral must be equal to the inelastic or absorptive scattering rate $nv\sigma_{\text{inel}}$, where v is the relative velocity and n is the number density. Now, if the wave function $\psi(\mathbf{r})$ is normalized to a plane wave of unit amplitude at large separation, then n reduces to unit number density. In such a situation, we thus have

$$\sigma_{\text{inel}} = \frac{2}{\hbar v}\int V_I |\psi(\mathbf{r})|^2 d^3\mathbf{r}. \tag{2.25}$$

2.2 Partial-wave expansion

For a spherically symmetric potential, the scattering wave function can be most conveniently expanded in partial waves. Each of the partial waves has a definite angular momentum of relative motion between the two particles. The quantized states of relative angular momentum of two free particles form a complete orthonormal basis in which the expansion is sought. To develop a partial-wave description of scattering phenomena, we first need to write the time-independent Schrödinger equation (2.4) in spherical coordinates $\mathbf{r} = (r, \theta, \phi) \equiv (r, \hat{r})$. In these coordinates, the operator ∇^2 takes the form

$$\nabla^2 = \frac{1}{r}\frac{\partial^2}{\partial r^2}r - \frac{1}{r^2}\hat{L}^2(\theta, \phi), \tag{2.26}$$

where

$$\hat{L}^2(\theta, \phi) = -\frac{1}{\sin\theta}\frac{\partial}{\partial\theta}\sin\theta\frac{\partial}{\partial\theta} - \frac{1}{\sin^2\theta}\frac{\partial^2}{\partial\phi^2}. \tag{2.27}$$

Note that here one may use $\frac{1}{r}\frac{\partial^2}{\partial r^2}r = \frac{1}{r^2}\frac{\partial}{\partial r}r^2\frac{\partial}{\partial r}$. The wave function $\psi_{\mathbf{k}}(\mathbf{r})$ can be expanded in the orthonormal eigenfunctions of $\hat{L}^2(\theta, \phi)$, which are the spherical harmonics $Y_{\ell m_\ell}(\theta, \phi)$, where ℓ and m_ℓ denote the quantum numbers associated with the angular motion in θ and ϕ coordinates, respectively. To derive an explicit expression for the expansion of the plane wave in terms of spherical harmonics, we write

$$e^{\mathbf{k}\cdot\mathbf{r}} = \sum_\ell c_\ell(kr)P_\ell(\hat{k}\cdot\hat{r}), \tag{2.28}$$

where $c_\ell(kr)$ is an expansion coefficient that depends on $kr = |\mathbf{k}||\mathbf{r}|$ and $P_\ell(\cos\theta')$ is the Legendre polynomial. Let $\hat{k} \equiv (\theta_k, \phi_k)$ and $\hat{r} \equiv (\theta', \phi')$. We consider a coordinate system where $\mathbf{k}$ points toward the z-direction. Then, $\theta_k = 0$. So, $\hat{k}\cdot\hat{r} = \cos\theta'\cos\theta_k + \sin\theta_k\sin\theta'\cos(\phi_k - \phi') = \cos\theta'$. Next, to calculate $c_\ell(r)$, we multiply both sides of the above equation by $P_{\ell'}(\cos\theta')$ and then integrate over θ' to obtain

$$\frac{2}{2\ell+1}c_\ell(r) = \int_0^\pi \sin\theta' d\theta' \exp[ikr\cos\theta']P_\ell(\cos\theta'). \tag{2.29}$$

Let $\alpha = \cos\theta$; then, the above integral can be expressed as

$$\int_0^\pi \sin\theta' d\theta' \exp[ikr\cos\theta'] = \int_{-1}^1 d\alpha \exp[ikr\alpha]P_\ell(\alpha). \tag{2.30}$$

The above integration can be carried out analytically [2]. The result is

$$\int_{-1}^1 d\alpha \exp[ikr\alpha]P_\ell(\alpha) = i^\ell\sqrt{2\pi}\frac{J_{\ell+\frac{1}{2}}(kr)}{\sqrt{kr}}, \tag{2.31}$$

where $J_{\ell+\frac{1}{2}}$ is the Bessel function of order $\ell + \frac{1}{2}$. Since the spherical Bessel function $j_\ell(x)$ is given by $j_\ell(x) = \sqrt{\frac{\pi}{2}}\frac{J_{\ell+\frac{1}{2}}(x)}{x}$, we have

$$c_\ell(kr) = i^\ell(2\ell+1)j_\ell(kr). \tag{2.32}$$

Thus, we finally obtain the formula

$$\phi_{\mathbf{k}}(\mathbf{r}) = \frac{1}{(2\pi)^{3/2}}e^{\mathbf{k}\cdot\mathbf{r}} = \frac{1}{(2\pi)^{3/2}}\sum_{\ell=0}^{\ell} i^\ell j_\ell(kr)(2\ell+1)P_\ell(\cos\theta'). \tag{2.33}$$

Note that

$$P_\ell(\cos\theta') = \frac{4\pi}{2\ell+1}\sum_{m_\ell=-\ell}^{\ell} Y_{\ell,m_\ell}(\theta, \phi)Y^*_{\ell,m_\ell}(\theta_k, \phi_k), \tag{2.34}$$

where $(\theta_k, \phi_k) \equiv \hat{k}$ and $(\theta, \phi) = \hat{r}$. So, we can write

$$\phi_{\mathbf{k}}(\mathbf{r}) = 4\pi \frac{1}{(2\pi)^{3/2}} \sum_{\ell=0}^{\ell} i^{\ell} j_{\ell}(kr) \sum_{m_\ell=-\ell}^{\ell} Y_{\ell, m_\ell}(\theta, \phi) Y^{*}_{\ell, m_\ell}(\theta_k, \phi_k). \tag{2.35}$$

In a similar fashion, we seek an expansion of the 3D scattering wave function in the following form:

$$\psi_{\mathbf{k}}(\mathbf{r}) = \frac{1}{(2\pi)^{3/2}} 4\pi \sum_{\ell=0}^{\infty} i^{\ell} \sum_{m_\ell=-\ell}^{\ell} \psi_{\ell, m_\ell}(kr) Y_{\ell, m_\ell}(\theta, \phi) Y^{*}_{\ell, m_\ell}(\theta_k, \phi_k), \tag{2.36}$$

where $\psi_{\ell,m_\ell}(r)$ represents a partial-wave radial wave function to be determined. Now, substituting equations (2.36), (2.33), and (2.34) into equation (2.12) yields

$$\begin{aligned} f_{\mathbf{k}', \mathbf{k}} = &- \frac{2\mu}{\hbar^2} \sum_{\ell,\ell'} \sum_{m_\ell, m_{\ell'}} \int r^2 dr j_\ell(kr) V(r) \psi_{\ell', m_{\ell'}}(kr) \\ &\times 4\pi \int\int \sin\theta d\theta d\phi Y^{*}_{\ell, m_{ell}}(\theta, \phi) Y_{\ell', m_{\ell'}}(\theta, \phi) Y_{\ell, m_{ell}}(\hat{k}') Y^{*}_{\ell, m_\ell}(\hat{k}). \end{aligned} \tag{2.37}$$

The angular integral in the above equation is zero unless $\ell = \ell'$ and $m_{\ell'} = m_\ell$. This means that upon scattering off a spherically symmetric potential, the plane of rotation cannot change. In the case of a spin-dependent potential or an anisotropic potential, the plane of rotation may change, and different partial waves become coupled. Scattering by an anisotropic potential is a case of multichannel scattering, where different partial-wave indexes are used to characterize different channels. We will discuss anisotropic scattering in the next chapter.

It is convenient to define the z-axis of the laboratory frame as lying along the direction $\hat{k}$ of the incident momentum $\mathbf{k}$. For this chosen laboratory frame, the incident relative angular momentum is perpendicular to the z-axis, meaning that $m_\ell = 0$, $Y_{\ell,m_\ell}(\hat{k}) \equiv Y_{l,0}(0, \phi_k) = \sqrt{(2l+1)/4\pi}$ and $Y_{\ell,m_\ell}(\hat{k}') \equiv Y_{\ell,0}(\hat{k}') = \sqrt{(2l+1)/4\pi}\, P_\ell(\cos\theta')$; also,

$$\psi_{\mathbf{k}}(\mathbf{r}) = \frac{1}{(2\pi)^{3/2}} \frac{1}{4\pi} \sum_{\ell=0}^{\infty} i^{\ell} (2\ell + 1) \psi_\ell(kr) P_\ell(\cos\theta), \tag{2.38}$$

where $\psi_\ell(kr) = \psi_{\ell,0}(kr)$. Thus, the scattering amplitude has the expansion

$$f_{\mathbf{k}', \mathbf{k}} = \sum_{\ell} (2\ell + 1) f_l(k) P_\ell(\cos\theta_{k'}), \tag{2.39}$$

where

$$f_\ell(k) = -\int r^2 dr j_\ell((kr)) U(r) \psi_\ell(kr) \tag{2.40}$$

with

$$U(r) = \frac{2\mu}{\hbar^2} V(r). \tag{2.41}$$

Substituting equations (2.33), (2.39), and (2.40) into equation (2.11), we obtain

$$\psi_{\mathbf{k}}(\mathbf{r}) \rightarrow \sum_{\ell}(2\ell + 1)i^{\ell}\left[j_{\ell}(kr)P_{\ell}(\cos\theta) + f_{\ell}(k)\frac{e^{i(kr-\ell\pi/2)}}{r}P_{\ell}(\cos\theta_{k'})\right]. \tag{2.42}$$

In the previous subsection, we noted that $\hat{k}' = \hat{r}$, implying that $\theta_{k'} = \theta$. Comparing the above expression with equation (2.38), one gets

$$\psi_{\ell}(kr \rightarrow \infty) \longrightarrow j_{\ell}(kr) + f_{\ell}(k)\frac{e^{i(kr-\ell\pi/2)}}{r}. \tag{2.43}$$

Substitution of the asymptotic form of $j_{\ell}(kr)$, namely

$$j_{\ell}(kr \rightarrow \infty) \longrightarrow \frac{1}{2ikr}[e^{i(kr-\ell\pi/2)} - e^{-i(kr-\ell\pi/2)}], \tag{2.44}$$

into equation (2.43) yields

$$\begin{aligned}\psi_{\ell}(kr \rightarrow \infty) \longrightarrow & \frac{1}{2ikr}[-e^{-i(kr-\ell\pi/2)} + \{1 + 2ikf_{\ell}(k)\}e^{i(kr-\ell\pi/2)}] \\ &= \frac{1}{2ikr}[-e^{-i(kr-\ell\pi/2)} + S_{\ell}(k)e^{i(kr-\ell\pi/2)}],\end{aligned} \tag{2.45}$$

where $S_{\ell}(k) = 1 + 2ikf_{\ell}(k)$ is the ℓth partial-wave **S** matrix element. In terms of the partial-wave phase shift $\delta_{\ell}(k)$, this element is given by $S_{\ell}(k) = \exp[2i\delta_{\ell}(k)]$ and $f_{\ell}(k) = e^{i\delta_{\ell}}\sin\delta_{\ell}/k$.

So far, we have expressed the partial-wave asymptotic radial wave function in two equivalent forms: one given by equation (2.44) as the superposition of the spherical Bessel function and an outgoing spherical wave and the other given by equation (2.45) as the superposition of an ingoing spherical wave $\exp[-(kr - \ell\pi/2)]/(kr)$ and an outgoing spherical wave $\exp[(kr - \ell\pi/2)]/(kr)$. The superposition coefficient in both equations is expressed in terms of the partial-wave scattering amplitude $f_{\ell}(k)$. There are different forms of the asymptotic radial wave functions, since the superposition of any two linearly independent solutions of a time-independent Schrödinger equation is also a solution. Therefore, it is important to find at least a pair of linearly independent base solutions that can then be superposed to obtain all other desirable solutions.

Substituting equation (2.36) into the Schrödinger equation (2.4) and using the result

$$\hat{L}^2(\theta, \phi)Y_{\ell m_{\ell}}(\theta, \phi) = \ell(\ell + 1)Y_{\ell m_{\ell}}(\theta, \phi) \tag{2.46}$$

and the orthonormality of spherical harmonics, one obtains

$$\left[-\frac{\hbar^2}{2\mu r}\frac{d^2}{dr^2}r + \frac{\hbar^2}{2\mu}\frac{\ell(\ell + 1)}{r^2} + V(r)\right]\psi_{\ell m_{\ell}}(kr) = E\psi_{\ell m_{\ell}}(kr). \tag{2.47}$$

Since the potential is isotropic, the radial wave function $\psi_{\ell m_\ell}(kr)$ does not depend on m_ℓ and hence, for simplicity, one can suppress the index m_ℓ. Henceforth, we replace $\psi_{\ell m_\ell}(kr)$ by $\psi_\ell(kr)$. The repulsive potential $\frac{\hbar^2}{2\mu}\frac{\ell(\ell+1)}{r^2}$ is called the centrifugal barrier. One can define an effective potential

$$V_{\text{eff}}^{(\ell)} = \frac{\hbar^2}{2\mu}\frac{\ell(\ell+1)}{r^2} + V(r). \tag{2.48}$$

Now, if the potential $V(r)$ goes to zero faster than $1/r^2$ as $r \to \infty$, then in the asymptotic regime, one can neglect $V(r)$ for $\ell \neq 0$. For $\ell = 0$, one can ignore the potential in the asymptotic regime only when $|V(r)| \ll E$. In this way, neglecting the potential in the asymptotic regime, one is left with the partial-wave Schrödinger equation for a free particle. This partial-wave free-particle Schrödinger equation can be cast into a Bessel equation of order $\ell + 1/2$ using the transformation $\psi_\ell(kr) = \bar{\psi}_{\ell+1/2}(kr)/\sqrt{kr}$. The mathematical method used to reduce the free-particle Schrödinger equation into a Bessel equation and to obtain its analytical solutions is available in any standard book on mathematical physics; for example, see Arfken and Weber [3]. The two linearly independent base solutions of the Bessel equation are the Bessel and Neumann functions $J_n(kr)$ and $N_n(kr)$ of order n. Accordingly, the two base solutions of the ℓ-wave free-particle Schrödinger equation are the spherical Bessel and spherical Neumann functions $j_\ell(kr)$ and $n_\ell(kr)$, respectively, which are related to $J_{\ell+1/2}$ and $N_{\ell+1/2}$ by

$$j_\ell(kr) = J_{\ell+1/2}(kr)/\sqrt{kr} \tag{2.49}$$

and

$$n_\ell(kr) = N_{\ell+1/2}(kr)/\sqrt{kr}. \tag{2.50}$$

These functions have the asymptotic boundary conditions

$$j_\ell(x \to 0) \sim x^\ell \tag{2.51}$$

$$n_\ell(x \to 0) \sim x^{-\ell-1} \tag{2.52}$$

$$j_\ell(x \to \infty) \sim \sin(x - \ell\pi/2)/x \tag{2.53}$$

$$n_\ell(x \to \infty) \sim -\cos(x - \ell\pi/2)/x. \tag{2.54}$$

They can be superposed to obtain another pair of linearly independent base functions, namely

$$h_\ell^{\pm}(x) = j_\ell(x) \pm i n_\ell(x), \tag{2.55}$$

which are called spherical Hankel functions of the first (+) and second (−) kind*. In the limit $x \to \infty$ they behave as

$$h_\ell^{\pm}(x \to \infty) \sim \mp i \frac{e^{\pm i(x-\ell\pi/2)}}{x}, \tag{2.56}$$

where (+) and (-) denote outgoing and ingoing spherical waves, respectively. The scattering wave function $\psi_\ell(kr)$ for large r can be expressed using different combinations of these free-particle partial-wave base wave functions. Accordingly, one can define different partial-wave scattering matrix elements, as we discuss below.

Partial-wave $\hat{\mathbf{t}}$-matrix element

Next, one can introduce a partial-wave $\hat{\mathbf{t}}$-matrix element $t_\ell(k)$ or $t_\ell(E)$. To establish a connection between $f_\ell(k)$ and $t_\ell(E)$, we first need to define the energy-normalized partial-wave basis $|E, \ell, m_\ell\rangle$, normalized in the following manner:

$$\langle E', \ell', m_{\ell'}|E, \ell, m_\ell\rangle = \delta(E - E')\delta_{\ell, \ell'}\delta_{m_\ell, m_{\ell'}}. \tag{2.57}$$

The coordinate representation of $|E, \ell, m\rangle$ is

$$\langle \mathbf{r}|E, \ell, m_\ell\rangle = \phi_{E\ell}(r) Y_{\ell, m_\ell}(\hat{r}), \tag{2.58}$$

where $\phi_{E\ell}(r)$ is a partial-wave radial basis function that satisfies the normalization condition

$$\int r^2 dr \phi^*_{E'\ell}(r)\phi_{E\ell}(r) = \delta(E - E'). \tag{2.59}$$

To express $\phi_{E\ell}(r)$ in terms of $j_\ell(kr)$, we need to know how $j_\ell(kr)$ is normalized. For this, we make use of the normalization of the plane wave as expressed in equation (2.35). Thus, we have

$$\delta^3(\mathbf{k} - \mathbf{k}') = \frac{(4\pi)^2}{(2\pi)^3}\sum_\ell \int r^2 dr j_\ell^2(r) \sum_{m_\ell} Y_{\ell, m_\ell}(\hat{k}') Y^*_{\ell, m_\ell}(\hat{k}). \tag{2.60}$$

Now, since

$$\delta^3(\mathbf{k} - \mathbf{k}') = \frac{1}{k^2}\delta(k - k')\frac{1}{\sin\theta}\delta(\theta - \theta')\delta(\phi - \phi') \tag{2.61}$$

and

$$\sum_\ell \sum_{m_\ell} Y_{\ell, m_\ell}(\hat{k}') Y^*_{\ell, m_\ell}(\hat{k}) = \frac{1}{\sin\theta}\delta(\theta - \theta')\delta(\phi - \phi'), \tag{2.62}$$

we have

$$\frac{(4\pi)^2}{(2\pi)^3}\int r^2 dr j_\ell^2(r) = \frac{1}{k^2}\delta(k - k'). \tag{2.63}$$

Using the property

$$\delta(f(x) - f(x')) = \left| \left(\frac{\partial f(x)}{\partial x}\right)^{-1} \right| \delta(x - x'), \tag{2.64}$$

one can write

$$\delta(k - k') = \frac{\partial E}{\partial k} \quad \delta(E - E') = \frac{\hbar^2 k}{\mu}\delta(E - E'). \tag{2.65}$$

Therefore,

$$\frac{\mu k}{\hbar^2}\frac{(4\pi)^2}{(2\pi)^3}\int r^2 dr j_\ell^2(r) = \delta(E - E'). \tag{2.66}$$

Now, comparing the above equation with equation (2.59), one can readily identify that

$$\phi_{E\ell}(r) = \sqrt{\frac{\mu k}{\hbar^2}\frac{(4\pi)^2}{(2\pi)^3}} j_\ell(kr) = \sqrt{\frac{2\mu k}{\pi\hbar^2}} j_\ell(kr) = \sqrt{\frac{k}{\pi E}} k j_\ell(kr). \tag{2.67}$$

In order to calculate the on-the-energy-shell $\hat{\mathbf{T}}$-matrix element, it is necessary to expand the plane wave in terms of the energy-normalized basis $|E, \ell, m_\ell\rangle$. Therefore, we can write

$$|\mathbf{k}\rangle = \sum_{E',\ell',m_{\ell'}} |E', \ell', m_{\ell'}\rangle\langle E', \ell', m_{\ell'}|\mathbf{k}\rangle. \tag{2.68}$$

One can calculate

$$\langle E', \ell', m_{\ell'}|\mathbf{k}\rangle = \int d^3\mathbf{r}\phi_{E'\ell'}(r) Y^*_{\ell', m_{\ell'}}(\hat{r})\phi_{\mathbf{k}}(\mathbf{r}). \tag{2.69}$$

Substituting the expansion of $\phi_{\mathbf{k}}(\mathbf{r})$ as given by equation (2.35) into the above equation and carrying out the integration over $\mathbf{r}$, one obtains

$$\langle E', \ell', m_{\ell'}|\mathbf{k}\rangle = \frac{4\pi}{(2\pi)^{3/2}} i^\ell \sqrt{\frac{\pi E}{k}} \frac{1}{k} Y_{\ell,m_\ell}(\hat{k})\delta_{\ell,\ell'}\delta_{m_\ell,m_{\ell'}}\delta(E - E') \tag{2.70}$$

Substituting this into equation (2.68), we get

$$|\mathbf{k}\rangle = \frac{4\pi}{(2\pi)^{3/2}}\sum_\ell\sum_{m_\ell} i^\ell \sqrt{\frac{\pi E}{k}} \frac{1}{k} Y_{\ell, m_\ell}(\hat{k})|E, \ell, m_\ell\rangle. \tag{2.71}$$

Thus, the on-the-energy-shell $\hat{\mathbf{T}}$-matrix element can then be conveniently expanded in the following form:

$$T_{\mathbf{k}', \mathbf{k}} = \sum_{\ell,\ell'}\sum_{m_\ell,m_{\ell'}} \langle E, \ell' m_{\ell'}|\hat{\mathbf{T}}(E + 0^+)|E, \ell, m_\ell\rangle \frac{\pi\hbar^2}{2\mu k} Y_{\ell, m_\ell}(\hat{k}) Y_{\ell', m_{\ell'}}(\hat{k}'). \tag{2.72}$$

Recall that the matrix $\hat{\mathbf{T}}(E + 0^+)$ as defined by equation (1.50) is basically given by the interaction operator $\hat{V}$. Now, if the interaction is spherically symmetric, then

$$\langle E, \ell' m_{\ell'}|\hat{\mathbf{T}}(E + 0^+)|E, \ell, m_\ell\rangle = \frac{(4\pi)^2}{(2\pi)^3}\langle E, \ell'|\hat{\mathbf{T}}(E + 0^+)|E, \ell\rangle\delta_{\ell\ell'}\delta_{m_\ell m_{\ell'}}. \tag{2.73}$$

This means that

$$T_{\mathbf{k}',\mathbf{k}} = \frac{4\pi}{(2\pi)^3}\sum_{\ell} T_\ell(E)(2\ell+1)\frac{\pi\hbar^2}{2\mu k}P_\ell(\cos\theta_{k'}). \tag{2.74}$$

If we now make use of equations (2.17) and (2.39), as well as the above equation, we obtain

$$f_\ell(k) = -\frac{\pi T_\ell(E)}{k} = -\frac{t_\ell(k)}{k}. \tag{2.75}$$

Partial-wave Ŝ-matrix element

Now, to establish a relation between $f_\ell(k)$ or $t_\ell(k)$ and the partial-wave Ŝ-matrix element $S_\ell(k)$ for an isotropic scattering, we make use of equation (1.58), which gives a relationship between $S_{\mathbf{p}'\mathbf{p}}$ and $T_{\mathbf{p}'\mathbf{p}}$. To this end, we first expand $S_{\mathbf{k}'\mathbf{k}}$ into the following form:

$$S_{\mathbf{k}'\mathbf{k}} = \frac{1}{4\pi}\sum_{\ell}(2\ell+1)S_\ell(k)\frac{1}{k^2}\delta(k-k')P_\ell(\cos\theta_{k'}). \tag{2.76}$$

Here, as before, we have assumed that the z-axis of the laboratory fixed coordinate system is aligned with the incident momentum $\mathbf{k}$. Furthermore, in this chosen laboratory frame, making use of equations (2.61) and (2.62), we can write

$$\delta^3(\mathbf{k}-\mathbf{k}') = \frac{1}{4\pi}\sum_{\ell}(2\ell+1)\frac{1}{k^2}\delta(k-k')P_\ell(\cos\theta_{k'}). \tag{2.77}$$

Substituting equations (2.39), (2.74) and the above two equations into equation (1.58), we obtain

$$\boxed{S_\ell(k) = 1 - 2it_\ell(k) = 1 + 2ikf_\ell(k)}. \tag{2.78}$$

Substituting equation (2.75) into equation (2.45), the asymmetric boundary condition of $\psi_\ell(kr)$ can be expressed in terms of $t_\ell(k)$ as given by

$$\psi_\ell(kr\to\infty) \longrightarrow \frac{1}{kr}[\sin(kr-\ell\pi/2) - t_\ell(k)e^{i(kr-\ell\pi/2)}]. \tag{2.79}$$

In terms of the partial-wave Ŝ-matrix element $S_\ell(k)$, the asymptotic boundary condition of $\psi_\ell(kr)$ is given by

$$\psi_\ell(kr\to\infty) \longrightarrow \frac{1}{2ikr}[-e^{-i(kr-\ell\pi/2)} + S_\ell(k)e^{i(kr-\ell\pi/2)}]. \tag{2.80}$$

In terms of the spherical Hankel functions $h^{\pm}(kr)$, the asymptotic boundary condition is given by

$$\psi_\ell(kr\to\infty) \longrightarrow h^{-}(kr) + S_\ell(k)h^{+}(kr). \tag{2.81}$$

The partial-wave boundary condition can also be expressed in terms of the $\hat{\mathbf{K}}$-matrix element $K_\ell(k)$ using the relation (1.52) between $\hat{\mathbf{S}}$ and $\hat{\mathbf{K}}$. For single-channel scattering, equation (1.52) provides

$$K_\ell(k) = -i\frac{S_\ell(k) - 1}{S_\ell(k) + 1}. \tag{2.82}$$

The substitution of this relation into equations (2.80) and (2.81) leads to the boundary conditions

$$\psi_\ell(kr \to \infty) \longrightarrow \frac{S_\ell(k) + 1}{2kr}[\sin(kr - \ell\pi/2) + K_\ell(k)\cos(kr - \ell\pi/2)] \tag{2.83}$$

and

$$\psi_\ell(kr \to \infty) \longrightarrow \frac{S_\ell(k) + 1}{2}[j_\ell(kr) - K_\ell(k)n_\ell(kr)], \tag{2.84}$$

respectively.

Phase shift

Since the $\hat{\mathbf{S}}$-matrix is unitary for elastic scattering, $S_\ell(k)$ can be expressed in the form

$$S_\ell(k) = e^{2i\delta_\ell(k)} = \frac{\cot\delta_\ell(k) + i}{\cot\delta_\ell(k) - i}, \tag{2.85}$$

where $\delta_\ell(k)$ is called the ℓ-wave phase shift. In terms of $\delta_\ell(k)$, we have

$$t_\ell(k) = -e^{i\delta_\ell(k)}\sin\delta_\ell(k) = -\frac{1}{\cot\delta_\ell(k) - i} \tag{2.86}$$

$$f_\ell(k) = \frac{e^{i\delta_\ell(k)}\sin\delta_\ell(k)}{k} = \frac{1}{k\cot\delta_\ell(k) - ik} \tag{2.87}$$

$$K_\ell(k) = \tan\delta_\ell(k). \tag{2.88}$$

The total scattering cross section σ can be evaluated using equation (2.39). This is given by

$$\begin{aligned}\sigma &= \int d\Omega_{k'}|f_{\mathbf{k}',\mathbf{k}}|^2 = \int d\Omega_{k'}\left|\sum_\ell(2\ell + 1)f_\ell(k)P_\ell(\cos\theta_k)\right|^2 \\ &= 4\pi\sum_\ell(2\ell + 1)|f_\ell(k)|^2\end{aligned} \tag{2.89}$$

where $d\Omega_{k'} = \sin\theta_{k'}d\theta_{k'}d\phi_{k'}$. Thus, $\sigma = \sum_\ell \sigma_\ell$, where

$$\sigma_\ell = 4\pi(2\ell + 1)|f_\ell(k)|^2 = 4\pi(2\ell + 1)\frac{\sin^2\delta_\ell(k)}{k^2} \tag{2.90}$$

is the total ℓ-wave scattering cross section. The partial-wave differential cross section is given by

$$\frac{d\sigma_\ell(\Omega_{k'})}{d\Omega_{k'}} = (2\ell + 1)^2 \frac{\sin^2\delta(k)}{k^2}[P_\ell(\cos\theta_{k'})]^2. \tag{2.91}$$

Before ending this section, we briefly discuss how the scattering cross section is affected by the indistinguishability or exchange symmetry of the particles. In the COM frame, exchanging two particles amounts to changing the spherical coordinates $\theta \to \pi - \theta$ and $\phi \to \phi + \pi$, implying an inversion of the Cartesian coordinates. Under inversion, the spherical harmonics change: $Y_{\ell m_\ell} \to (-1)^\ell Y_{\ell m_\ell}$. Under exchange, the wave function of two bosons is symmetric, while that of two fermions is antisymmetric. Therefore, for odd partial waves, the wave function of the two bosons vanishes, while for even partial waves, the wave function of two fermions vanishes. As a result, only even partial waves are allowed for scattering between bosons, and only odd partial waves are allowed for fermion–fermion scattering. Because of the symmetrization of the wave functions of two indistinguishable particles, the total cross section for scattering between two spinless bosons or fermions has to be multiplied by a factor of two. For scattering between particles with spin in the absence of any external field, the total cross section has to be multiplied by the spin degeneracy factor. Scattering between particles with spin in an external magnetic field is a topic of multichannel scattering that is discussed in the next chapter.

2.3 Partial-wave Green's function

Let $\tilde{\psi}_{k\ell}(r) = A_k r\psi_\ell(r)$, where A_k is a normalization constant such that

$$\int dr\tilde{\psi}^*_{k\ell}(r)\tilde{\psi}_{k'\ell}(r) = \delta(E - E'). \tag{2.92}$$

Using the method of partial-wave expansion as already discussed, the time-independent Schrödinger equation for $\tilde{\psi}_{k\ell}(r)$ takes the simple one-dimensional (albeit one-sided, since $r \geqslant 0$) form

$$\left[-\frac{\hbar^2}{2\mu}\frac{d^2}{dr^2} + \frac{\hbar^2}{2\mu}\frac{\ell(\ell + 1)}{r^2} + V(r)\right]\tilde{\psi}_{k\ell}(r) = E\tilde{\psi}_{k\ell}(r). \tag{2.93}$$

One can rewrite the partial-wave radial Schrödinger equation in the form

$$\hat{L}_\ell(r)\tilde{\psi}_{k\ell}(r) = 0, \tag{2.94}$$

where

$$\hat{L}_\ell(r) = -\frac{d^2}{dr^2} + \frac{\ell(\ell + 1)}{r^2} + U(r) - k^2. \tag{2.95}$$

The ℓth-wave Green's function $G_\ell(r, r')$ of equation (2.94) is a symmetric function of the two radial coordinates r and r', defined as the solution of the inhomogeneous equation

$$\hat{L}_\ell(r)G_\ell(r, r') = \frac{2\mu}{\hbar^2}\delta(r - r'). \tag{2.96}$$

Let $\tilde{\psi}^{(1)}_{k\ell}(r)$ and $\tilde{\psi}^{(2)}_{k\ell}(r)$ be the two linearly independent solutions of equation (2.94). We choose the function $\tilde{\psi}^{(1)}_{k\ell}(r)$ as the one defined by its boundary condition $\tilde{\psi}^{(1)}_{k\ell}(r = 0) = 0$ at $r = 0$, while $\tilde{\psi}^{(2)}_{k\ell}(r)$ is defined by its boundary condition at $r \to \infty$. The function $\tilde{\psi}^{(1)}_{k\ell}(r)$ that vanishes at $r = 0$ is called the regular solution, while the function $\tilde{\psi}^{(2)}_{k\ell}(r)$, which is defined only by its long-range behavior, is called the irregular solution. For free-particle scattering, $(U(r) = 0)$, $\tilde{\psi}^{(1)}_{k\ell}(r)$ and $\tilde{\psi}^{(2)}_{k\ell}(r)$ reduce to the energy-normalized free wave functions

$$\phi^{(1)}_{k\ell}(r) = \sqrt{\frac{2\mu}{\pi\hbar^2 k}}\, krj_\ell(kr) \tag{2.97}$$

and

$$\phi^{(2)}_{k\ell}(r) = \sqrt{\frac{2\mu}{\pi\hbar^2 k}}\, krn_\ell(kr), \tag{2.98}$$

respectively. The Green's function is constructed in the following way:

$$\begin{aligned} G_\ell(r, r') &= -\pi\tilde{\psi}^{(1)}_{k\ell}(r)\tilde{\psi}^{(2)}_{k\ell}(r'), \quad r < r' \\ &= -\pi\tilde{\psi}^{(1)}_{k\ell}(r')\tilde{\psi}^{(2)}_{k\ell}(r) \quad r > r'. \end{aligned} \tag{2.99}$$

We will now prove that the Green's function as defined by the above equation indeed satisfies equation (2.96). The Green's function of equation (2.99) is finite and continuous, but its derivative is discontinuous at $r = r'$. The first derivative of the Green's function $\frac{d}{dr}G_\ell(r, r')$ exhibits a step function at $r = r'$. This means that $\frac{d^2}{dr^2}G_\ell(r, r')$ is infinite at $r = r'$, while its integration over an infinitesimally small radius around $r = r'$ should be finite. By integrating both sides of equation (2.96) from $r = r' - \Delta$ to $r = r' + \Delta$, where Δ is infinitesimally small, one obtains

$$\tilde{\psi}^{(1)}_{k\ell}(r)\frac{d}{dr}\tilde{\psi}^{(2)}_{k\ell} - \tilde{\psi}^{(2)}_{k\ell}(r)\frac{d}{dr}\tilde{\psi}^{(1)}_{k\ell} = \frac{2\mu}{\hbar^2}. \tag{2.100}$$

Here, we have used the fact that $[U(r) - k^2]G_\ell(r, r')$ is finite at $r = r'$, and therefore its integration vanishes in the limit $\Delta \to 0$. The left-hand side of the above equation is the Wronskian $W(r)$ of equation (2.94). The fact that $W(r)$ is independent of r is a consequence of the linear independence of the two functions. This proves that the Green's function, as defined in equation (2.99), satisfies equation (2.96).

Depending on the asymptotic boundary condition imposed on $\tilde{\psi}^{(2)}_{k\ell}(r)$, one can define three types of Green's functions: $G^{\pm}_{\ell}(r, r')$ and $G^{0}_{\ell}(r, r')$. Here, the superscript $\pm$ denotes whether the outgoing (+) or ingoing (-) asymptotic boundary condition is imposed on $\phi_{k\ell}(r)$, while the superscript '0' implies a standing-wave boundary condition. Thus, the Green's function $G^{0}_{\ell}(r, r')$ for a free particle is given by

$$
\begin{aligned}
G^{0}_{\ell}(r, r') &= -\frac{2\mu}{\hbar^2} k r^2 j_\ell(kr) n_\ell(kr'), \quad r < r' \\
&= -\frac{2\mu}{\hbar^2} k r^2 j_\ell(kr') n_\ell(kr), \quad r > r'.
\end{aligned} \tag{2.101}
$$

Equation (2.101) can be conveniently expressed in the following form:

$$
G^{0}_{\ell}(r, r') = -\frac{2\mu}{\hbar^2} k r^2 j_\ell(kr_<) n_\ell(kr_>), \tag{2.102}
$$

where $r_<$ ($r_>$) is either r or r', whichever is smaller (larger) than the other.

The partial-wave scattering wave function $\psi_\ell(r)$ can be expressed in the form of a partial-wave Lippmann–Schwinger equation with the help of a partial-wave Green's function in the following form:

$$
\psi_\ell(r) = j_\ell(kr) + \int_0^\infty dr' G^{0}_{\ell}(r, r') V(r') \psi_\ell(r'), \tag{2.103}
$$

which satisfies the standing-wave boundary condition in the limit $r \to \infty$. Similarly, using the other two free-particle Green's functions, one can write the wave functions that satisfy outgoing and ingoing asymptotic boundary conditions.

The free-particle 3D Green's function $G^{+}_{0}(\mathbf{r}, \mathbf{r}')$ given in equation (2.7) can be expanded in terms of the free-particle ($U(r) = 0$) partial-wave Green's functions $G^{+}_{\ell}(r, r')$. $G_0(\mathbf{r}, \mathbf{r}')$ satisfies the equation

$$
\left[-\frac{\hbar^2}{2\mu} \nabla^2 - E\right] G^{+}_{0}(\mathbf{r}, \mathbf{r}') = \delta^3(\mathbf{r} - \mathbf{r}'). \tag{2.104}
$$

Defining

$$
G_k(\mathbf{r} - \mathbf{r}') = \frac{\hbar^2}{2\mu} G^{+}_{0}(\mathbf{r} - \mathbf{r}') = -\frac{1}{4\pi} \frac{\exp ik|\mathbf{r} - \mathbf{r}'|}{|\mathbf{r} - \mathbf{r}'|}, \tag{2.105}
$$

we can rewrite equation (2.104) in the form

$$
[-\nabla^2 - k^2] G_k(\mathbf{r} - \mathbf{r}') = \delta^3(\mathbf{r} - \mathbf{r}'). \tag{2.106}
$$

From here onward, we only consider the outgoing boundary condition, omitting the superscript.

2.4 Wigner threshold laws

For single-channel scattering by a central potential, the wave number or energy dependence of the partial-wave scattering amplitude $f_\ell(k)$ and phase shift $\delta_\ell(k)$ follows some definite laws near the threshold of the potential, i.e. in the limit $k \to 0$ or $E \to 0$. Here, the limit $E \to 0$ implies that E is much smaller than the absolute magnitude of the effective potential $Ve_{\text{eff}} = |V(r) + \hbar^2\ell(\ell+1)/(2\mu r^2)$ for large r, such that $kr \to 0$. In what follows, we will derive these laws, which are known as Wigner threshold laws.

Derivation of the Wigner threshold laws

To derive the Wigner threshold laws, we start with the expression for $f_\ell(k)$ from equation (2.40), which is

$$f_\ell(k) = -\int r^2 dr j_\ell((kr))U(r)\psi_\ell(r) = -\pi\frac{T_\ell(k)}{k}, \tag{2.107}$$

where $T_\ell(k)$ is given by

$$T_\ell(k) = \int dr \phi_{k\ell}^{(1)}(r)V(r)\widetilde{\psi}_{k\ell}(r). \tag{2.108}$$

The general on-the-energy-shell **T**-matrix element $T_{\ell m_\ell,\ell' m_{\ell'}}(k)$ for the interaction Hamiltonian operator $\hat{V}$ is given by

$$T_{\ell m_\ell,\, \ell' m_{\ell'}}(k) = \langle E, \ell, m_\ell|\hat{V}|E, \ell', m_{\ell'}; +\rangle, \tag{2.109}$$

where $|E, \ell, m_\ell\rangle$ is the energy-normalized free ($\hat{V} = 0$) state, and $|E, \ell, m_\ell; +\rangle$ represents an energy-normalized state with $\hat{V} \neq 0$. For an isotropic interaction Hamiltonian, $T_{\ell m_\ell,\ell' m_{\ell'}}(k)$ vanishes unless $\ell = \ell'$ and is independent of the magnetic quantum number m_ℓ; thus, it reduces to $T_\ell(k)$.

The energy-normalized wave function $\widetilde{\psi}_{k\ell}(r)$ can be calculated using the integral equation (2.103) and is given by

$$\widetilde{\psi}_{k\ell}(r) = \phi_{k\ell}^{(1)}(r) + \int_0^\infty dr' G_\ell^0(r, r')V(r')\widetilde{\psi}_{k\ell}(r').$$

The Green's function $G_\ell^0(r, r')$ is given by equation (2.102). Using the boundary conditions

$$j_\ell(x \to 0) \sim x^\ell, \tag{2.110}$$

$$n_\ell(x \to 0) \sim -x^{-\ell-1}, \tag{2.111}$$

it can be readily proven that $G_\ell^0(r, r')$ is independent of k in the limit $k \to 0$. So, the k-dependence of $\widetilde{\psi}_{k\ell}(r)$ for $kr \to 0$ is the same as that of $\phi_{k\ell}^{(1)}(kr \to 0)$. If $V(r)$ goes to zero for $r \geqslant r_0$, where r_0 stands for the range of the potential, or if $V(r)$ is exponentially bounded, then for $kr_0 \ll 1$ one can deduce from equation (2.40) that

$$f_\ell(k) \simeq -a_\ell k^{2\ell}, \tag{2.112}$$

where a_ℓ is a constant known as the scattering length. Since $f_\ell(k) = \frac{1}{k\cot\delta_\ell - ik} = e^{i\delta_\ell}\sin\delta_\ell/k$, $f_\ell(k)$ can be approximated only when $|\cot\delta_\ell(k)| \gg 1$, or in other words, $|\delta_\ell(k)| \ll \pi/2$. In such a case, one can approximate $e^{i\delta_\ell} \simeq 1$ and $\sin\delta_\ell(k) \simeq \delta_\ell(k)$ and write

$$\delta_\ell(k) \simeq -a_\ell k^{2\ell+1}. \tag{2.113}$$

Relation (2.112) or (2.113) is one of the Wigner threshold laws that applies to scattering by a short-range potential. Note that, since $f_\ell(k)$ has the dimension of length, $f_\ell(k)$ is reduced to a constant $-a_0$ only for $\ell = 0$; that is, for the s-wave only. In this case, a_0 has the dimension of length. For $\ell > 0$, although a_ℓ is known as the scattering length, it has different dimensions from those of length; for instance, for $\ell = 1$, it has the dimension of volume, and for $\ell = 2$, the dimension of a_ℓ is length to the power of five.

If the potential has the power-law behavior $V(r) = C_n/r^n$ in the limit $r \to \infty$, where C_n is a constant, then in order to derive the low-energy behavior of the phase shift, one has to use the following bounds [4]:

$$|xj_\ell(x)| \leqslant \left(\frac{x}{1+x}\right)^{\ell+1} \tag{2.114}$$

$$|xn_\ell(x)| \leqslant \left(\frac{x}{1+x}\right)^{-\ell}. \tag{2.115}$$

Suppose that $V(r)$ assumes a power-law form for $r \geqslant r_0$ with $kr_0 \ll 1$. Let us split the integral into $\int_0^\infty \cdots dx = \int_0^{x_0} \cdots dx + \int_{x_0}^\infty \cdots dx$, where $x_0 = kr_0$. The contribution from the part $\int_0^{x_0} \cdots dx$ is quite small, since the regular wave functions (physical solutions) oscillate strongly at short separations and go to zero as $r \to 0$. This means that at low energy, the scattering properties are primarily determined by the long-range behavior of the potential, while the details of the short-range part of the potential are not significant. Using the bounds given above, one can then write

$$|f_\ell(k)| \leqslant AC_n k^{n-3}\int_{x_0}^\infty x^{2+2\ell-n}dx\left(\frac{1}{1+x}\right)^{2\ell}, \tag{2.116}$$

where A is a constant. If $2 + 2\ell - n < -1$ or $\ell < (n-3)/2$, then upon carrying out the integration, one again obtains

$$f_\ell(k) \simeq -a_\ell k^{2\ell}, \quad \ell < (n-3)/2 \tag{2.117}$$

or $\delta_\ell(k) \simeq -a_\ell k^{2\ell+1}$ for $\delta_\ell(k) \ll \pi/2$. If $\ell \geqslant (n-3)/2$, then performing the integration on the rhs of equation (2.116) by parts, considering $x^{2+2\ell-n}$ as the first part and $1/(1+x)^{2\ell}$ as the second part, one arrives at the result

$$f_\ell(k) \sim k^{n-3}, \quad \ell \geqslant (n-3)/2, \tag{2.118}$$

which implies that

$$\delta_\ell(k) \sim k^{n-1}, \quad \ell \geqslant (n-3)/2. \tag{2.119}$$

Scattering length

We have already defined the scattering length a_ℓ using the Wigner threshold law, as expressed in relation (2.112) or equivalently (2.113). This formulation applies to short-range potentials or potentials that change according to $\sim 1/r^n$ as $r \to \infty$, where $n \geqslant 3$ and $\ell < (n-3)/2$. Equation (2.112) shows that the scattering amplitude $f_0(k)$ for $\ell = 0$ (an s-wave) is a constant (independent of k) and is equal to the negative of the s-wave scattering length $a_s \equiv a_0$ in the limit $k \to 0$. This means that at a very low energy, the total scattering amplitude is almost equal to $-a_s$, since the contribution to the amplitude from higher partial waves is vanishingly small, as dictated by the Wigner threshold laws. This fact has important consequences for the mean-field description of dilute quantum gases at low energy, as we will discuss in subsequent chapters.

The s-wave scattering length a_s has a simple geometrical interpretation. For $r > r_0$, the s-wave scattering wave function $\psi_0(r) \sim \sin(kr + \delta_0)$, which, in the limit $k \to 0$, may be approximated as

$$\psi_0(r) \simeq \sin(kr - ka_s) \simeq k(r - a_s), \tag{2.120}$$

which is a straight line with a_s being the intercept of the straight line on the x-axis (r-axis). In the weak potential limit, a_s is negative for an attractive potential; accordingly, the intercept occurs on the negative side of the x-axis. Similarly, for a repulsive potential, a_s is positive, and so the straight line makes the intercept on the positive side of the x-axis. However, for a sufficiently strong attractive potential, a_s may be large and positive, indicating the presence of a scattering resonance that is closely related to a bound state in the system. We will discuss the possible connections between resonances and bound states in chapter 4.

Effective range expansion

According to the Wigner threshold laws, at low energy, $\delta_\ell(k) \simeq -a_\ell k^{2\ell+1}$ for a short-range potential is an odd function of k in the limit $k \to 0$. Consequently, the low-energy **K**-matrix element

$$K_\ell(k) = \tan \delta_\ell \tag{2.121}$$

is also an odd function of k. Therefore, the function $k^{-2\ell-1}\tan\delta_\ell$ or the function $k^{2\ell+1}\cot\delta_\ell$ must be an even function of k. In the low-energy limit, $K_\ell(k)\simeq kf_\ell(k)$. Keeping the next higher-order term in the expansion of $k^{-2\ell}f_\ell(k)$, we may write

$$k^{-2\ell}f_\ell(k)\simeq -a_\ell - b_\ell k^2, \tag{2.122}$$

where b_ℓ is a constant. Thus, we can write

$$k^{2\ell+1}\cot\delta_\ell = \frac{1}{k^{-2\ell}f_\ell(k)} = -\frac{1}{a_\ell}+\frac{1}{2}r_{\mathrm{eff},\,\ell}k^2, \tag{2.123}$$

where $r_{\mathrm{eff},\ell}=2b_\ell/a_\ell^2$ is called the ℓ-wave effective range. This effective range expansion is most relevant and useful for the s-wave. We henceforth consider the effective range expansion for the s-wave only and suppress the index ℓ.

The physical meaning of the effective range can be understood in the following way. Let $\psi(k, r)$ be the solution of the s-wave Schrödinger equation with potential $V(r)$ and wave number k. Upon multiplying the Schrödinger equation for $\psi(k, r)$ by $\psi(0, r)$, multiplying the Schrödinger equation for $\psi(0, r)$ by $\psi(k, r)$, and subtracting the resulting latter equation from the former one, one obtains

$$\frac{d}{dr}W[\psi(0, r), \psi(k, r)] = -k^2\psi(0, r)\psi(k, r), \tag{2.124}$$

where $W[X, Y]$ is the Wronskian between the two functions $X(r)$ and $Y(r)$. In the case of large r, $\psi(k, r)$ has the asymptotic form

$$\psi(k, r\to\infty) = C\sin[kr+\delta_0(k)] = \phi(k, r), \tag{2.125}$$

where C is a constant. We now construct a similar Wronskian equation for the functions $\phi(k, r)$ and $\phi(0, r)$:

$$\frac{d}{dr}W[\phi(0, r), \phi(k, r)] = -k^2\phi(0, r)\phi(k, r). \tag{2.126}$$

Subtracting equation (2.126) from (2.124) and integrating the resulting equation over r from $r=0$ to $r=\infty$, we obtain

$$W[[\phi(0, r), \phi(k, r)]_{r=0}]_{r=0} = -k^2\int_0^\infty dr[\psi(0, r)\psi(k, r) - \phi(0, r)\phi(k, r)], \tag{2.127}$$

where we have used the boundary condition $\psi(k, r=0)=0$ for all k. Assuming $C=1/\sin\delta_0$, for $kr\ll 1$ the wave function $\phi(k, r)$ reduces to

$$\phi(k, r) = 1 + rk\cot\delta_0. \tag{2.128}$$

Taking the limit $k\to 0$ and using the Wigner threshold law, we can further write

$$\phi(0, r) = 1 - \frac{r}{a_s}. \tag{2.129}$$

Thus, we have

$$W[[\phi(0, r), \phi(k, r)]_{r=0}]_{r=0} = k \cot \delta_0 + \frac{1}{a_s}. \tag{2.130}$$

Therefore, we can write

$$k \cot \delta_0 = -\frac{1}{a_s} + k^2 \int_0^\infty dr[\phi(0, r)\phi(k, r) - \psi(0, r)\psi(k, r)]. \tag{2.131}$$

By making a low-energy expansion of the integral, keeping only the leading-order term, and comparing the resulting equation with equation (2.123), we obtain

$$r_{\text{eff}} = 2 \int_0^\infty dr[\phi(0, r)^2 - \psi(0, r)^2], \tag{2.132}$$

where $r_{\text{eff}} \equiv r_{\text{eff},0}$. Now, if the potential vanishes for $r > r_0$ then we can restrict the integration on the rhs extending up to r_0, since for $r > r_0$, we have $\psi(k, r) = \phi(k, r)$. The effective range expansion was first derived by Schwinger [5] in the context of proton–neutron scattering. A simpler derivation of the expansion was given by Bethe [6]. The effective range expansion is also called the shape-independent approximation because it does not depend on the shape of the potential. It is particularly useful if the potential is strong, such that the integral at low energy is insensitive to k, as long as k remains much smaller than a characteristic wave number that determines the strength of the potential. For this reason, the effective range expansion is widely applied in low-energy nuclear collisions such as neutron–proton scattering, where the two-body potential is strong.

2.5 Scattering by a short-range potential

A short-range potential $V(r)$ is one that either goes to zero exponentially, has a faster than exponential falloff, or becomes zero after a finite separation $r = r_0$ known as the range of the potential. A paradigmatic model short-range potential is the rectangular well or barrier used to illustrate some of the basic features of low-energy scattering by a short-range potential and schematically shown in figures 2.1 and 2.2.

Rectangular well and barrier

Consider the potential

$$V(r) = \pm V_0, \quad r \leqslant r_0, \tag{2.133}$$

$$=0 \quad r > r_0, \tag{2.134}$$

Let us define two wave numbers

$$k_0 = \frac{\sqrt{2m(E \mp V_0)}}{\hbar}, \quad k = \frac{\sqrt{2mE}}{\hbar}, \tag{2.135}$$

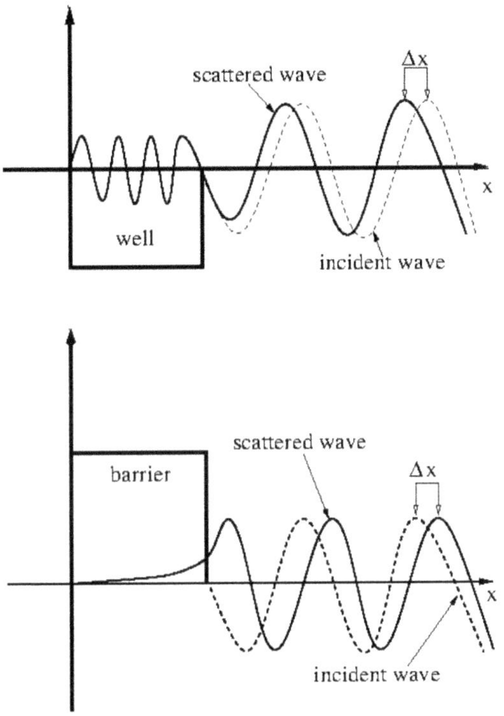

Figure 2.1. Schematic showing the scattering of waves by a rectangular well and a barrier for a finite collision energy. Here, $x = kr$ and $\Delta x = k(r_i - r_f)$ is the phase shift, where r_i and r_f refer to the position of a crest or a trough of the incident and scattered sinusoidal waves, respectively, in the asymptotically large r regime. A wave scattered by a weak potential well has a positive phase shift, while one scattered by a barrier has a negative phase shift. In other words, a weak potential well (equivalently, an attractive potential) pulls the incident wave toward the origin, while a weak potential barrier (or a repulsive potential) pushes the incident wave away from the origin.

which correspond to the inner-range and outer-range wave numbers, respectively. Here, we consider scattering for the s-wave, that is, for $\ell = 0$:

$$\phi_0^{\mathrm{out}}(r) = A[\sin(kr) + \tan\delta_0 \cos(kr)], \quad r > r_0 \tag{2.136}$$

$$\phi_0^{\mathrm{in}}(r) = B\sin(k_0 r), \quad r \leqslant r_0. \tag{2.137}$$

Since both ϕ_0^{out} and ϕ_0^{in} must be the same and continuous at the boundary $r = r_0$, we have

$$\left\{\frac{1}{\phi_0^{\mathrm{in}}}\frac{d\phi_0^{\mathrm{in}}}{dr}\right\}_{r=r_0} = \left\{\frac{1}{\phi_0^{\mathrm{out}}}\frac{d\phi_0^{\mathrm{out}}}{dr}\right\}_{r=r_0}. \tag{2.138}$$

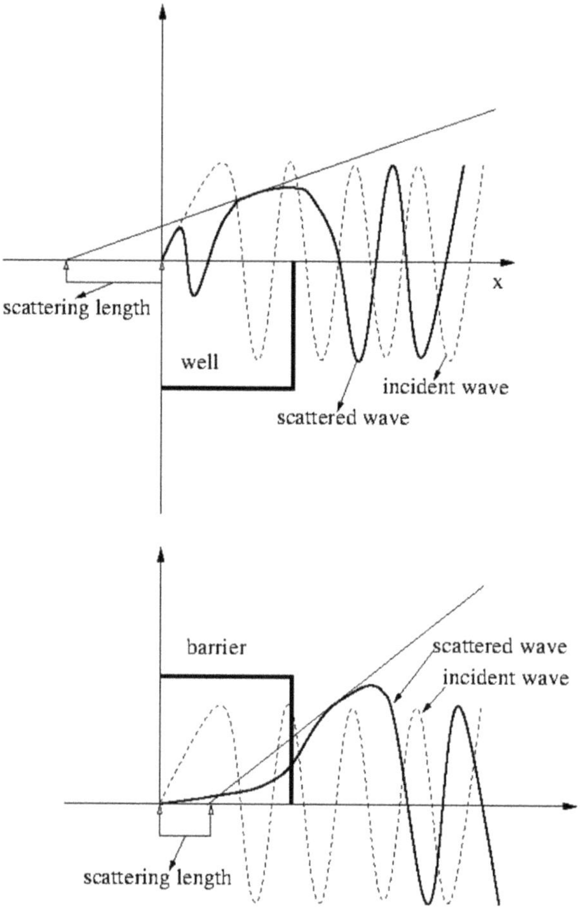

Figure 2.2. Schematic showing s-wave scattering by a rectangular well and a barrier in the zero-energy limit. Also illustrated is the geometric significance of the scattering length. For a weak potential well (barrier), the scattering length is negative (positive).

The above equation can be conveniently put into the form

$$\left\{\frac{d}{dr}ln\phi_0^{\text{in}}\right\}_{r=r_0} = \left\{\frac{d}{dr}ln\phi_0^{\text{out}}\right\}_{r=r_0}; \tag{2.139}$$

that is, the logarithmic derivatives of the inner and outer wave functions must be the same at the boundary of the well or barrier. Equation (2.139) provides

$$k_0 \cot(k_0 r_0) = \frac{k[\cot(kr_0) - \tan\delta_0]}{1 + \tan\delta_0 \cot(kr_0)}. \tag{2.140}$$

The above equation gives

$$\cot\delta_0 = \frac{k + k_0 \cot(kr_0)\cot(k_0 r_0)}{k\cot(kr_0) - k_0 \cot(k_0 r_0)}, \tag{2.141}$$

which yields

$$\delta_0 = kr_0 - \cot^{-1}\left\{\frac{k_0}{k}\cot(k_0 r_0)\right\}. \tag{2.142}$$

To interpret the results, let us first distinguish between weak and strong potentials. For this purpose, we first define a characteristic energy scale $E_c = \hbar^2/(2\mu r_0^2)$. If $V_0 < E_c$, then we call it a weak potential; otherwise, it is a strong potential. The low-energy limit of scattering generally implies $kr_0 \ll 1$ and $E \ll V_0$, which in turn provide

$$k_0 \simeq \kappa_0 + \frac{k^2}{2\kappa_0}, \tag{2.143}$$

where $\kappa_0 = \sqrt{\mp 2\, mV_0}/\hbar$. Obviously, for a potential well, κ_0 is real, while for a potential barrier, it is purely imaginary. Low-energy scattering due to a weak potential well occurs under the condition $|\kappa_0| r_0 < \pi/2$.

Here, we discuss scattering due to a weak potential only. Scattering due to a strong potential for which $|\kappa_0| r_0 \geqslant \pi/2$ leads to scattering resonances and the appearance of bound states, which we will discuss in chapter 4. To comprehend some of the qualitative behavior of $\delta_0(k)$ as given by (2.141) due to a weak potential in the low-energy limit, we first consider the case of a potential well for which $k_0 > k$. For a very weak potential well, the denominator of (2.141) is positive, while the numerator is always positive, meaning that $\cot\delta_0 > 0$ or $\delta_0 > 0$. We next consider the case of a potential barrier, for which k_0 is purely imaginary in the low-energy limit. Thus, $k_0\cot(k_0 r_0) = |k_0|\coth(k_0 r_0)$ is real. Since, in this case, $k > |k_0|$, we have $\cot\delta_0 < 0$, implying that $\delta_0 < 0$. Thus, we infer that for scattering due to a weak potential in the low-energy limit, δ_0 is positive or negative if the potential is a well or a barrier, respectively; in other words, δ_0 is positive or negative if the potential is attractive or repulsive, respectively.

For a potential well, by using the limiting values $\lim_{k\to 0} k\cot(kr_0) \simeq 1/r_0$ and $k_0 \simeq \kappa_0$, one obtains

$$\lim_{k\to 0} k\cot\delta_0(k) = -\frac{1}{a_s}, \tag{2.144}$$

where a_s is called the s-wave scattering length, which is given by

$$-\frac{1}{a_s} = \frac{\kappa_0\cot(\kappa_0 r_0)}{1 - \kappa_0 r_0\cot(\kappa_0 r_0)}. \tag{2.145}$$

In the limit $|\kappa_0| r_0 \ll 1$, a_s is negative and its magnitude is quite small, reflecting the fact that the potential is very weak. On the other hand, for a potential barrier, we have $\kappa_0\cot(\kappa_0 r_0) = |\kappa_0|\coth(|\kappa_0| r_0)$ and a_s becomes positive. Thus, for a weak potential, a_s is positive (negative) for a repulsive (attractive) potential. The s-wave scattering amplitude is given by

$$f_0(k) = \frac{1}{k\cot\delta_0(k) - ik}. \tag{2.146}$$

Substituting equation (2.144) into the above equation, one obtains

$$f_0 k = -\frac{a_s}{1 + ika_s}. \tag{2.147}$$

In the limit $|ka_s| \ll 1$, $f_0 \simeq -a_s$ and the s-wave scattering cross section is $\sigma_0 \simeq 4\pi a_s^2$.

Before ending this subsection, we mention that the procedure above allows one to deduce the phase shift δ_ℓ for any partial wave ℓ from the logarithmic derivative boundary condition

$$\frac{d}{dr}\ln\,[rj_l(k_0 r)]_{r=r_0} = \frac{d}{dr}\ln\,[rj_l - rn_l \tan\delta_l]_{r=r_0}. \tag{2.148}$$

Rigid sphere

Let us consider scattering by a rigid sphere of radius R_0, which is basically scattering by a potential of the form

$$\begin{aligned} V(r) &= \infty, \quad r \leqslant R_0 \\ &= 0 \quad r > R_0. \end{aligned} \tag{2.149}$$

Obviously, the boundary condition for the wave function $\psi_\ell(r) = 0$ at $r = R_0$, while for $r > R_0$, it becomes a free wave, which may be written in the form

$$\psi_\ell(r) = A[j_\ell(kr) - K_\ell(k)n_\ell(kr)], \tag{2.150}$$

where A is a constant and $K_\ell(k)$ is a **K**-matrix element. Since $\psi_\ell(R_0) = 0$, we must have $j_\ell(kR_0) - K_\ell(k)n_\ell(kR_0) = 0$, which implies

$$K_\ell(k) = \tan\delta_\ell(k) = \frac{j_\ell(kR_0)}{n_\ell(kR_0)}. \tag{2.151}$$

For $kR_0 \ll 1$, using the limiting values of the functions $j_\ell(x)$ and $n_\ell(x)$ in the limit $k \to 0$, we obtain

$$\tan\delta_\ell(k) \simeq -\frac{(kR_0)^{2\ell+1}}{(2\ell+1)!!(2\ell-1)!!}, \tag{2.152}$$

which shows that $a_s = R_0$. In the limit $k \to 0$, the s-wave scattering amplitude is most significant, and the total scattering cross section is indeed almost equal to $4\pi a_s^2$. For scattering by a rigid sphere of radius R_0, the total low-energy cross section is approximately equal to $4\pi R_0^2$; thus, the low-energy scattering due to a short-range repulsive potential is equivalent to the scattering produced by a rigid sphere of radius a. This means that the inner details of the potential for $r < a_s$ do not matter much in low-energy scattering by a short-range potential.

For rigid-sphere scattering, as $k \to \infty$, $K_\ell(k)$ is given by

$$\tan\delta_\ell(k) \simeq \tan(kR_0 - \ell\pi/2), \tag{2.153}$$

which shows that $\delta_\ell(k)$ diverges for $k \to \infty$.

Zero-range potentials

Zero-range potentials are model potentials of the form

$$V_{zr}(\mathbf{r}) = g\delta^3(\mathbf{r}). \tag{2.154}$$

Here, g is the strength of the potential, which is usually independent of energy, and $\delta^3(\mathbf{r})$ stands for the 3D delta function. This potential is also called the contact potential, as it is finite only at $r = 0$. At this potential, the Schrödinger equation reads

$$[\nabla^2 + k^2]\psi(\mathbf{r}) = \frac{2\mu g}{\hbar^2}\psi(\mathbf{r})_{\mathbf{r}\to 0}\delta^3(\mathbf{r}). \tag{2.155}$$

Clearly, the quantity $\alpha = \frac{2\mu g}{\hbar^2}$ has the dimension of length. Using a Green's function and the Lippmann–Schwinger equation, we can write the scattering solution

$$\psi(\mathbf{r}) = \frac{1}{(2\pi)^{3/2}}e^{i\mathbf{k}\cdot\mathbf{r}} - \frac{1}{4\pi}\alpha\frac{e^{ikr}}{r}\psi(\mathbf{r})_{\mathbf{r}\to 0}. \tag{2.156}$$

Under the first-order Born approximation, one can replace $\psi(\mathbf{r})$ on the rhs of the above equation by the plane wave $\frac{1}{(2\pi)^{3/2}}e^{i\mathbf{k}\cdot\mathbf{r}}$, resulting in

$$\psi(\mathbf{r}) = \frac{1}{(2\pi)^{3/2}}\left[e^{i\mathbf{k}\cdot\mathbf{r}} - \frac{1}{4\pi}\alpha\frac{e^{ikr}}{r}\right]. \tag{2.157}$$

Comparing this equation with the asymptotic boundary condition of equation (2.20), we can associate the quantity $-\frac{1}{4\pi}\alpha$ with the scattering amplitude $f_{\mathbf{k}',\mathbf{k}}$. Since g is a constant by assumption, this first-order Born-approximated scattering amplitude is also a constant. This is only possible at very low energies, where s-wave scattering is dominant according to the Wigner threshold laws. Therefore, $-\alpha/4\pi$ can be replaced by the s-wave scattering amplitude $-a_s$. Thus, we obtain

$$g = \frac{4\pi\hbar^2 a_s}{2\mu}. \tag{2.158}$$

For a delta-function-type contact potential, the higher-order terms in the Born approximation diverge. This divergence problem can be overcome if one first expands the wave function $\psi(\mathbf{r})$ around $\mathbf{r} = 0$ up to the first order in the following way:

$$\psi(\mathbf{r}) \simeq (1 + \mathbf{r}\cdot\boldsymbol{\nabla})\psi(\mathbf{r}) \tag{2.159}$$

and then takes the limit $\mathbf{r} \to 0$. If g is isotropic (which is the usual case), then the wave function is also isotropic; in which case, $\psi(\mathbf{r})$ can be replaced by $\frac{1}{\partial r}\mathbf{r}\psi(\mathbf{r})$. So, an improved or regularized zero-range potential can be defined as

$$V_{zr}^{reg}(\mathbf{r}) = g\delta^3(\mathbf{r})\frac{1}{\partial r}r\cdot \tag{2.160}$$

One can calculate all the Born terms if $k\,|a_s| < 1$. Therefore, the scattering amplitude is given by

$$f(k) = -\frac{a_s}{1 + ika_s}. \tag{2.161}$$

Earlier, we derived the partial-wave scattering amplitude in terms of the partial-wave phase shift. In terms of the s-wave phase shift δ_0, the s-wave scattering amplitude is given by

$$f_0(k) = \frac{1}{k \cot \delta_0 - ik}. \tag{2.162}$$

If the effective range vanishes, then in the limit $k \to 0$, we have $k \cot \delta_0 = -\frac{1}{a_s}$. The perturbatively calculated scattering amplitude $f(k)$ becomes equal to $f_0(k)$. Thus, at sufficiently low energies, where the Winger threshold laws apply, one can resort to a zero-range potential of the form

$$V_{\text{zr}} = \frac{4\pi\hbar^2 a_s}{2\mu}\delta^3(\mathbf{r}) \tag{2.163}$$

or its regularized form

$$V_{\text{zr}} = \frac{4\pi\hbar^2 a_s}{2\mu}\delta^3(\mathbf{r})\frac{\partial}{\partial r}r\cdot \tag{2.164}$$

In the above analysis, we have shown that if there exists a zero-range potential, then its scattering amplitude is essentially that of the s-wave and therefore a constant. Here, we prove that if the total scattering amplitude is approximated by $-a_s$, then the corresponding potential is a zero-range potential. The fact that the total scattering amplitude is a constant implies that the scattering occurs deep in the Wigner threshold law regime, where one can ignore all partial-wave scattering except for the s-wave amplitude. If one equates this constant scattering amplitude with the first-order Born-approximated amplitude at low energy then it follows that

$$T^{(1)}(q) = -\frac{1}{4\pi}\frac{2\mu}{\hbar^2}\lim_{\mathbf{q}\to 0}\int d^3\mathbf{r}\, V(r)e^{i\mathbf{q}\cdot\mathbf{r}} = -a_s. \tag{2.165}$$

The above equation shows that a_s is essentially related to the Fourier transform of the potential under the first-order Born approximation. So, the potential $V(r)$ is related to the inverse Fourier transform:

$$V(r) = \frac{4\pi\hbar^2 a_s}{2\mu}\frac{1}{(2\pi)^3}\int d^3\mathbf{q}e^{-i\mathbf{q}\cdot\mathbf{r}} = \frac{4\pi\hbar^2 a_s}{2\mu}\delta^3\mathbf{r}. \tag{2.166}$$

Zero-range potentials are also known as contact potentials or Fermi's pseudo-potentials. The mean interaction potential (also called the mean-field potential) of a dilute system of particles undergoing rigid- or hard-sphere interactions can be approximated by a regularized zero-range potential [7, 8]. For a dilute system of rigid-sphere interacting particles, the interparticle separation is so large that their interaction is effectively determined by the asymptotic boundary condition [7]. In this case, the hard-sphere radius can be extended to $r = 0$, leading to the regularized zero-range potential operator as given by equation (2.164).

2.6 Classical theory of collisions

In the classical theory of collisions, one has to calculate the classical trajectories ($\mathbf{r}(t)$) or orbits ($\theta(r)$) of a projectile particle under the influence of the target's force field. This can be done by applying classical Hamiltonian or Lagrangian dynamics, subject to certain conservation constraints. We assume that for $r \to \infty$ and $t \to \pm\infty$, the force field vanishes, implying that the trajectories at $t \to \pm\infty$ are rectilinear. However, for a Coulomb potential, this assumption is not strictly valid, because the Coulomb potential affects the motion even at a large distance. In the case of nonreactive and elastic collisions, the total energy, i.e. the magnitude of the linear and angular momenta, must be conserved (figure 2.3).

The classical theory of collisions is described in terms of the impact parameter b defined by

$$b = \frac{L}{p} = \frac{|\mathbf{r} \times \mathbf{p}|}{p} = r \sin \alpha, \tag{2.167}$$

where α is the angle between $\mathbf{p}$ and $\mathbf{r}$ and $\mathbf{L} = \mathbf{r} \times \mathbf{p}$ is the angular momentum of the projectile particle that is being scattered by a scattering center (target), which is the origin of the coordinate system as shown in figure 2.3. For $r \to \infty$, the velocity of the scattered projectile is directed along $\hat{r}$, and so α is the angle of scattering. Note that by 'scattering center' we mean the center of force that is responsible for

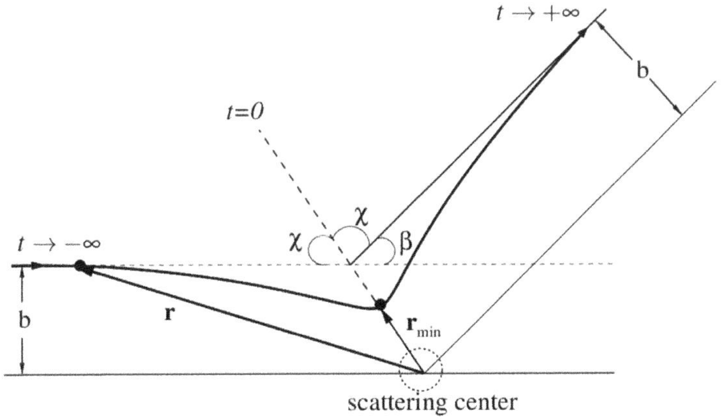

Figure 2.3. Geometry of the classical scattering of a projectile by a target. The center of force, referred to as the scattering center, is the origin of the coordinate frame. The time $t = 0$ is the instant at which the radial distance $r_{\min}$ of the projectile is at a minimum. Asymptotically, at $t \to \pm\infty$, the projectile moves freely in a rectilinear path. b is the impact parameter. $\beta = \pi - 2\chi$ is the deflection angle, where χ is the angle subtended by the incident or outgoing asymptotic rectilinear trajectory with $\mathbf{r}_{\min}$.

deflecting the path or trajectory of the projectile particle. Here, **p** and **r** refer to the momentum and position vector, respectively, of the particle at $t \to \pm\infty$, and $t = 0$ is the instant at which the magnitude of the radial vector is at a minimum (i.e. $r_{\min} = r(t = 0)$) and the radial velocity vanishes. It is assumed that at short distances $r \leqslant r_{\min}$, the force is repulsive. From the geometry in figure 2.3, it is clear that b is the perpendicular distance between the two parallel lines—one along the asymptotic ($t = \pm\infty$) path of the particle's motion and the other passing through the origin of the force. For a fixed target, the conservation of angular momentum implies that b is constant, since p is conserved and $b = L/p$. This means that if one draws a line passing through the origin and parallel to the outgoing ($t = \infty$) rectilinear path, the perpendicular distance between these two lines equals b. As the projectile approaches very close to the target, the radial speed of the projectile is slowed down if the force exerted by the target is repulsive, and the angular momentum starts to increase. As a result, there is a minimum distance of approach where the radial speed vanishes and the particle is deflected away from the target. Let $\mathbf{r}_{\min}$ denote this minimum position vector. Also, let $\mathbf{v}_{\text{in}}$ and $\mathbf{v}_{\text{out}}$ denote the incoming ($t \to -\infty$) and outgoing ($t \to \infty$) velocity vectors, respectively, and let χ represent the angle between $\mathbf{r}_{\min}$ and $-\mathbf{v}_{\text{in}}$. Due to symmetry, the velocity $\mathbf{v}_{\text{out}}$ also makes the same angle χ with $\mathbf{r}_{\min}$. So, from the geometry, it is clear that the deflection angle β between $\mathbf{v}_{\text{in}}$ and $\mathbf{v}_{\text{out}}$ is given by

$$\beta(p, b) = \pi - 2\chi. \tag{2.168}$$

The purpose here is to calculate this angle $\beta(p, b)$ as a function of $p = mv$ and b for elastic collisions, where $v = |\mathbf{v}_{\text{in}}| = |\mathbf{v}_{\text{out}}|$. Note that the deflection angle is not the same as the scattering angle, because β can range from $-\infty$ to π, while the scattering angle α, by definition, lies between 0 and π. Therefore, α and β are related by [9]

$$\beta = \pm\alpha - 2n\pi, \tag{2.169}$$

where n is a positive integer, so that $\pi \geqslant \alpha \geqslant 0$.

Let us consider that the force field is spherically symmetric; that is, we consider scattering due to a central field. In three dimensions, the momentum **p** has three components: the radial component $p_r = m\dot{r}$ and two angular components $p_\theta = r^{-1}I\dot{\theta}$ and $p_\phi = (r \sin\theta)^{-1} I\dot{\phi}$, where $I = mr^2$ is the moment of inertia and m is the mass of the particle. Therefore, the Hamiltonian is given by

$$H = \frac{1}{2m}\mathbf{p}^2 + V(r) = \frac{1}{2}m\left(\frac{dr}{dt}\right)^2 + \frac{L^2}{2I} + V(r) = E, \tag{2.170}$$

where $V(r)$ is the potential energy and

$$L^2 = I^2(\dot{\theta}^2 + \sin\theta^{-2}\dot{\phi}^2) \tag{2.171}$$

is the square of the angular momentum. Since the problem has cylindrical symmetry, the angle ϕ remains constant throughout the scattering process. This means that the initial and final rectilinear trajectories of the particle lie in the same plane that contains the z-axis. If we assume that $V(r \to \pm\infty) = 0$, then E must equal the free-particle energy $E = p^2/(2m)$. We can thus write

$$m\frac{dr}{dt} = \left[p^2 - 2mV(r) - \frac{L^2}{r^2}\right]^{1/2}. \tag{2.172}$$

The minimum distance of approach $r_{\min}$ is given by $\frac{dr}{dt}\,|_{r=r_{\min}} = 0$, implying that

$$p^2 - 2mV(r_{\min}) - \frac{L^2}{r_{\min}^2} = 0, \tag{2.173}$$

which can also be written as

$$\frac{V(r_{\min})}{E} + \frac{b^2}{r_{\min}^2} = 1. \tag{2.174}$$

If the potential is repulsive, then the above equation can have solutions for $b \geqslant 0$, but for an attractive potential, the above equation can have solutions only for $b \neq 0$ and $\frac{b^2}{r_{\min}^2} > |\frac{V(r_{\min})}{E}|$.

We consider the z-axis to be aligned with the velocity or momentum of the incident projectile. Since $\dot{\phi} = 0$ for a spherically symmetric potential, we have $\alpha \equiv \theta$, and the angular momentum L is given by

$$L = mr^2\frac{d\theta}{dt}, \tag{2.175}$$

from which one can express the differential dt as

$$dt = \frac{mr^2}{L}d\theta \tag{2.176}$$

Similarly, from equation (2.172), we have

$$dt = \frac{mdr}{\left[p^2 - 2mV(r) - \frac{L^2}{r^2}\right]^{1/2}}. \tag{2.177}$$

Thus, one can write the orbit equation in the form

$$d\theta = \frac{Ldr}{r^2\left[p^2 - 2mV(r) - \frac{L^2}{r^2}\right]^{1/2}}. \tag{2.178}$$

Since $b = L/p$, one can express the above equation in terms of b as given by

$$d\theta = \frac{bdr}{r^2\left[1 - 2mV(r)/p^2 - \frac{b^2}{r^2}\right]^{1/2}}. \tag{2.179}$$

Therefore, the angle χ between $\mathbf{r}_{\min}$ and $\mathbf{r}(t=\infty)$ is given by

$$\chi = \int_0^{\chi} d\theta = \int_{r_{\min}}^{\infty} \frac{bdr}{r^2\left[1 - 2mV(r)/p^2 - \frac{b^2}{r^2}\right]^{1/2}}. \tag{2.180}$$

If a potential is repulsive everywhere, that is, if $V(r)$ is a monotonically decreasing function of r, such as a repulsive Coulomb potential, then it is obvious that $V(r) < V(r_{\mathbf{min}})$ for $r > r_{\min}$. Consequently, the integral on the rhs of equation (2.180)

$$\int_{r_{\min}}^{\infty} \frac{bdr}{r^2\left[1 - 2mV(r)/p^2 - \frac{b^2}{r^2}\right]^{1/2}} \leqslant \int_{r_{\min}}^{\infty} \frac{bdr}{r^2\left[1 - 2mV(r_{\mathbf{min}})/p^2 - \frac{b^2}{r^2}\right]^{1/2}}. \tag{2.181}$$

Now, substituting $V(r_{\mathbf{min}})$ given by equation (2.174) into the above inequality, the rhs of the inequality can be expressed as

$$\int_{r_{\min}}^{\infty} \frac{dr}{r\left[\frac{r^2}{r_{\min}^2} - 1\right]^{1/2}}. \tag{2.182}$$

Letting $x = r/r_{\min}$, the above integral reduces to

$$\int_1^{\infty} \frac{dx}{x\sqrt{x^2 - 1}} = \frac{\pi}{2}. \tag{2.183}$$

Thus, we find that for such a repulsive potential, $\chi \leqslant \pi/2$, and therefore $0 \leqslant \beta \leqslant \alpha$. For a strongly attractive potential, the projectile can make several orbits around the scattering center at close range before entering the asymptotic regime.

An alternative way to calculate χ is to use the Hamilton–Jacobi equations in terms of the angle θ_r and the action variables $J \equiv L$. The corresponding generating functions $S(r, \theta_r)$ are given by

$$\frac{1}{2m}\left(\frac{dS}{dr}\right)^2 + \frac{J^2}{2mr^2} + V(r) = E \tag{2.184}$$

and

$$\frac{\partial S}{\partial \theta_r} = J. \tag{2.185}$$

These two equations yield

$$S(r, \theta_r) = \int dr\left[2mE - \frac{J^2}{r^2} - 2mV\right]^{1/2} + J\theta_r. \tag{2.186}$$

Differentiating the above equation with respect to J, one obtains

$$\theta_r = \int dr \frac{J}{r^2\left[2mE - \frac{J^2}{r^2} - 2mV\right]^{1/2}}. \tag{2.187}$$

Upon substituting 2 $mE = p^2$ and $J = pb$ into this result, one can readily obtain the same expression as that in equation (2.180).

Let us recall the definition of the differential scattering cross section $\sigma(\theta)$. The quantity $\sigma(\theta)d\Omega$ is the ratio of the number of particles scattered into the solid angle $d\Omega$ per unit time to the incident flux of particles. If $\mathcal{N}$ denotes the incident flux of particles, then the number of particles scattered within the annular area element $2\pi bdb$ per unit time is $2\pi bdb\mathcal{N}$. Since this area element subtends a solid angle of $2\pi \sin\theta$ at the origin, we can write

$$\sigma(\theta)2\pi \sin\theta|d\theta|\mathcal{N} = 2\pi b|db|\mathcal{N}. \tag{2.188}$$

Therefore, we have

$$\sigma(\theta) = \frac{1}{\sin\theta} b \left| \frac{db}{d\theta} \right|. \tag{2.189}$$

The absolute magnitude of $\frac{db}{d\theta}$ is necessary, since $\sigma \geqslant 0$. Thus, in order to derive the functional dependence of σ on θ, it is necessary to express the impact parameter in terms of θ. This can be done by inverting the expression for $\theta = \pi - 2\chi$ in terms of b, where the angle χ can be evaluated using the integral in equation (2.180).

2.7 Coulomb scattering

2.7.1 Classical expression for $\sigma(\theta)$

The classical form of the differential Coulomb scattering cross section can be obtained by applying the method described in the preceding section. For a Coulomb potential of the form $V(r) = C/r$, and writing $u_0 = 1/r_{\min}$, from equation (2.174) we have

$$u_0^2 + \frac{C}{Eb^2}u_0 = \frac{1}{b^2}, \tag{2.190}$$

which yields

$$u_0 = \frac{1}{b}\sqrt{1 + \frac{C^2}{4E^2b^2}} - \frac{C}{2Eb^2}; \tag{2.191}$$

that is,

$$r_{\min} = \frac{b^2}{\sqrt{b^2 + \frac{C^2}{4E^2}} - \frac{C}{2E}}. \tag{2.192}$$

The above equation shows that for $E \gg |C/2b|$, $r_{\min} \simeq b$, and for $E \ll |C/2b|$, $r_{\min} \simeq |C|/(2E)$, if $C > 0$ (repulsive potential). For a Coulomb potential, the integration in equation (2.180) can be carried out analytically [10], resulting in

$$\chi = \cot^{-1}\frac{C}{2Eb}. \tag{2.193}$$

Therefore, we have

$$\theta = \pi - 2\cot^{-1}\frac{C}{2Eb} \tag{2.194}$$

and thus

$$b = \frac{C}{2E}\cot(\theta/2). \tag{2.195}$$

Substituting this into equation (2.189), one obtains

$$\sigma(\theta) = \left[\frac{C}{4E\sin^2(\theta/2)}\right]^2. \tag{2.196}$$

2.7.2 Quantum mechanical expression for $\sigma(\theta)$

The quantum theory of collisions views the collision phenomenon as the deflection of an incoming Schrödinger wave into an outgoing one. The asymptotic boundary conditions that apply to the scattered waves depend on the nature of the potential. We have already discussed the boundary conditions for non-Coulomb potentials. The boundary conditions for Coulomb collisions are quite different from those already discussed.

Let us first understand how the asymptotic boundary condition depends on the nature of the potential [1]. To analyze the asymptotic nature of the ℓ-partial-wave wave function $\chi_\ell = r\psi_\ell(r)$ due to a central potential $V(r)$, we make an ansatz for the asymptotic wave function:

$$\chi_\ell(r \to \infty) \sim Ae^{\pm ikr}\exp\left[\int_0^r f(r')dr'\right], \tag{2.197}$$

where $f(r)$ is an unknown function to be determined, and A is a constant. Substituting this ansatz into the Schrödinger equation, we obtain

$$\left[-\frac{d^2}{dr^2} + \frac{\ell(\ell+1)}{r^2} + U(r)\right]\chi_\ell(r) = k^2\chi_\ell(r). \tag{2.198}$$

If $U(r) = \frac{2\mu}{\hbar^2}V(r)$, one obtains

$$\frac{df}{dr} + f^2 \pm 2ikf = U(r) + \frac{\ell(\ell+1)}{r^2}. \tag{2.199}$$

If $U(r)$ behaves according to $1/r^n$ for $n > 0$, then the last term ($\pm 2ikf$) is the leading term. If $n > 1$, then $\chi_\ell(r)$ reduces to the form $Ae^{\pm ikr}$ for $r \to \infty$; that is, the asymptotic wave function is sinusoidal. If $U(r)$ is an exponentially decreasing function in the asymptotic regime, then both the first and the third terms on the lhs of the above equation are the leading terms, and again the asymptotic wave function behaves sinusoidally. If $V(r) = C/r$ (i.e. we are considering a Coulomb potential), then the asymptotic wave function takes the form

$$\chi_\ell(r) \sim Ae^{\pm i[kr - \nu \ln(kr)]}, \tag{2.200}$$

where $\nu = \frac{2\mu C}{\hbar^2 k}$. Thus, the Coulomb wave function does not behave sinusoidally, even at asymptotically large separations for all partial waves.

Let us consider a partial-wave analysis of scattering between two charged particles 1 and 2 with charges q_1 and q_2, respectively. We assume that the two particles scatter only under the influence of the mutual Coulomb interaction potential $V = C/r$, where $C = q_1 q_2/(4\pi\varepsilon)$. The ℓth partial-wave Schrödinger equation for relative motion can be written as

$$\frac{1}{r}\frac{d^2}{dr^2}r\psi_\ell\left[k^2 - \frac{2\nu k}{r} - \frac{\ell(\ell+1)}{r^2}\right]\psi_\ell(r) = 0, \tag{2.201}$$

where $\nu = \mu C/(\hbar^2 k)$. Substituting $\psi_\ell(r) = r^\ell e^{ikr}\chi_\ell(r)$, we have

$$r\frac{d^2\chi_\ell}{dr^2} + [2ikr + 2(\ell+1)]\frac{d\chi_\ell}{dr} + [2ik(\ell+1) - 2\nu k]\chi_{ell} = 0. \tag{2.202}$$

This is an equation for a confluent hypergeometric function, and its solution that is regular at $r = 0$ is given by

$$\chi_\ell = \mathcal{N}_\ell F(\ell + 1 + i\nu, 2\ell + 2, -2ikr), \tag{2.203}$$

where $\mathcal{N}_\ell$ is a constant chosen so that the total wave function $\Psi(\mathbf{r}) = \sum_\ell \psi_\ell P_\ell(\cos\theta)$ coincides with that expressed in parabolic coordinates.

Let us now derive the Coulomb scattering wave function in parabolic coordinates $\xi = r - z = r(1 - \cos\theta)$, $\eta = r + z = r(1 + cos\theta)$, and ϕ. Using the ansatz $\Psi(\mathbf{r}) = e^{ikz}F(\xi, \eta, \phi)$ in the time-independent Schrödinger equation, we have

$$\xi\frac{d^2F}{d\xi^2} + (1 - ik\xi)\frac{dF}{d\xi} - \nu kF = 0, \tag{2.204}$$

which does not involve ϕ since the potential is spherically symmetric. Comparing this equation with the standard form of the confluent hypergeometric function

$$q\frac{d^2f}{dq^2} + (b - z)\frac{df}{dq} - af, \tag{2.205}$$

we have $q = ik\xi$, $b = 1$, $f = ikF$, and $a = -i\nu$. Thus, the solution is given by

$$F(\xi, \eta) = KF(-i\nu, 1, ik\xi), \tag{2.206}$$

where K is a constant. In the limit $r \to \infty$, $\Psi(\xi, \eta, z)$ behaves like

$$\Psi \underset{r\to\infty}{\longrightarrow} \frac{Ce^{\nu\pi/2}}{\Gamma(1 + i\nu)}[U(r - z)e^{i(kz)+\nu \ln k(r-z)} + r^{-1}f(\theta)e^{i(kr - \nu \ln 2kr)}], \tag{2.207}$$

where $U(r - z) = 1 + \nu^2/[ik(r - z)]$ and

$$f(\theta) = \frac{\Gamma(1 + i\nu)}{i\Gamma(-i\nu)} \frac{e^{-i\nu \ln[\sin^2(\theta/2)]}}{2k\sin^2(\theta/2)}. \tag{2.208}$$

The term associated with $f(\theta)$ on the rhs of equation (2.207) is the scattered wave, which differs from that for a non-Coulomb potential by the factor $e^{-i\nu \ln(2kr)}$. The first term on the rhs corresponds to the incident plane wave. The factor $U(r - z)$ goes to zero for large r. Both the incident and scattered waves are different from those for non-Coulomb potentials by the logarithmic phase factors. The different scattering cross section is given by

$$\begin{aligned} \sigma = |f(\theta)|^2 &= \left[\frac{\nu}{2k\sin^2(\theta/2)}\right]^2 \\ &= \left(\frac{\mu q_1 q_2}{2\hbar^2 k^2}\right)^2 \csc^4(\theta/2). \end{aligned} \tag{2.209}$$

Replacing $\hbar k/\mu$ with the speed v allows one to express

$$\sigma = \left(\frac{q_1 q_2}{2\mu v^2}\right)^2 \csc^4(\theta/2), \tag{2.210}$$

which coincides with Rutherford's formula for classical Coulomb scattering. So, for a Coulomb potential, the total differential scattering cross section has the same form in both classical and quantum scattering theories.

2.8 High vs low-energy and classical vs quantum collisions

Quantum theory is the most general theory for describing natural phenomena. One can go from the quantum world to the classical world under some limiting conditions, but not the other way around. The Ehrenfest theorem provides a prescription for this type of limiting procedure. Whether the classical approximation applies in describing a collision event depends on how the length scale of variation of the target's force experienced by the incoming projectile compares with the de Broglie wavelength, $\lambda_{dB} = h/p$, associated with the wave function of the projectile. Here, p is the magnitude of the asymptotic momentum of the projectile. A measure of the length scale of the variation of the force or potential is the minimum distance of approach, r_{min}, as defined in the previous section. If $\lambda_{dBG} \ll r_{min}$, then the target sees the motion of the projectile more as the classical trajectory of a particle than as wave motion. Since λ_{dB} is inversely proportional to the momentum or velocity, the question arises as to whether classical collision theory implies the high-energy approximation of quantum collision theory.

For a non-Coulomb potential, it is true that classical theory applies to the relatively high-energy regime, while in the low-energy limit, classical theory breaks down. However, for a Coulomb potential, the low-energy limit does not necessarily imply that classical theory breaks down. This can be understood from a comparison between the length scale associated with the Coulomb force in classical scattering and λ_{dB}. For simplicity of analysis, let us consider scattering by a repulsive Coulomb potential ($C > 0$). Since the radial velocity vanishes at $r = r_{min}$, one may consider r_{min} as the length scale of significant variation of the Coulomb force. As equation (2.192) shows, for a repulsive Coulomb potential, r_{min} is lowest for $b = 0$, or equivalently $L = 0$, and it is given by $r_{min} = C/E = 2\,mC/p^2$. Obviously, this length scale increases much faster than λ_{dB} as p decreases. Thus, at low energy, Coulomb scattering can be treated using classical theory. In contrast, for a short-range repulsive potential or a potential with a long-range repulsive tail of the form C_n/r^n with $n > 2$, the corresponding r_{min} always decreases faster than λ_{dB} as $p \to 0$. Therefore, the ratio λ_{dB}/r_{min} determines whether the classical theory of collisions breaks down or not. Classically, the range of a potential may be given by the maximum impact factor b_{max}, above which no particle is deflected or scattered. Since a Coulomb potential has infinite range, it allows scattering with large impact parameters even at very low energy. This can be verified from the dependence of b and σ on E, as given by equations (2.195) and (2.196). As $E \to 0$, σ diverges for all θ and also b diverges for all θ except $\theta \to \pi$. This means that at low energy, the projectile never comes close to the target.

For an attractive potential, as we mentioned earlier, r_{min} classically exists only when $b \neq 0$ or $L \neq 0$. For a potential that is everywhere attractive, the scattering for $L = 0$ cannot be explained by classical theory. Quantum mechanically, for an attractive potential, discrete bound states exist at negative energies. Classically, an attractive central potential can lead to closed orbits at low energy by balancing the attractive radial force with the centrifugal force associated with a definite angular momentum. If a projectile that has a collision energy close to the energy of the closed orbit is incident and has an impact parameter that closely corresponds to the angular momentum of the closed orbit, then the projectile spirals around the center of the force before reaching the minimum distance of approach, from where it is deflected to the asymptotic regime [9]. As a result, arrival in the asymptotic regime is delayed compared to the time it would have taken in the case of a repulsive potential of similar strength. In the context of quantum scattering, this is resonant scattering due to the existence of a bound state, as we will discuss in chapter 4.

To better understand the validity regime of classical theory in the context of binary collisions, one can define a quantum length scale λ_q associated with the long-range part $V_L(r)$ of the potential $V(r)$, as follows:

$$|V_L(r = \lambda_q)| = \frac{1}{2\mu}\left(\frac{h}{\lambda_q}\right)^2. \tag{2.211}$$

If $V_L(r)$ is given by an inverse power law r, as shown below

$$V_L(r) = C_n/r^n, \tag{2.212}$$

where $n > 2$ is an integer, then

$$\lambda_q = \left(\frac{2\mu|C_n|}{h^2}\right)^{(n-2)}. \tag{2.213}$$

Since the quantum expression for the collision energy is $E = (1/2m)(h/\lambda_{\text{dB}})^2$, it is clear that if $\lambda_{\text{dB}} \geqslant \lambda_q$, the classical description of the collision cannot be valid, since the wavelength of the Schrödinger wave is comparable to or greater than the significant spatial dimension of the variation of potential. Therefore, for such potentials, low-energy collisions are essentially quantum mechanical. For an attractive proton–electron Coulomb potential, λ_q is equal to half the Bohr radius s_0. For a potential that is attractive everywhere, such as an attractive Coulomb force, a bound state with $L = 0$ can exist only due to quantum fluctuations that prevent the particle from collapsing at the center of the force. The validity of the classical theory of collisions also rests on the quantum fluctuation or uncertainty about the position of the particle—if this uncertainty over the long term is comparable to or larger than r_{min} or λ_q, then classical theory breaks down. Heisenberg's uncertainty principle of quantum mechanics states that the position and momentum of a quantum particle cannot both be precisely measured. As a result, the radius of an orbit and its angular velocity cannot both be precisely determined. In quantum collision theory, aspects of these quantum fluctuations can be incorporated by considering the propagation of a wave packet through the potential. Wave packets spread as they propagate, but for classical theory to be applicable, this spread should be negligible.

Exercise

- By drawing a schematic diagram of classical scattering of a particle by a target, introduce the concept of impact parameter b. From classical-quantum correspondence, show that $b = \sqrt{\ell(\ell+1)}\,\lambda_{dB}/(2\pi)$, where ℓ is an integer and λ_{dB} is the de Broglie wavelength.
- Define differential and total scattering cross sections. What is scattering rate?
- Show that for two particles interacting via a central potential, the Hamiltonian is separable in relative and center-of-mass coordinates.
- Writing down the Schroedinger equation of a particle moving in a potential $V(\mathrm{r}, t)$, derive the equation of continuity.
- Discuss the normalization procedure of plane waves in terms of Dirac delta function.
- What is a minimum uncertainty wave packet? Discuss how to construct a wave packet using plane waves as basis functions.
- What are the open and closed channels of scattering?
- Define **S**, **K** and **T** matrices of scattering. Write down Lippmann-Schwinger equation in terms of free-particle Green operator.
- Show that for scattering by a conservative force, the total elastic scattering cross section is proportional to the imaginary part of the forward scattering amplitude

References

[1] Schiff L I 1968 *Quantum Mechanics* (New York, NY: McGraw-Hill)

[2] Gradshteyn I S and Tyzhik I M 2007 *Table of Integrals, Series and Products* (Academic)

[3] Arfken G B and Weber J J 2005 *Mathematical Methods for Physicists* (Elsevier)

[4] Taylor J R 2006 *Scattering Theory: The Quantum Theory of Nonrelativistic Collisions* (Dover)

[5] Schwinger J 1947 A variational principle for scattering problems *Phys. Rev.* **72** 742

[6] Bethe H A 1949 Theory of the effective range in nuclear scattering *Phys. Rev.* **76** 38–50

[7] Huang K and Yang C N 1957 Quantum-mechanical many-body problem with hard-sphere interaction *Phys. Rev.* **105** 767–75

[8] Lee T D, Huang K and Yang C N 1957 Eigenvalues and eigenfunctions of a bose system of hard spheres and its low-temperature properties *Phys. Rev.* **106** 1135–45

[9] Newton R G 1982 *Scattering Theory of Waves and Particles* (Springer)

[10] Goldstein H 2011 *Classical Mechanics* (Pearson)

IOP Publishing

Low Energy Collision Physics
Applications to cold atoms
Bimalendu Deb

Chapter 3

Multichannel scattering

In the previous chapter, we discussed theoretical methods in single-channel scattering. In this chapter, we present formal theoretical methods in multichannel scattering that are applicable for calculating not only two-particle elastic and inelastic scattering amplitudes but also for calculating quantum many-body reactive scattering rates [1]. Historically, the theory of multichannel scattering was primarily developed for applications in nuclear scattering, including nuclear reactions. Early attempts to understand resonances in low-energy nuclear collisions [2–4] stimulated interest in the multichannel aspects of scattering phenomena. The R-matrix method of Wigner and Eisenbud [5] paved the way for numerical calculations of multichannel quantum many-body scattering and reaction rates [6]. In this method, the scattering or reaction zone is divided into two parts: an inner region and an outer region. In the outer region, different channels of reaction or scattering products, which may interact only via long-range forces, are defined. The compound nucleus, or collision complex, is formed inside the inner region where short-range forces dominate. The R-matrix is defined in terms of the logarithmic derivative of the multichannel wave function at the boundary surface between the two regions. From the explicit energy dependence of the R-matrix, one can deduce the position and width of isolated resonances. This method can be used to compute the scattering rates of any collision system, including molecular reactions [7]. Once the R-matrix is calculated, one can express the K- and S-matrices in terms of the R-matrix. Apart from the R-matrix method, various other multichannel scattering methods, such as close coupling [8] and multichannel quantum defect theory [9–12], have been developed over the years for applications to atomic collisions in specific physical situations.

Here, we discuss the formal theory of the multichannel scattering method for calculating the K-matrix. In the context of atomic collisions, one usually deals with two- and three-body reactive or nonreactive phenomena. By 'reactive phenomena,' we mean phenomena in which the chemical compositions of the products differ from

doi:10.1088/978-0-7503-3959-9ch3 © IOP Publishing Ltd 2025. All rights, including for text and data mining (TDM), artificial intelligence (AI) training, and similar technologies, are reserved.

those of the reactants. In nonreactive scattering phenomena, the initial and final angular momentum states (including spin states) of the colliding particles may change. In a nonreactive inelastic collision, the total kinetic energy of the scattered particles necessarily differs from that of the incoming colliding particles. In this chapter, we focus on two-particle multichannel scattering only and do not describe any rearrangement reactive scattering. In the two-particle case, most of the methods of single-channel scattering discussed in the previous chapter can be generalized for multichannel scattering by considering the two-particle wave function to be a multicomponent one. In chapter 1, we briefly introduced the meaning of a 'channel' of a scattering event, stating simply what constitutes a channel. For the scattering of particles with spin, the scattering is essentially multichannel. By this, we mean that the interparticle potential depends on the different spin configurations of the particles, and hence different spin channels exist.

3.1 Two-particle scattering with spin

Spin is an internal angular momentum of a particle, having no relation to its position and momentum variables, unlike its orbital angular momentum defined by $\mathbf{r} \times \mathbf{p}$. Both spin and orbital angular momenta are generators of rotation, follow similar frame transformation laws, and can be added together. For a composite particle like an atom or a molecule, the term 'orbital angular momentum' generally means its total electronic orbital angular momentum, unless otherwise specified. Similarly, the spin of a composite particle denotes the total spin of all its constituent particles.

To start with, we keep our discussion quite general by considering two particles 1 and 2 that possess angular momenta $\mathbf{J}_1$ and $\mathbf{J}_2$, respectively, where $\mathbf{J}_n$ may refer to the total spin, the total spin plus the orbital angular momentum, or just the total orbital angular momentum of the nth particle if it has no spin. The total internal angular momentum of the two particles is $\mathbf{J}_i = \mathbf{J}_1 + \mathbf{J}_2$, and their total angular momentum is $\mathbf{J} = \mathbf{J}_i + \mathbf{L}$, where $\mathbf{L}$ is the angular momentum of the relative motion of the two particles. In the case of asymptotically large separation, there is no interaction between $\mathbf{L}$ and $\mathbf{J}_i$. As a result, the eigenstate $|J, M_J\rangle$ of both J^2 and the projection J_z of $\mathbf{J}$ on the z-axis is product separable in the states $|J_i, M_{J_i}\rangle$ and $|\ell, m_\ell\rangle$, which are the eigenstates of the pairs of operators (J_i^2, J_{iz}) and (L^2, L_z), respectively. Since $\mathbf{J}_i$ can assume values ranging from $|J_1 - J_2|$ to $J_1 + J_2$, there exist $(2J_2 + 1)$ or $(2J_1 + 1)$ internal angular momenta if $J_1 > J_2$ or $J_2 > J_1$, respectively. A channel for a two-particle scattering is defined by a set of all the angular momenta, including internal and relative motional ones, that characterizes the joint state of the two particles when they are far apart (practically non-interacting).

In order to enumerate the possible number of channels, we must first distinguish between two cases: (1) the absence of any external field and (2) the presence of an external magnetic or electric field.

In the absence of any external field

In the absence of any external field, the threshold of the channel corresponding to the total internal angular momentum $\mathbf{J}_i$ depends only on the value of $\mathbf{J}_i$ and not on

J_{iz}. So, a channel with a particular value of $\mathbf{J}_i$, say J_i, is $(2J_i + 1)$-fold degenerate. These channels may have different thresholds if there are spin–spin or spin–orbit interactions. Otherwise, all these channels have the same threshold with no interchannel coupling. In the absence of any spin–spin or spin–orbit interaction, one can perform a single-channel calculation for each J_i separately, since different J_is are uncoupled. In the absence of an external field, J is a good quantum number, implying that it is a conserved quantity. So, in this case, one can carry out scattering calculations for different J values separately. The scattering between particles with spin in the absence of any external field becomes essentially multichannel when spin–spin interaction, spin–orbit interaction, or any other type of anisotropic interaction is present.

In the presence of an external field

At the outset, we distinguish between two types of fields: static and dynamic fields. If the particles possess a permanent electric or magnetic dipole moment, then an external electric or magnetic field can directly interact with each of the particles via electric or magnetic dipole interaction, respectively. Therefore, the channel thresholds change due to the action of an external field. Furthermore, an external field can lift the degeneracies of the system, leading to an increase in the number of nondegenerate channels. If the field is static, then the net effect of the field is to shift (via, for example, Zeeman or Stark shifts) the internal states of the colliding particles without any real transition between states. In contrast, a dynamic field may induce real transitions in the system if the field is at resonance or near resonance with the system's transition frequency. If the colliding particles are polarizable but do not possess any permanent electric or magnetic dipole moment, then an external electric or magnetic field can induce an electric or magnetic dipole moment in the particles, respectively. These induced dipoles can interact with each other via higher-order dipolar interactions. Thus, an external field not only lifts degeneracies and causes shifts, but it can also modify the scattering forces in a significant manner. In the presence of an external field, the scattering is essentially anisotropic and multichannel. The channels in this case are usually chosen in such a way that the interaction potential matrix in the long range becomes diagonal in the channel basis. The technical details of the methods of multichannel scattering will be elaborated on in subsequent chapters with specific examples and applications. Here, we focus on the formal theoretical methods of multichannel scattering only.

3.2 Formal theory of multichannel scattering

Here, we present a general and formal mathematical theory of two-particle multichannel scattering by generalizing the method of single-channel partial-wave scattering outlined in the previous chapter to a multichannel situation. We assume that the interparticle potential is a function of relative coordinates only, enabling us to work only with the relative coordinates in the center-of-mass (COM) frame. In the asymptotic limit of separation, the angular momenta of relative motion, described by a pair of partial-wave quantum numbers (ℓm_ℓ), are good quantum numbers, apart from other quantum numbers that characterize the internal (electronic and nuclear) angular momentum states of the two particles at large separation. A channel state $|\mu\rangle$ describes an angular momentum state, where μ collectively denotes all the quantum numbers that characterize an asymptotic angular state of two colliding particles. The channel states are assumed to be separable in internal and relative angular momentum states. A multichannel scattering state $|F^{\mu}\rangle$ can be expanded in terms of channel states as given by

$$|F^{\mu}\rangle = \sum_{\nu}\phi_{\nu\mu}|\nu\rangle, \tag{3.1}$$

whose coordinate representation $F^{\mu}(\mathbf{r}) = \langle \mathbf{r}|F^{\mu}\rangle$ is given by

$$F^{\mu}(\mathbf{r}) = \frac{1}{r}\sum_{\nu}\phi_{\nu\mu}(r)\chi_{\nu}(\tau), \tag{3.2}$$

where $\chi_{\nu}(\tau)$ is an angular momentum wave function and τ collectively denotes all the angular ($\hat{r} \equiv (\theta, \phi)$) and spinor ($\sigma$) variables. The spinor variables may include electronic (σ_e) and nuclear (σ_n) spinor variables for all the electrons and nuclei of the system of two particles (which may consist of a pair of atoms or molecules). Here, $\phi_{\nu\mu}(r)$ denotes a radial wave function characterized by two channel indices μ and ν. Let N be the total number of channels. In the channel basis states, one can define an N-component radial wave function $\bar{\Phi}^{\mu}(r)$ in vector form:

$$\bar{\Phi}^{\mu}(r) = \begin{pmatrix} \phi_{1\mu}(r) \\ \phi_{2\mu}(r) \\ \vdots \\ \phi_{N\mu} \end{pmatrix}. \tag{3.3}$$

These vectors then form the matrix

$$\bar{\bar{\Phi}}(r) = [\bar{\Phi}^{1}(r), \bar{\Phi}^{2}(r), \cdots, \bar{\Phi}^{N}(r)], \tag{3.4}$$

which contains complete information about the multichannel radial wave function.

The interaction potential in a multichannel scattering problem is a matrix $\mathbf{V}(r)$ in the channel basis. The matrix element $V_{\nu\mu}$ is given by

$$V_{\nu\mu}(r) = \int d\tau \chi_{\nu}^{*}(\tau)\mathbf{V}(\tau, r)\chi_{\mu}(\tau), \tag{3.5}$$

where $\mathbf{V}(\tau, r)$ is the interaction that may depend on angular and spinor variables τ in addition to the radial coordinate r. The multichannel counterpart of the time-independent single-channel partial-wave Schrödinger equation (2.93) takes the matrix form

$$\frac{\hbar^2}{2\mu}\left[-\mathbf{1}_N\frac{d^2}{dr^2} + \frac{\Lambda^2}{r^2}\right]\bar{\bar{\Phi}}(r) + \mathbf{V}(r)\bar{\bar{\Phi}}(r) = E\bar{\bar{\Phi}}(r), \tag{3.6}$$

where $\mathbf{1}_N$ denotes the $N \times N$ identity matrix and Λ^2 is a diagonal matrix given by

$$\Lambda^2 = \begin{pmatrix} \ell_1(\ell_1+1) & 0 & \cdots & 0 \\ 0 & \ell_2(\ell_2+1) & \cdots & 0 \\ \vdots & \vdots & \ddots & \vdots \\ 0 & 0 & \cdots & \ell_N(\ell_N+1) \end{pmatrix}, \tag{3.7}$$

with ℓ_μ being the partial-wave quantum number of the μth channel. The off-diagonal matrix elements of $\mathbf{V}(r)$ are interchannel coupling terms, while the diagonal elements are the channel potentials. In the limit $r \to \infty$, the diagonal matrix element $V_{\mu\mu}(r)$ becomes a constant $E_{\rm th}^{\mu} = V_{\mu\mu}(r \to \infty)$, which is called the threshold energy of channel μ. Let us enumerate the channels in ascending order according to the threshold values of the channels, meaning that $E_{\rm th}^{(1)} < E_{\rm th}^{(2)} < \cdots < E_{\rm th}^{(N)}$.

There are three cases in multichannel scattering phenomena: (i) all N channels are open, (ii) all N channels are closed, and (iii) N_c channels are closed and $N_o = N - N_c$ channels are open. In case (i), it is possible to treat any of the channels as an incident channel, meaning that an asymptotic incident state may be prepared in this channel, while all channels may be treated as scattered channels. All scattering matrices are $N \times N$ matrices in this case. In case (ii), the states of all channels vanish at asymptotically large distances, implying the existence of a multichannel bound state with no scattering or continuum state. The third case, where some channels are closed and the remaining channels are open, is a tricky one and requires careful treatment, as we will elaborate in subsequent chapters. In this case, the physical scattering matrices are $N_o \times N_o$ matrices that are to be derived by eliminating the effects of the closed channels from the analytically consistent $N \times N$ scattering problem.

3.2.1 Case (i): all channels are open

Let us first consider case (i); that is, all channels are open, implying that $E > E_{\rm th}^{(N)}$. Then, the asymptotic kinetic energy of the νth channel is given by $E_p^{(\nu)} = \frac{p_\nu^2}{2\mu} = E - E_{\rm th}^{(\nu)}$, where p_ν denotes the magnitude of the νth channel's momentum. In this case, all the scattering matrices are $N \times N$ matrices. Redefining the potential matrix $\tilde{\mathbf{V}}(r) = \mathbf{V}(r) - \bar{\bar{E}}_p$, where

$$\bar{\bar{E}}_p = \text{diag.}\left[E_p^{(1)}, E_p^{(2)}, \cdots E_p^{(N)}\right] \tag{3.8}$$

is a diagonal matrix, one can rewrite equation (3.6) in the following form:

$$\frac{\hbar^2}{2\mu}\left[-\mathbf{1}_N \frac{d^2}{dr^2} + \frac{\Lambda^2}{r^2}\right]\bar{\bar{\Phi}}(r) + \tilde{\mathbf{V}}(r)\bar{\bar{\Phi}}(r) = \bar{\bar{E}}_p \bar{\bar{\Phi}}(r). \tag{3.9}$$

In discussing the theory of single-channel scattering in the previous chapter, we first introduced the **S**-matrix, then the **T**-matrix, and finally the **K**-matrix. Here, in introducing the multichannel counterparts of these matrices, we will first introduce the multichannel **K**-matrix, then the **S**-matrix, and finally the **T**-matrix. We will show the connection of these matrices with the multichannel scattering amplitude matrix. The multichannel versions of the spherical Bessel and Neumann functions are the matrices

$$\bar{\bar{\mathbf{J}}} = \text{diag.}[rj_{\ell_1}(k_1 r),\ rj_{\ell_2}(k_2 r),\ \cdots rj_{\ell_N}(k_N r)] \tag{3.10}$$

$$\bar{\bar{\mathbf{N}}} = \text{diag.}[rn_{\ell_1}(k_1 r),\ rn_{\ell_2}(k_2 r),\ \cdots rn_{\ell_N}(k_N r)]. \tag{3.11}$$

The mathematical form of the asymptotic boundary condition of the multichannel wave function $\bar{\bar{\Phi}}(r)$ is the same as in the partial-wave single-channel scattering given by equation (2.84), but j_ℓ and n_ℓ are replaced by $\bar{\bar{\mathbf{J}}}$ and $\bar{\bar{\mathbf{N}}}$, respectively; and K_ℓ is replaced by the matrix **K**. Therefore, the boundary condition is given by

$$\bar{\bar{\Phi}}(r \to \infty) \sim \bar{\bar{\mathbf{J}}} - \bar{\bar{\mathbf{N}}}\mathbf{K}, \tag{3.12}$$

which also implies that

$$\phi_{\nu\mu}(r \to \infty) \sim rj_{\ell_\nu}(k_\nu r)\delta_{\nu\mu} - rn_{\ell_\nu}(k_\nu r)K_{\nu\mu}. \tag{3.13}$$

The multichannel **S**-matrix is given by

$$\mathbf{S} = (\mathbf{1} + i\mathbf{K})(\mathbf{1} - i\mathbf{K})^{-1}. \tag{3.14}$$

To relate the multichannel **S**-matrix elements to the multichannel scattering amplitudes, one needs to introduce another matrix **F** whose diagonal element $F_{\nu\nu}$ is the νth channel's scattering amplitude, while the off-diagonal element $F_{\nu\mu}$ $(\nu \neq \mu)$ corresponds to the amplitude of the transition from the $|\nu\rangle$ channel states to the $|\mu\rangle$ channel states. Since the scattering amplitude in the single-channel case involves the wave number, to relate **S** to **F** we also need to introduce a multichannel wave vector matrix

$$\mathbf{P} = \text{diag.}[k_1,\ k_2,\ \cdots k_N], \tag{3.15}$$

where $k_\nu = p_\nu/\hbar$. The multichannel counterpart of equation (2.78) is given by

$$\mathbf{S} = \mathbf{1} + 2i\mathbf{P}^{1/2}\mathbf{F}\mathbf{P}^{1/2}. \tag{3.16}$$

The multichannel **T**-matrix is related to the **S**-matrix by

$$\mathbf{S} = \mathbf{1} - 2i\mathbf{T}. \tag{3.17}$$

Thus,

$$\mathbf{T} = -\mathbf{P}^{1/2}\mathbf{F}\mathbf{P}^{1/2}. \tag{3.18}$$

In terms of the **S**-matrix element, the asymptotic boundary condition is given by

$$\phi_{\nu\mu}(r \to \infty) \sim rh^{-}_{\ell_\nu}(k_\nu r)\delta_{\nu\mu} + rh^{+}_{\ell_\nu}(k_\nu r)S_{\nu\mu}. \tag{3.19}$$

The same can also be expressed in terms of the **T**-matrix element in the following form:

$$\phi_{\nu\mu}(r \to \infty) \sim rj_{\ell_\nu}(k_\nu r)\delta_{\nu\mu} - rh^{+}_{\ell_\nu}(k_\nu r)T_{\nu\mu}. \tag{3.20}$$

In case (ii), where all N channels are closed, there is no scattering, and only multichannel bound states are possible. The problem in this case is to devise an appropriate numerical method to compute a multicomponent bound state. We do not discuss this case further.

3.2.2 Case (iii): some channels are open, others are closed

The most interesting case is case (iii), which consists of N_o open channels and $N_c = N - N_o$ closed channels. Here, we give a brief outline of the method used to deal with this case. The details of the method and its applications in cold collisions will be discussed in chapters 3 and 4.

As mentioned earlier, the physical scattering properties can be described in terms of $N_o \times N_o$ scattering matrices. However, mathematically, one can formulate the problem as a generalized $N \times N$ multichannel scattering problem by using the analyticity of the closed-channel wave functions at negative energies or at purely imaginary wave numbers. One then obtains solutions in the form of multicomponent wave functions. The closed-channel components of these wave functions diverge at asymptotically large separations. Since these wave functions are obtained by solving coupled second-order linear differential equations (Schrödinger equations), any linear superposition of the wave functions is also a mathematically valid solution. To obtain a desirable linear superposition in the asymptotic regime, one applies a transformation **A** to the $N \times N$ matrix $\bar{\bar{\Phi}}(r)$ to obtain an $N \times N_o$ submatrix $\bar{\bar{\Psi}}(r)$. Obviously, the dimensions of the transformation matrix **A** are $N \times N_o$. The matrix **A** is chosen in such a way that the coefficients of all the closed-channel states (diverging components) are zero in the asymptotic regime, leaving $\bar{\bar{\Psi}}(r)$ as a physically acceptable solution. The asymptotic analysis of the sub-matrix $N_o \times N_o \bar{\bar{\eta}}(r)$ of $\bar{\bar{\Psi}}(r)$ yields multichannel scattering properties, while the $N_c \times N_o$ submatrix $\bar{\bar{\chi}}(r)$ represents the closed-channel wave function in matrix form, whose elements all asymptotically decay in space. Closed channels affect the scattering properties in a nontrivial way, leading to scattering resonances, as we discuss in the next chapter.

Over the years, many theoretical and numerical methods have been developed to treat multichannel scattering and associated effects such as multichannel resonances, inelastic collisions, and reactive collisions. We will discuss some of these methods that are relevant in the context of cold collisions in the next chapter.

3.2.3 Multichannel scattering cross section

The multichannel scattering matrices are derived by asymptotic analysis of the $N_o \times N_o$ wave function matrix, as we discussed above. Suppose all N_o channels have different thresholds; this means all the channel momenta are different. Then the elastic collision can happen only if the incident and outgoing channels are the same.

On the other hand, if a particle is incident in one channel and scatters into another channel, then this event contributes to inelastic scattering. The multichannel scattering method is quite general, since one can calculate all possible types of inelastic scattering cross sections, including rearrangement collisions, nuclear reactions, and chemical reactions. Let i represent an incident or incoming channel, while all other channels $j \neq i$ are outgoing channels. The incident wave function $\psi_{ii}(\mathbf{r})$ in channel i is a plane wave. Let us recall the asymptotic form of the wave function as discussed in chapter 2. After scattering, for large r, the wave function can be expressed as a spherical wave, as given by

$$\psi_{ii}(\mathbf{r}) \underset{r\to\infty}{\longrightarrow} \frac{\sqrt{\pi}}{ik_i r}\sum_{\ell}\sqrt{2\ell+1}\,Y_{\ell,0}\left[-e^{-i(k_i r-\ell\pi/2)} + e^{i(k_i r-\ell\pi/2)}S_{ii}^{\ell}\right]. \tag{3.21}$$

Asymptotically, the incident plane wave also behaves as a spherical wave:

$$e^{i\mathbf{k}_i\cdot\mathbf{r}} \sim \frac{\sqrt{\pi}}{ik_i r}\sum_{\ell}\sqrt{2\ell+1}\,Y_{\ell,0}[-e^{-i(k_i r-\ell\pi/2)} + e^{i(k_i r-\ell\pi/2)}]. \tag{3.22}$$

Therefore, the scattered part of the wave function $\psi_{\mu\mu}^{\text{scat}}(\mathbf{r})$ is given by

$$\begin{aligned}\psi_{ii}^{\text{scat}}(\mathbf{r}) &= \psi_{ii}(\mathbf{r}) - e^{i\mathbf{k}_i\cdot\mathbf{r}} \\ &\underset{r\to\infty}{\longrightarrow} \frac{\sqrt{\pi}}{ik_i r}\sum_{\ell}\sqrt{2\ell+1}\,Y_{\ell,0}(S_{ii}^{\ell} - 1)e^{i(k_i r-\ell\pi/2)}.\end{aligned} \tag{3.23}$$

Now, recall the definition of the differential scattering cross section: it is the ratio of the number of particles scattered per second into the solid angle $d\Omega$ to the incident flux of particles passing perpendicular to a plane of unit area per second. The current density, as defined by equation (1.17), is an outgoing or diverging flux or probability current density; that is, the number of particles passing perpendicularly through the unit area per second. For an unnormalized plane wave, it is simply given by $\hbar k_i/\mu = v_i$, where v_i is the relative velocity. Replacing the expression $\phi(\mathbf{r}, t)$ in equation (1.17) with $\psi_{ii}^{\text{scat}}(\mathbf{r})$ as given by equation (3.23), one obtains the scattered current density $\mathbf{j}_{\text{scat}}$ given by

$$\mathbf{j}_{\text{scat}} = \pi v_i \sum_{\ell}(2\ell+1)|S_{ii}^{\ell} - 1|^2/(k_i^2 r^2)|Y_{\ell,0}|^2 \tag{3.24}$$

The flux of scattered particles going outward through the surface of a large sphere of radius r within a solid angle $d\Omega$ is given by $r^2 d\Omega\mathbf{j}_{\text{scat}}$. Therefore, the differential elastic scattering cross section is given by

$$\sigma^{\text{el}}(\theta, \phi)_i = \pi\sum_{\ell}(2\ell+1)|S_{ii}^{\ell} - 1|^2/(k_i^2)|Y_{\ell,0}|^2 \tag{3.25}$$

The total elastic scattering cross section is obtained by integrating the above quantity over all solid angles, resulting in

$$\sigma_i^{\text{el}} = \pi\sum_{\ell}(2\ell+1)|S_{ii}^{\ell} - 1|^2/k_i^2 \tag{3.26}$$

The above expression applies when there is only one incident channel. If there are multiple incident channels, then the total elastic scattering cross section is given by the weighted sum over all the channels, that is, by the expression $\sigma_{\text{tot}}^{\text{el}} = \sum_i w_i \sigma_i^{\text{el}}$, where w_i is a weight factor.

Next, we discuss how to calculate the inelastic scattering cross section using the multichannel scattering method. The component of the multichannel wave function that is scattered into an outgoing channel j when the wave is incident in channel $i \neq j$ is given by

$$\psi_{ji}(\mathbf{r}) \underset{r\to\infty}{\longrightarrow} \sim \frac{\sqrt{\pi}}{ikr}\sum_{\ell}\sqrt{2\ell+1}\, e^{i(k_j r-\ell\pi/2)} S_{ji}^{\ell}. \tag{3.27}$$

The flux of particles scattered into channel $j \neq i$ and passing through the surface of a large sphere of radius r within the solid angle $d\Omega$ is given by

$$\pi v_j \sum_{\ell}(2\ell+1)|S_{ji}^{\ell}|^2/(k_j^2)|Y_{\ell,0}|^2 d\Omega \tag{3.28}$$

Therefore, the total inelastic scattering cross section for the wave incident in channel i is given by

$$\sigma_i^{\text{inel}} = \sum_j \frac{v_j}{v_i}\sum_{\ell}(2\ell+1)\pi|S_{ji}^{\ell}|^2/(k_j^2) \tag{3.29}$$

If the wave is incident in several channels, then again one has to take a weighted sum over all the incident channels. An alternative way of expressing the inelastic scattering cross section is to use S_{ii}^{ℓ} only; that is, the diagonal S-matrix element of the incident channel only. The inelastic scattering must result from the depletion of waves in the incident channel. Suppose the waves in the ith channel propagate inward only, while the waves in all other channels propagate outward at the expense of the ith channel's wave amplitude. For a converging flux of particles, the motion of which is described by the wave function $\phi(\mathbf{r}, t)$, the current density is given by

$$\mathbf{j} = -\frac{\hbar}{2\mu}[\phi^*(\mathbf{r}, t)(-i\nabla)\phi(\mathbf{r}, t) - \phi(\mathbf{r}, t)(-i\nabla)\phi^*(\mathbf{r}, t)]. \tag{3.30}$$

Note the minus sign on the rhs—it is due to the fact that the wave is ingoing or converging. The total flux of particles entering channel i through the surface of a large sphere of radius r into the solid angle $d\Omega$ is equal to the current density of $\psi_{ii}^{\text{scat}}(\mathbf{r})$ for large r, multiplied by $r^2 d\Omega$. The result is

$$\frac{\pi v_i}{k_i^2}\sum_{\ell}(2\ell+1)(1-|S_{ii}^{\ell}|^2)|Y_{\ell,0}|^2 d\Omega \tag{3.31}$$

Thus, the total inelastic scattering cross section for the wave incident in channel i only is given by

$$\sigma_i^{\text{inel}} = \frac{\pi}{k_i^2}\sum_{\ell}(2\ell+1)(1-|S_{ii}^{\ell}|^2) \tag{3.32}$$

For single-channel scattering, $|S_{ii}^{\ell}|^2 = 1$, and so there is no inelastic part unless one considers single-channel scattering with a complex potential, as discussed in chapter 2.

The above expression shows that for multichannel scattering with more than one open channel, $|S_{ii} < 1$. Again, for multiple incident channels, one has to take a weighted sum over all the incident channels to obtain the total inelastic cross section.

It is worthwhile to examine how the optical theorem holds in the multichannel case. Using equations (3.26) and (3.32), one can calculate the total cross section σ_t^i for the wave incident in the ith channel. Explicitly, this yields

$$\sigma_t^i = \sigma_i^{\mathrm{el}} + \sigma_i^{\mathrm{inel}} = \frac{2\pi}{k_i^2}\sum_\ell (2\ell + 1)\left(1 - \mathrm{Re}[S_{ii}^\ell]\right). \tag{3.33}$$

The scattered wave function is given by equation (3.23), from which one can obtain the scattering amplitude $f_k^i(\theta)$ in the forward direction $\theta = 0$, as given by

$$f_k^i(0) = \frac{\sqrt{\pi}}{ik_i}\sum_\ell \sqrt{2\ell + 1}\, Y_{\ell,0}(0, \phi)(S_{ii}^\ell - 1). \tag{3.34}$$

By substituting $Y_{\ell,0}(0, \phi) = \sqrt{(2\ell + 1)/4\pi}$ and comparing the equations above, one immediately finds the relation

$$\sigma_t^i = \frac{4\pi}{k}\,\mathrm{Im}\left[f_k^i(0)\right], \tag{3.35}$$

which is the statement of the optical theorem. Thus, we have proven that the multichannel scattering method satisfies the optical theorem.

Exercise

- Discuss how coupled Schroedinger equations under partial wave expansion are expressed by a single equation in matrix form in a multichannel scattering problem.
- Discuss asymptotic boundary conditions imposed on a multichannel wave function when all channels are open and when some channels are open and the rest are closed.
- Define elastic and inelastic scattering cross sections for a multichannel problem.

References

[1] Barrett R F, Robson B A and Tobocman W 1983 Calculable methods for many-body scattering *Rev. Mod. Phys.* **55** 155–243

[2] Breit G and Wigner E 1936 Capture of slow neutrons *Phys. Rev.* **49** 519–31

[3] Breit G 1940 The interpretation of resonances in nuclear reactions *Phys. Rev.* **58** 506–37

[4] Wigner E P 1946 Resonance reactions and anomalous scattering *Phys. Rev.* **70** 15–33

[5] Wigner E P and Eisenbud L 1947 Higher angular momenta and long range interaction in resonance reactions *Phys. Rev.* **72** 29

[6] Lane A M and Thomas R G 1958 R-matrix theory of nuclear reactions *Rev. Mod. Phys.* **30** 257–353

[7] Smith F T 1962 Scattering matrix and chemical reaction rates *J. Chem. Phys* **36** 248–55

[8] Burke P G and Smith K 1962 The low-energy scattering of electrons and positrons by hydrogen atoms *Rev. Mod. Phys.* **34** 458–502
[9] Seaton M J 1966 Quantum defect theory II. Illustrative one-channel and two-channel problems *Proc. Phys. Soc.* **88** 815
[10] Seaton M J 1983 Quantum defect theory *Rep. Prog. Phys.* **46** 167
[11] Aymar M, Greene C H and Luc-Koenig E 1996 Multichannel rydberg spectroscopy of complex atoms *Rev. Mod. Phys.* **68** 1015–23
[12] Burke J P, Greene C H and Bohn J L 1998 Multichannel cold collisions: simple dependences on energy and magnetic field *Phys. Rev. Lett.* **81** 3355–8

IOP Publishing

Low Energy Collision Physics
Applications to cold atoms
Bimalendu Deb

Chapter 4

Resonances and bound states

Resonances manifest as the sharp peaks observed in graphs of scattering cross section versus energy. Resonances in binary collisions are closely related to two-body bound or quasi-bound states, which are the discrete quantum states whose amplitudes fall off (usually exponentially) in space as the separation $r \to \infty$. In the case of a single channel, a bound state is formed at a specific negative energy; that is, below the channel threshold. In the multichannel case, as mentioned in the preceding chapter, a bound state occurs when all the channels are closed; that is, below the threshold of the lowest channel. Under specific circumstances, a bound state may occur at zero energy; that is, right at the lowest threshold. Quasi-bound states are also discrete quantum states that are formed temporarily at a discrete positive energy (above the lowest threshold). Both resonant and bound states (including quasi-bound ones) can be described by a unified approach to cold collisions. The temporary binding of a quasi-bound state at a discrete energy enhances the scattering cross section at that energy. This occurs because the quasi-bound state eventually decays into the continuum of scattering states whose central energy equals the binding energy. The resonance width is determined by the coupling between the bound and continuum states.

Broadly, resonances can be classified into two categories: single-channel and multichannel resonances. Single-channel resonances are grouped into two types: shape resonance and zero-energy resonance. A shape resonance occurs at a higher partial wave due to the well-like shape of the effective potential, which is the sum of the two-body potential plus the centrifugal barrier. A zero-energy resonance may occur due to the existence of a bound state at zero or near-zero energy because there is a lot of interplay between the resonance and such low-energy bound states if the collision energy is lower than the binding energy of these bound states. Multichannel atomic resonances at low energy may be classified into two classes: Feshbach resonances (FRs) [1] and Fano resonances [2]. Both resonances occur due to

doi:10.1088/978-0-7503-3959-9ch4

© IOP Publishing Ltd 2025. All rights, including for text and data mining (TDM), artificial intelligence (AI) training, and similar technologies, are reserved.

continuum–bound couplings. Here, by 'bound' states, we generally mean quasi-bound states unless specified otherwise. While FRs may occur due to the existence of one or multiple bound states coupled to a single continuum or multiple continua of scattering states, Fano resonances necessarily involve at least two bound states and one continuum of scattering states. In Fano resonances, at least one bound state is strongly coupled to a continuum, leading to the continuum–bound dressed state, which is a superposition of the bound and continuum states. Such superposition states are known as Fano states. When a Fano state interacts with another bound state, the resulting scattering states exhibit an asymmetric structure with a sharp peak and a sharp dip. Thus, Fano resonances may be viewed as a particular case of FRs.

4.1 Energy dependence of phase shift near a resonance

Since the ℓth partial-wave scattering cross section $\sigma_\ell(k)$ can be expressed in terms of various scattering matrix elements that are functions of the partial-wave phase shift, understanding the energy dependence of the phase shift near a resonance is essential for describing the resonant behavior of a system. As we discussed in chapter 2, for single-channel scattering due to a central potential, $\sigma_\ell(k)$ is given by

$$\sigma_\ell(k) = \frac{4\pi}{k^2}(2\ell + 1)|t_\ell(k)|^2 = 4\pi(2\ell + 1)\frac{\sin^2\delta(k)}{k^2}. \tag{4.1}$$

Since a resonance means a sharp peak in the $\sigma_\ell(k)$ versus k (or energy) curve, $\delta_\ell(k)$ as a function of k should vary sharply through $(n + 1/2)\pi$, where n is an integer. In other words, $|t_\ell(k)|$ should reach its maximum value, namely unity, at the resonant energy $E_r = \hbar^2k^2/2\mu$. Since $S_\ell(k) = e^{2i\delta_\ell(k)}$, at the resonant energy, $S_\ell(k) = -1$. A plot of $\delta_\ell(k)$ as a function of k within the range from zero to π is shown in figure 4.1. Since $S_\ell(k)$ is a periodic function of δ_ℓ with a modulus of π, one may restrict the variation of δ_ℓ between zero and π.

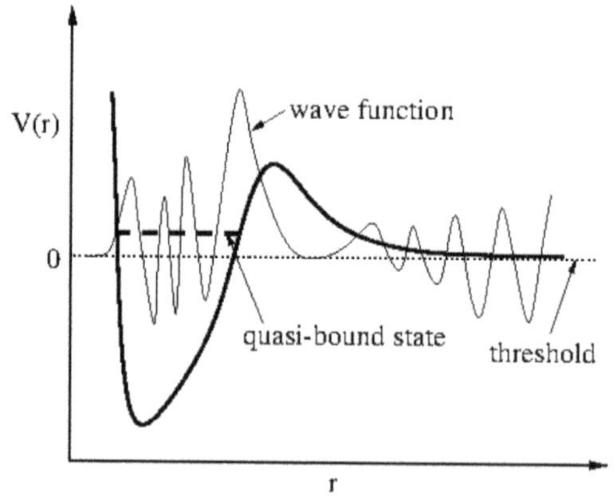

Figure 4.1. Shape resonance due to a potential barrier that supports a quasi-bound state.

4.2 Resonances, bound states, and analyticity of S-matrix elements

In this section, we discuss ℓ-wave resonances and bound states from the point of view of the analyticity of $S_\ell(E)$ or $S_\ell(k)$ as a function of complex E or complex k, respectively. A resonance occurs at a complex energy $E_r = E_0 - i\hbar\Gamma/2$, typically with $E_0 \gg \hbar\Gamma$, where Γ is the angular frequency width of the resonance. This means that resonances appear in the fourth quadrant of the complex energy plane, with the imaginary part being quite small. The inverse of width Γ^{-1} is the time delay in the detection of the resonantly scattered particles in comparison to the time of detection of the non-resonantly scattered ones. To understand the resonances as the complex poles of $S_\ell(E)$, we write the asymptotic wave function as given by equation (2.80):

$$\psi_\ell(r, E) \sim -e^{-i(kr-\ell\pi/2)} + S_\ell(E)e^{i(kr-\ell\pi/2)}, \tag{4.2}$$

where we have replaced the explicit k-dependence with explicit E-dependence for the present purpose. It is instructive to use the time-dependent form of the wave function,

$$\psi_\ell(r, t) = \frac{1}{2}\int dE\left[\psi_\ell(r, E)e^{-iEt/\hbar} + \psi_\ell^*(r, E)e^{iEt/\hbar}\right]. \tag{4.3}$$

The implicit assumption behind this time-dependent wave function is that we are actually dealing with a wave packet and not just a plane wave. A plane wave at energy E is an approximation of a wave packet which is sharply peaked at E. The first part on the rhs of the above equation corresponds to the positive-frequency part, while the second corresponds to the negative-frequency part in the language of field theory. Considering the positive-frequency component only, that is $\psi_\ell(r, E)$, one can infer that $\psi_\ell(r, E)$ must be analytic in the upper half-plane of the complex energy plane when only resonances occur. This implies that $S_\ell(E)$ may have poles only in the lower half-plane. Otherwise, any pole of $S_\ell(E)$ in the upper half-plane implies an exponential rise of the wave function with time. So, when we calculate the energy integration, the contour for the positive (negative) frequency component must be closed in the lower (upper) half-plane. Since

$$S_\ell(E) = \frac{\cot\delta_\ell + i}{\cot\delta_\ell - i}, \tag{4.4}$$

a pole at $E_r = E_0 - i\hbar\Gamma/2$ implies that

$$\cot\delta_\ell(E)_{E=E_r} = i. \tag{4.5}$$

In terms of k, the resonance occurs at $k_r = k_0 - ik_i$, which is related to E_r by $E_r = \hbar^2k_r^2/2\mu$. Thus, we have $\Gamma = 2\hbar k_0k_i/\mu$. Equations (4.2) and (4.3) imply that, at the resonance, the wave function diverges according to $e^{ik_0 + k_ir}$ at large separation. This is an artifact of the plane-wave approximation of the time-independent wave packet. The wave function does not diverge if a wave packet is used instead of a plane wave. However, a resonance clearly implies a solution that decays over time. This implies that a resonance is an excited state of the system that decays. In reality, a resonance occurs due to the temporal binding of the particles; thus, a resonance is associated with a quasi-bound state of the system.

A bound state is a negative-energy solution of the Schrödinger equation and is bounded in space. Typically, at large separations, a bound state $\psi_\ell(k)$ behaves as $\psi_\ell(k) \sim e^{-\kappa r}$. The spatial part of the wave function that oscillates in time as $e^{-iEt/\hbar}$ is $e^{i(kr-\ell\pi/2)}$ as $r \to \infty$. In order for this wave function to behave as $e^{-\kappa r}$ at large separation, we must set $k = i\kappa$ and $E = -\hbar^2\kappa^2/2\mu = -E_b$, where E_b is the magnitude of the binding energy. A general state at large separation may also be expressed as a superposition of ingoing (or incident) and outgoing waves as given by

$$\psi_\ell(r) \sim -\xi_\ell(k)e^{-i(kr-\ell\pi/2)} + \xi_\ell^*(k)(E)e^{i(kr-\ell\pi/2)}, \tag{4.6}$$

where $\xi_\ell(k)$ is called a Jost function, which has the property $\xi_\ell^*(k) = \xi_\ell(-k)$. Since, for a bound state, there is no ingoing wave, a bound state occurs when $\xi_\ell(k) = 0$ for $k = i\kappa$. Comparing equation (4.6) with equation (4.2), one can write

$$S_\ell(E) \equiv S_\ell(k) = \frac{\xi_\ell^*(k)}{\xi_\ell(k)}. \tag{4.7}$$

Thus, a bound state is given by a pole of the **S**-matrix element with the pole lying on the imaginary axis in the upper half of the complex k-plane. The wave functions of an s-wave resonant scattering state and a near-zero energy bound state are closely related, as we discuss below.

4.3 Resonances and bound states for a rectangular potential well

For a rectangular potential well, we explicitly calculated the s-wave phase shift δ_0 in section 2.5. The quantity $\cot\delta_0$ is given by equation (2.141). To find the resonances and bound-state energies for a rectangular potential, one has to solve the equation $\cot\delta_0(k) = i$, that is, the equation

$$k + k_0 \cot(kr_0)\cot(k_0r_0) = i[k \cot(kr_0) - k_0 \cot(k_0r_0)]. \tag{4.8}$$

This equation can be recast into the form

$$[1 - i\cot(kr_0)][k + ik_0 \cot(k_0r_0)] = 0. \tag{4.9}$$

The first factor of $1 - i\cot(kr_0)$ on the lhs of the above equation does not involve the potential depth V_0. We know that the scattering resonances occur at a complex k, with the imaginary part being negative. Thus, we can see that this factor cannot become zero at a complex k. This implies that

$$k_0 \cot(k_0r_0) = ik. \tag{4.10}$$

Here, $k = k_r + ik_i$ is a complex variable, with k_r and k_i being the real and imaginary parts of k. The above equation (4.8) has to be solved numerically as a transcendental equation. The resonances occur when $k_r \neq 0$, $k_i < 0$, and $k_i \ll k_r$; that is, when the poles of the scattering amplitude or the **S**-matrix element appear close to the real axis and the imaginary part of the poles is quite small. The bound states occur when

$k_r = 0$ and $k_i > 0$; that is, when the poles appear on the positive imaginary axis. The energy of a resonant pole is given by $E = E_r - i\hbar\Gamma = \hbar^2k^2/2\mu$, where $E_r = \hbar^2(k_r^2 - k_i^2)/2\mu$ is the resonant energy and $\Gamma/2 = \hbar k_r k_i/\mu$ is the half-width of the resonance. These features of the resonances and bound states follow from the analytic properties of the **S**-matrix elements. All the resonant and bound-state poles of the rectangular potential well were evaluated and analyzed by Nussenzveig [3].

As mentioned earlier, when an s-wave resonance occurs, δ_0 as a function of real energy goes through $\pi/2$. It then follows from equation (2.144) that a_s diverges at the resonance. If we define the dimensionless strength s_p of a rectangular potential as $s_p = \kappa_0 r_0$, where $\kappa_0 = \sqrt{2\mu V_0}/\hbar$, the minimum strength of the potential for the occurrence of the first resonance is

$$s_p = \pi/2. \tag{4.11}$$

This shows that a weak rectangular potential, for which $s_p \ll 1$, cannot give rise to any resonance. As s_p increases from a small to a large value across $\pi/2$, a_s goes from a small negative to a large negative value, and at $s_p = \pi/2$ it diverges. As s_p is increased slightly above $\pi/2$, a_s becomes large positive. With a further increase in s_p, a_s decreases from a large positive to a small positive value or zero. Thus, a large positive a_s is a signature of the occurrence of resonance. Resonances are closely related to bound states supported by the potential. As we mentioned in chapter 2, the s-wave scattering-wave function can be approximated as a straight line near the range of the potential. a_s is the intercept of this straight line with the x-axis. As the depth V_0 of a rectangular potential well of range r_0 is gradually increased, the intercept on the negative side of the x-axis increases until the straight line bends toward the positive side of the x-axis, making the intercept large positive. Exactly at the resonance, a_s diverges, implying that the straight line becomes parallel to the x-axis. If there exists a bound state close to the threshold, then its wave function for $r > r_0$ must decay with r. A bound state with energy $E_b = -\hbar^2\kappa^2/2\mu$ has the form $\psi_b(r) \sim Ae^{-\kappa r}$ for $r > r_0$, where A is a constant. In the limit $\kappa \to 0$, one can approximate it as

$$\psi_b(r) \sim A(1 - \kappa r). \tag{4.12}$$

Now, upon comparing this with the straight-line form of the scattering-wave function $\psi_0(r) \sim -ka_s(1 - r/a_s)$, one finds that $\kappa = 1/a_s$. Thus, the binding energy E_b of a near-zero energy s-wave bound state is related to a_s by

$$E_b \simeq -\frac{\hbar^2}{2\mu a_s^2}. \tag{4.13}$$

Zero-energy resonances

A resonance may occur right at the threshold of a potential; that is, at $E = 0$ when $s_p = (2n + 1)\pi/2$, i.e. where the strength of the potential is an odd integral multiple of $\pi/2$. In such a case, $\delta_\ell(0) = (2n + 1)\pi/2$, and the scattering amplitude diverges at $E = 0$, resulting in ℓ-wave zero-energy resonances. This can be readily understood for the case of $\ell = 0$ for $n = 0$. Substituting equation (2.143) and $s_p = \kappa_0 r_0 = \pi/2$ into equation (2.142) and taking the limit $k \to 0$, one finds that $\delta_0(0) = \pi/2$. Zero-energy resonances are associated with the zero-energy bound states. However, an s-wave zero-energy bound state is unstable and therefore physically not realizable. Therefore, an s-wave bound state may occur only at a nonzero negative energy. However, zero-energy bound states for higher partial waves are stable because the centrifugal barrier can protect such bound states from being disintegrated.

Let us now consider that $s_p \neq (2n + 1)\pi/2$. In this case, as k decreases, $\delta_0(k)$ must rapidly pass through $(n + 1/2)\pi$ near $k = 0$ for the occurrence of a resonance at near-zero energy, with the condition that $E \neq 0$. Here, the integer n is the number of bound states that the potential can support. The value of n depends on the strength of the potential—the larger the strength, the larger the value of n. The value $\delta_0(k)$ at $k = 0$ can be calculated from equation (2.142) as follows:

$$\delta_0(0) = n\pi, \quad s_p \neq (2n + 1)\pi/2. \tag{4.14}$$

The above equation is a statement of Levinson's theorem. For higher partial waves, Levinson's theorem states that

$$\delta_\ell(0) = n_\ell \pi, \quad s_p \neq (2n + 1)\pi/2, \tag{4.15}$$

where n_ℓ is the number of bound states of the ℓ partial wave.

If the potential is relatively strong but not strong enough to support the first near-zero-energy bound state, then the phase shift does not reach the resonant value $\pi/2$ but is close to it and eventually goes to zero as $k \to 0$.

4.4 Shape resonances

Shape resonances are single-channel resonances that occur due to the formation of a quasi-bound state above the threshold of a potential. These quasi-bound states are temporarily supported by a well-like structure that forms due to the shape of the potential, as shown in figure 4.1. For two ground-state atoms, this kind of resonance may arise for higher partial-wave scattering. The effective potential for two ground-state atoms at long range is

$$V_{\text{eff}}(r) = \frac{\hbar^2 \ell(\ell + 1)}{2\mu r^2} - \frac{C_6}{r^6}, \tag{4.16}$$

where $C_6 > 0$ is a van der Waals coefficient. Obviously, for $\ell \neq 0$, the effective potential passes through a maximum at the point

$$r_{\max} = \left(\frac{6\mu C_6}{\hbar^2 \ell(\ell + 1)} \right)^{1/4}, \tag{4.17}$$

leading to a barrier called the centrifugal barrier. The value of the potential at the top of the barrier is $V_{\max} = V_{\text{eff}}(r_{\max})$, which is larger for smaller values of C_6 and smaller values of μ for a fixed value of ℓ. If $V_{\max}$ is sufficiently large compared to the vibrational-level spacing near the dissociation threshold, then the potential can support a quasi-bound vibrational state above the threshold. When the collision energy coincides with the energy of this quasi-bound state, a shape resonance occurs.

4.5 Feshbach resonances

FRs are multichannel resonances. They occur if a closed channel can support a bound state, the existence of which manifests as a resonance in the lower open channels. So, in order to calculate an FR, one has to resort to a multichannel scattering method. One of the well-known multichannel resonance scattering methods is the Feshbach method. In this method, the bound states of one or more channels are calculated by first assuming the channels are uncoupled; then, the couplings of these bound states with other channel states (which include both continuum as well as bound states) are introduced. Projection operator techniques are used to project the total wave function into bound and continuum states. Here, we resort to a simplified two-channel Feshbach method to illustrate the essential features of an FR.

4.5.1 Two-channel model

Here, we consider a two-channel model for s-wave FR in the electronic ground-state manifold of an atom. In reality, many channels may be coupled. In modeling a multichannel s-wave FR using a two-channel model, it is assumed that the upper channel is closed and the lower one is open, while the effects of all the closed channels are effectively embodied by the single closed channel. Since we are interested only in s-wave resonance, the collision energy is close to the threshold of the lowest nondegenerate open channel or the lowest degenerate open channels. Whatever the number of open channels involved in an actual s-wave FR, all these open channels can be reliably represented by a single open channel in the two-channel model.

The two-channel model is schematically depicted in figure 4.2. The two channels are coupled via an interaction term V. The thresholds of the two channels, as well as V, may be varied by an external field. At an FR, the two atoms form a quasi-bound state in the closed channel because the asymptotic collision energy is close to the energy of a true bound state supported by the closed channel when $V = 0$. As we are interested in modeling only one isolated FR, it is also assumed that the closed channel can support only one bound state because closely lying multiple bound states may lead to overlapping resonances. In a magnetic Feshbach resonance (MFR) (to be discussed in the next chapter), the bound-state energy can be varied by tuning the strength of the applied magnetic field.

Let $|o\rangle$ and $|c\rangle$ represent the s-wave open- and closed-channel internal (spin or electronic) states, respectively. The Hamiltonian of a two-channel model in the center-of-mass (COM) frame can be written as

$$\hat{H} = -\frac{\hbar^2}{2\mu}\frac{1}{r}\frac{\partial^2}{\partial r^2}r\mathbf{1}_2 + V_o(r)|o\rangle\langle o| + V_c(r)|c\rangle\langle c| + V(r)|c\rangle\langle o| + V^*(r)|o\rangle\langle c|, \quad (4.18)$$

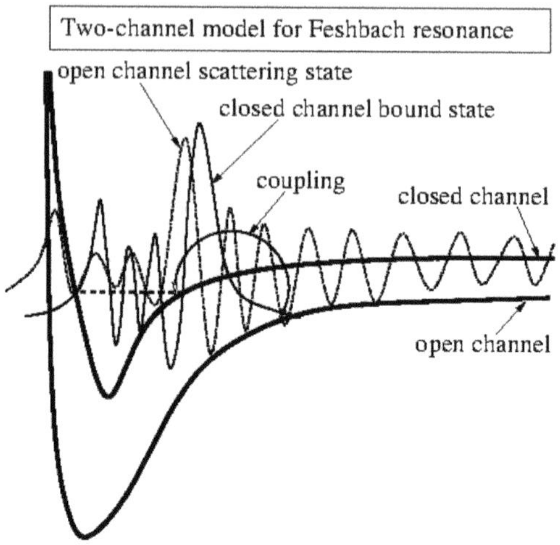

Figure 4.2. Schematic showing a two-channel model of FR.

where $\mathbf{1}_2 = |o\rangle\langle o| + |c\rangle\langle c|$ is a 2×2 identity matrix and $V_o(r)$ and $V_c(r)$ are the open- and closed-channel potential terms, respectively. Let the wave function be

$$|\psi\rangle = \frac{1}{r}[\phi(r)|o\rangle + \chi(r)|c\rangle], \tag{4.19}$$

which satisfies the time-independent Schrödinger equation $H|\psi\rangle = E|\psi\rangle$. Here, $\phi(r)/r$ and $\chi(r)/r$ represent the open- and closed-channel components of the wave function, respectively. Substituting equations (4.18) and 4.19 into the Schrödinger equation, we obtain two coupled differential equations

$$\left[-\frac{\hbar^2}{2\mu}\frac{d^2}{dr^2} + V_o(r)\right]\phi(r) = E\phi(r) - V^*(r)\chi(r) \tag{4.20}$$

$$\left[-\frac{\hbar^2}{2\mu}\frac{d^2}{dr^2} + V_c(r)\right]\chi(r) = E\chi(r) - V(r)\phi(r). \tag{4.21}$$

When the interchannel coupling $V \to 0$, the above two equations reduce to two independent eigenvalue equations. Since, in the Feshbach method, it is assumed that the collision energy E is close to the binding energy of a bound state in the closed channel, we are interested in the bound-state solution of the closed channel. The two homogeneous equations in the absence of interchannel coupling are

$$\left[-\frac{\hbar^2}{2\mu}\frac{d^2}{dr^2} + V_o(r)\right]\phi^{(0)}(r) = E\phi(0)(r) \tag{4.22}$$

$$\left[-\frac{\hbar^2}{2\mu}\frac{d^2}{dr^2} + V_c(r)\right]\chi^{(0)}(r) = E_c\chi^{(0)}(r), \tag{4.23}$$

where $\phi^{(0)}(r)$ and $\chi^{(0)}(r)$ the continuum and bound-state solutions of the two homogeneous equations (4.22) and (4.23), respectively. Here, E_c is the binding energy. Using the Green's function technique, one can write

$$\phi(r) = A\left[\phi^{(\text{reg})}(r) + \int dr' \int G_0(r, r')V^*(r')\chi(r')\right], \tag{4.24}$$

where A is an arbitrary normalization constant and ϕ^{reg} is the regular solution of equation (4.22) in the asymptotic form $\phi^{\text{reg}}(r) \sim \sin(kr + \delta_0)$. Here, $G_0(r, r')$ is the Green's function given by

$$G_0(r, r') = -\pi\phi^{\text{reg}}(r_<)\phi^+(r_>), \tag{4.25}$$

where $\phi^+(r) = \phi^{\text{irr}}(r) + i\phi^{\text{reg}}(r)$, and $\phi^{\text{irr}}(r)$ is the irregular solution of equation (4.22), which asymptotically behaves as $\phi^{\text{irr}}(r) \sim \cos(kr + \delta_0)$. Moreover, $r_<$ ($r_>$) implies whichever of r or r' is smaller (greater) than the other. The bound-state solution can be expressed as

$$\chi(r) = \frac{1}{E - E_c}\chi^{(0)}(r)\int \chi^{(0)}(r')V(r')\phi(r')dr'. \tag{4.26}$$

Substituting equation (4.24) into equation (4.26), we obtain

$$\begin{aligned}\chi(r) = \frac{1}{E - E_c}\chi^{(0)}(r)\Big[&\int dr' V(r')\chi^{(0)}(r')\phi^{(0)}(r') \\ &+ \int dr' V(r')\chi^{(0)}(r')\int dr'' G_0(r', r'')V^*(r'')\chi(r'')dr''\Big].\end{aligned} \tag{4.27}$$

Multiplying both sides by $V(r''')\chi^{(0)}(r''')G_0(r''', r)V^*(r)$ and integrating over r''' and r, one obtains

$$\Sigma = \frac{1}{E - E_c}\Sigma_0[V_E + \Sigma], \tag{4.28}$$

where

$$\Sigma = \int dr' V(r')\chi^{(0)}(r')\int dr'' G_0(r', r'')V^*(r'')\chi(r'')dr'' \tag{4.29}$$

and

$$\Sigma_0 = \int dr' V(r')\chi^{(0)}(r')\int dr'' G_0(r', r'')V^*(r'')\chi^{(0)}(r'')dr''. \tag{4.30}$$

Therefore, we have

$$\Sigma = \frac{V_E\Sigma_0}{E - E_c - \Sigma_0}, \tag{4.31}$$

where

$$V_E = \int dr \phi^{(0)}(r)V(r)\chi^{(0)}(r) \tag{4.32}$$

is the free–bound coupling constant. Note that Σ_0 is a complex function which can be written as $\Sigma_0 = E_{\text{shift}} - i\Gamma/2$, where $E_{\text{shift}} = \text{Re}[\Sigma_0]$ is the shift of the bound state due to the free–bound coupling and $\Gamma/2 = -\text{Im}[\Sigma_0] = \pi|V_E|^2$. Substituting these into (4.27), we can express $\chi(r)$ in the form

$$\chi(r) = \frac{V_E}{E - E_c - E_{\text{shift}} + i\Gamma/2}\chi^{(0)}(r). \tag{4.33}$$

Substituting equation (4.33) into (4.24), we get

$$\phi(r) = A\left[\phi^{\text{reg}}(r) + \frac{V_E}{E - E_c - \Sigma_0}\int dr' G_0(r, r')V^*(r')\chi^{(0)}(r')\right]. \tag{4.34}$$

Upon substituting the Green's function $G_0(r, r')$ given by equation (4.25) into the above equation, using the asymptotic forms of $\phi^{\text{reg}}(r)$ and $\phi^{+}(r)$, and choosing $A = e^{i\delta_0}$, we obtain

$$\phi(r \to \infty) \sim \sin(kr) - T_0 e^{ikr} - \frac{e^{i\delta_0}}{\varepsilon + i}e^{i(kr+\delta_0)}, \tag{4.35}$$

where

$$\varepsilon = \frac{E - E_c - E_{\text{shift}}}{\Gamma/2} \tag{4.36}$$

and $T_0 = -\sin\delta_0 e^{i\delta_0}$. Now, equating the above equation with the standard boundary condition in terms of the **T**-matrix element given by

$$\phi(r \to \infty) \sim \sin(kr) - T(k)e^{ikr}, \tag{4.37}$$

where $T(k) = -\sin\delta(k)e^{i\delta(k)} = \frac{1}{-\cot\delta(k) + i}$, with $\delta(k)$ being the total phase shift, we obtain

$$T(k) = T_0(k) + T_r(k)e^{2i\delta_0}, \tag{4.38}$$

where $T_r(k)$ is the resonant part of the **T**-matrix element given by

$$T_r(k) = \frac{1}{\varepsilon + i}. \tag{4.39}$$

Introducing the resonant phase shift δ_r given by

$$-\cot\delta_r = \varepsilon, \tag{4.40}$$

one can write

$$\sin\delta e^{i\delta} = \sin\delta_0 e^{i\delta_0} + \sin\delta_r e^{2i\delta_0 + i\delta_r}. \tag{4.41}$$

The **S**-matrix element is given by

$$\begin{aligned} S = e^{2i\delta} = 1 - 2iT &= e^{i(\delta_0 + \delta_r)}[e^{-i(\delta_0 + \delta_r)} + 2i\sin\delta_0 e^{-i\delta_r} + 2i\sin\delta_r e^{i\delta_0}] \\ &= e^{2i(\delta_0 + \delta_r)}. \end{aligned} \tag{4.42}$$

Therefore, $\delta = \delta_0 + \delta_r$.

4.6 Fano resonances

Fano resonances were first described in the context of the spectroscopy of autoionization, where they were observed in doubly excited states of two-electron atoms such as He. These resonances occur when a continuum–bound dressed state belonging to one configuration is coupled to another bound state belonging to another configuration [2]. A continuum–bound dressed state is an eigenstate of a continuum–bound coupled system. The notion of such dressed states is relevant when the continuum states are strongly coupled to a bound state. A method for obtaining such an eigenstate was first described by Fano [2]. The transition processes that lead to an autoionizing Fano resonance are schematically shown in figure 4.3. In some sense, Fano resonances are closely related to FRs; both deal with continuum–bound coupled systems.

4.6.1 Fano state

An energy-normalization eigenstate of a system of a single bound state interacting with a continuum of states can be written in the form

$$|E\rangle_F = b_E|b\rangle + \int dE' C_E(E')|E'\rangle \tag{4.43}$$

where E is an energy eigenvalue and $|\psi_b\rangle$ denotes a unit-normalized bound state meaning that $\int d^3\mathbf{r}|\psi_b(\mathbf{r})|^2 = 1$, where $\psi_b(\mathbf{r}) = \langle\mathbf{r}|\psi_b\rangle$. Here,$|E'\rangle$ represents an energy-normalized bare continuum or scattering state, implying that $\langle E|E'\rangle = \delta(E - E')$. The state $|E\rangle_F$ is an energy-normalized continuum–bound dressed or superposition state known as a Fano state, ${}_F\langle E|E'\rangle_F = \delta(E - E')$. Here, b_E and $C_E(E')$ are coefficients to be determined by solving the time-independent Schrödinger equation $\hat{H}|E\rangle_F = E|E\rangle_F$, where $\hat{H}$ is the Hamiltonian that describes the interaction between the bound state and the continuum of states. A general continuum–bound Hamiltonian may be written as

$$\hat{H} = \int EdE|E\rangle\langle E| + E_b|b\rangle\langle b| + \int dE[\Lambda_E|E\rangle\langle b| + \text{h.c.}], \tag{4.44}$$

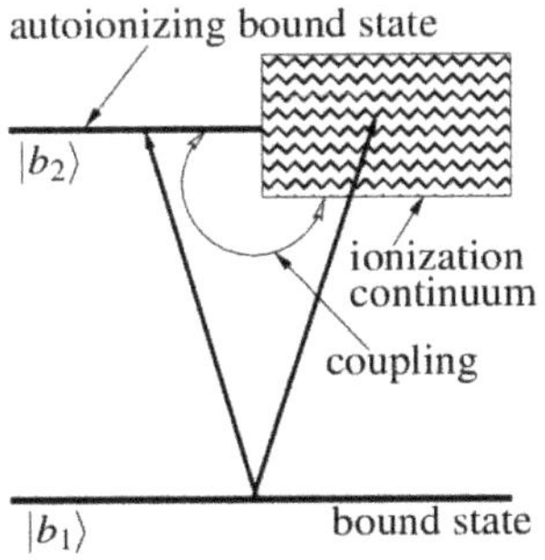

Figure 4.3. Schematic showing the levels and transition processes that lead to Fano resonance in autoionization.

where Λ_E is the coupling between the continuum state $|E\rangle$ and the bound state $|b\rangle$, and h.c. denotes the Hermitian conjugate. The scattering amplitude is determined by asymptotic analysis of the wave function $\Psi_E(\mathbf{r}) = \langle \mathbf{r}|\Psi_E\rangle$ as $r \to \infty$. Since $\psi_b(\mathbf{r}) = \langle \mathbf{r}|b\rangle$ vanishes as $r \to \infty$, the scattering amplitude is determined by the coefficient $C_E(E')$. However, $C_E(E')$ strongly depends on $\Lambda_{E'}$. As a result, Fano resonances occur at energies near the binding energy of $|b\rangle$. A detailed discussion of Fano resonances and their applications will be presented in the next chapter.

Substituting equations (4.43) and (4.44) into the Schrödinger equation $\hat{H}_{cb}|E\rangle_F = E|E\rangle_F$, one obtains

$$E_b b_E + \int dE' \Lambda^*_{E'} C_E(E') = E b_E \tag{4.45}$$

$$\Lambda_{E'} b_E + E' C_E(E') = E C_E(E'). \tag{4.46}$$

Since $|b\rangle$ is coupled to the continuum, we must impose the condition that as $\Lambda_E \to 0$, $|\Psi_E\rangle \to |E\rangle$. This implies that when $\Lambda_E = 0$, we have $C_E(E') = \delta(E - E')$. So, (4.65) can be rewritten as

$$C_E(E') = \delta(E - E') + \frac{\Lambda_{E'}}{E - E'} b_E = \delta(E - E') + \Lambda_{E'} G(E') b_E, \tag{4.47}$$

where $G(E') = 1/(E - E')$ is a Green's function. Assuming that E has a small positive imaginary part: $E \to E + i0^+$, we can write

$$G(E') = \mathscr{P}\frac{1}{E - E'} - i\pi\delta(E - E'), \tag{4.48}$$

where $\mathscr{P}$ denotes the principal value. Using the above relation and substituting equation (4.47) into equation (4.45), one gets

$$b_E = \frac{\Lambda^*_E}{E - E_b + i\Gamma/2 - E_{\text{shift}}}, \tag{4.49}$$

where

$$\Gamma = 2\pi|\Lambda_E|^2 \tag{4.50}$$

and

$$E_{\text{shift}} = \mathscr{P}\int dE' \frac{|\Lambda_{E'}|^2}{E' - E} \tag{4.51}$$

Now, substituting equation (4.49) into equation (4.47), one obtains the explicit analytical expressions for both b_E and $C_E(E')$. Thus, we have obtained the dressed or Fano state of an interacting system of a single bound state and a continuum of states. By expressing the Green's function $G_E(E')$ in real space (as done in chapter 2), one can express the Fano wave function in real space and thereby obtain all the scattering matrices by asymptotic analysis of the wave function, as shown in chapter 2. The analysis leads to resonances, the origin of which is similar in two-channel FR—when the collision energy coincides with the shifted bound-state energy $\tilde{E}_b = E_b + E_{\text{shift}}$, there is resonant enhancement of the scattering cross section of the Lorentzian profile with a width equal to Γ.

Now, in order to derive scattering matrices, we need to analyze the asymptotic form of the Fano wave function $\psi_E(\mathbf{r}) = \langle \mathbf{r}|E\rangle_F$ which is given by

$$\psi_E(\mathbf{r}) = \phi_E(\mathbf{r}) + b_E \psi_b(\mathbf{r}) + b_E \int dE' G(E') \Lambda_{E'} \phi_{E'}(\mathbf{r}), \tag{4.52}$$

where $\phi_E(\mathbf{r}) = \langle \mathbf{r}|E\rangle$ is the bare continuum state. As $r \to \infty$, $\phi_E(\mathbf{r}) \sim \sum_\ell \sqrt{2\ell+1}\, \sin(kr + \delta_\ell - \ell\pi/2) Y_{\ell,0}(\theta, \phi)$, $\psi_b(\mathbf{r}) \sim 0$ where $(\theta, \phi) \equiv \hat{r}$, and δ_ℓ is the ℓ-wave background phase shift due to the atom–atom central potential in the open channel. Thus, for large r, the integral on the rhs of the above equation can be expressed in the following form:

$$\begin{aligned} \int dE' G(E') \Lambda_{E'} \phi_{E'}(\mathbf{r}) \sim & \sum_\ell \sqrt{2\ell+1}\, Y_{\ell,0}(\theta, \phi) \frac{1}{i} \int_0^\infty \Lambda_{E'} k' dk' \frac{1}{k^2 - k'^2 \pm i\eta} \\ & \times [e^{i(k'r + \delta_\ell - \ell\pi/2)} - e^{-i(k'r + \delta_\ell - \ell\pi/2)}], \end{aligned} \tag{4.53}$$

where η is a small positive number, $\hbar^2 k^2/2\mu = E$, and $\hbar^2 k'^2/2\mu = E'$. So, we can further write

$$\begin{aligned} & i \int_0^\infty k' dk' \frac{1}{k^2 - k'^2 + i\varepsilon} [e^{ikr + \delta_\ell - \ell\pi/2} - e^{-(ikr + \delta_\ell - \ell\pi/2)}] \\ & = i \int_{-\infty}^\infty k' dk' \frac{e^{i(k'r + \delta_\ell - \ell\pi/2)}}{k'^2 - k^2 \pm i\eta}. \end{aligned} \tag{4.54}$$

Closing the contour of complex integration in the upper half-plane in the complex k' plane and performing the complex integration using the residue theorem, we obtain the result

$$-\pi \Lambda_E \exp[i(kr + \delta_\ell - \ell\pi/2)]. \tag{4.55}$$

Now, collecting all these terms together, we finally have the asymptotic form

$$\psi_E(\mathbf{r}) \sim \sum_\ell \sqrt{2\ell+1}\, \psi_\ell(r) Y_{\ell,0}(\theta, \phi), \tag{4.56}$$

where

$$\begin{aligned} \psi_\ell(r) & \sim \sin(kr + \delta_\ell - \ell\pi/2) - T_\ell e^{i(kr + \delta_\ell - \ell\pi/2)} \\ & = e^{-i\delta_\ell} \Big[\sin(kr - \ell\pi/2) - \big(T_\ell^0 + T_\ell e^{2i\delta_\ell}\big) e^{i(kr - \ell\pi/2)} \Big]. \end{aligned} \tag{4.57}$$

Here, $T_\ell^0 = -e^{i\delta_\ell} \sin\delta_\ell$ is the background T-matrix element and

$$T_\ell = \frac{\Gamma/2}{E - \tilde{E}_b + i\Gamma/2} \tag{4.58}$$

is the resonant part of the T-matrix element. Furthermore, $\tilde{E}_b = E_b + E_{\text{shift}}$. The total T-matrix element is $\tilde{T}_\ell = T_\ell^0 + T_\ell e^{2i\delta_\ell}$, and so the S-matrix element is given by

$$S_\ell = 1 - 2i\widetilde{T}_\ell = \frac{E - \widetilde{E}_b - i\Gamma/2}{E - \widetilde{E}_b + i\Gamma/2} e^{2i\delta_\ell}, \tag{4.59}$$

while the resonant phase shift δ_r is given by

$$-\cot\delta_r = \frac{E - \widetilde{E}_b}{\Gamma/2}. \tag{4.60}$$

We can identify the quantity on the rhs of the above equation with the parameter ε as introduced in the two-channel FR model in the previous section. Thus, we see that both the Feshbach and Fano methods of treating a continuum coupled to a bound state result in the same type of resonance.

Next, we discuss how a Fano state can lead to quantum interference in continuum–bound systems. Quantum interference means interference between or among transition pathways from one quantum state to another. It is analogous to optical interference in a double-slit experiment where two optical paths interfere when the two paths have a definite phase.

4.6.2 Doubly dressed Fano state: quantum interference

When the Fano state $|E\rangle_F$ is coupled to another bound state of a different configuration, as in the autoionization of atoms, a different kind of resonance occurs that has an asymmetric profile with a peak and a dip in the scattering cross section. This kind of asymmetric resonance is known as Fano resonance. This originates from quantum interference between two transition pathways—one is the direct pathway from the second bound state $|b'\rangle$ to the continuum, that is, $|b'\rangle \rightarrow |E\rangle$, while the other is the indirect pathway from $|b'\rangle$ to $|E\rangle$ via the first bound state $|b\rangle$, that is, $|b'\rangle \rightarrow |b\rangle \rightarrow |E\rangle$. Let us see how this happens.

When $|E\rangle_F$ is coupled to $|b'\rangle$, we can define a doubly dressed state

$$|E\rangle_{FF} = \int \tilde{C}_E(E')dE'|E'\rangle_F + \tilde{b}_E|b'\rangle \tag{4.61}$$

where $\tilde{C}_E(E')$ and $\tilde{b}_E$ are unknown coefficients that can be determined by following the same procedure as in the derivation of $|E\rangle_F$ and replacing $|E\rangle$ with $|E\rangle_F$. Suppose an interaction Hamiltonian V' couples $|b'\rangle$ to $|E'\rangle_F$, and the corresponding coupling $\tilde{\Lambda}_E$ is

$$\tilde{\Lambda}_{E\prime} = \langle b'|V'|E\rangle_F = b_{E\prime}\hbar g + \int dE'' C_{E\prime}(E'')\Lambda'_{E''} \tag{4.62}$$

where $g = \langle b'|V'|b\rangle/\hbar$ is the bound–bound coupling and $\Lambda'_{E''} = \langle b'|V'|E''\rangle$ is the coupling between the second bound state and the bare continuum $|E''\rangle$. Then, the Hamiltonian of the system can be written in the form

$$H = E_{b\prime}|b'\rangle\langle b'| + \int dE'\tilde{\Lambda}_{E\prime}|b'\rangle_F\langle E'| \tag{4.63}$$

where $E_{b'}$ is the binding energy of $|b'\rangle$. From the Schrödinger equation, we have

$$\tilde{C}_E(E') = \delta(E - E') + \frac{\tilde{b}_E\tilde{\Lambda}_{E'}}{E - E'} \tag{4.64}$$

$$(E - E_{b'})\tilde{b}_E - \tilde{b}_E\int dE'\frac{\widetilde{\Lambda}^*_{E'}\widetilde{\Lambda}_{E'}}{E - E'} = \widetilde{\Lambda}^*_E, \tag{4.65}$$

Following the same procedure as that used to derive the Fano state $|E\rangle_F$ above, we can readily write the result

$$\tilde{b}_E = \frac{\widetilde{\Lambda}^*_E}{E - E_{b'} + i\tilde{\Gamma}/2 - \widetilde{E}_{\text{shift}}}, \tag{4.66}$$

where

$$\tilde{\Gamma} = 2\pi|\tilde{\Lambda}_E|^2 \tag{4.67}$$

and

$$\tilde{E}_{\text{shift}} = \mathscr{P}\int dE'\frac{|\tilde{\Lambda}_{E\prime}|^2}{E - E'} \tag{4.68}$$

We now substitute equation (4.47) into equation (4.62) to obtain

$$\widetilde{\Lambda}_E = \Lambda'_E + b_E[\hbar g + \beta - i\zeta], \tag{4.69}$$

where

$$\beta = \mathscr{P}\int dE'\frac{\Lambda'_{E'}\Lambda_{E'}}{E - E'} \tag{4.70}$$

and

$$\zeta = \pi\Lambda'_E\Lambda_E. \tag{4.71}$$

Introducing the parameter

$$q = \frac{g + \beta}{\zeta}, \tag{4.72}$$

which is known as the Fano q-parameter, we can write

$$\widetilde{\Lambda}_E = \Lambda'_E + b_E\zeta(q - i). \tag{4.73}$$

Furthermore, by introducing the parameter $\varepsilon = (E - \tilde{E}_b)/(\Gamma/2)$, we can express b_E from equation (4.49) in the following form:

$$b_E = \frac{1}{\pi \Lambda_E(\varepsilon + i)}. \tag{4.74}$$

Substituting equations (4.74) and (4.71) into equation (4.73), we obtain

$$\widetilde{\Lambda}_E = \Lambda'_E \left[1 + \frac{q - i}{\varepsilon + i}\right] = \frac{q + \varepsilon}{\varepsilon + i} \Lambda'_E. \tag{4.75}$$

Thus, we obtain

$$\tilde{\Gamma} = 2\pi |\tilde{\Lambda}_E|^2/\hbar = \frac{(q + \varepsilon)^2}{\varepsilon^2 + 1} \Gamma' \tag{4.76}$$

where

$$\Gamma' = 2\pi |\Lambda'_E|^2/\hbar \tag{4.77}$$

and

$$\tilde{E}_{\text{shift}} = \mathscr{P} \int dE' \frac{|\tilde{\Lambda}_{E'}|^2}{E - E'} \tag{4.78}$$

The evaluation of $\tilde{E}_{\text{shift}}$ requires the calculation of double energy integrals.

We now discuss how this doubly dressed state $|E\rangle_{FF}$ leads to quantum interference in resonances. We have already noted that the singly dressed state $|E\rangle_F$ can give rise to Lorentzian-type resonance, as in FR. When two bound states belonging to two configurations are coupled to the same continuum of states, the state $|E\rangle_{FF}$ is formed. Following the same method as that used in the case of $|E\rangle_F$, one can calculate the scattering matrices associated with $|E\rangle_{FF}$ by making an asymptotic analysis of the wave function $\tilde{\Psi}_E(\mathbf{r}) = \langle \mathbf{r}|E\rangle_{FF}$. Substituting equation (4.64) into equation (4.61), the asymptotic form of $\Psi_{FF}(\mathbf{r})$ is given by

$$\widetilde{\Psi}_E(\mathbf{r}) \xrightarrow[r \to \infty]{} \Psi_E(\mathbf{r}) + \int dE' \frac{\widetilde{b}_E \widetilde{\Lambda}_{E'}}{E - E'} \Psi_{E'}(\mathbf{r}), \tag{4.79}$$

where $\Psi_E(\mathbf{r})$ is the Fano wave function defined in equation (4.52). We have already obtained the asymptotic form of $\Psi_E(\mathbf{r})$. Resorting to partial-wave decomposition $\tilde{\Psi}_E(\mathbf{r}) \sim \sum_\ell \sqrt{2\ell + 1}\, \tilde{\psi}_\ell(r) Y_{\ell,0}(\theta, \phi)$, the asymptotic form of $\tilde{\psi}_\ell(r)$ is given by

$$\begin{aligned} \widetilde{\psi}_\ell(r) &\sim \sin(kr + \delta_\ell - \ell\pi/2 + \delta_r(k)) + \int 2k'dk' \frac{\widetilde{b}_E \widetilde{\Lambda}_{E'}}{k^2 - k'^2} \\ &\times \sin(k'r + \delta_\ell - \ell\pi/2 + \delta_r(k')), \end{aligned} \tag{4.80}$$

where δ_r is given by equation (4.60). We have seen earlier that $\tilde{b}_E$, $\tilde{\Lambda}_{E'}$, and T_ℓ all have poles only in the lower half-plane. Therefore, they are analytic in the upper half-plane. The quantity $\frac{1}{k^2 - k'^2 \pm i\eta}$ can have poles in both the upper and lower half-planes. Now, as before, by closing the contour of complex integration in the upper half-plane, we can explicitly calculate the above integral. The result of the integration is

$$\widetilde{\psi}_\ell(r) \sim \sin(kr + \delta_\ell - \ell\pi/2 + \delta_r(k)) - T'_\ell\, e^{i(kr+\delta_\ell - \ell\pi/2+\delta_r(k))}, \tag{4.81}$$

where

$$T'_\ell = \frac{\widetilde{\Gamma}/2}{E - E' - \widetilde{E}_{\text{shift}} + i\widetilde{\Gamma}/2}. \tag{4.82}$$

Note that the element T'_ℓ appears due to the involvement of the second bound state $|b'\rangle$, and recall that T_ℓ is due to the first bound state $|b\rangle$. We will now derive the total T-matrix element by writing the asymptotic form $\widetilde{\psi}_\ell$ in the following form:

$$\begin{aligned}\widetilde{\psi}_\ell &\sim \frac{1}{2i}e^{-i\delta_\ell}[-e^{-i(kr-\ell\pi/2)} + e^{2i\delta_\ell}e^{i(kr-\ell\pi/2)}]\\ &+ \frac{1}{2i}e^{-i\delta_\ell}\{-2iT'_\ell e^{2i\delta_\ell}e^{i(kr-\ell\pi/2)}\}\\ &= \frac{1}{2i}e^{-i\delta_\ell}[-e^{-i(kr-\ell\pi/2)} + S_\ell e^{-i(kr-\ell\pi/2)}],\end{aligned} \tag{4.83}$$

where

$$S_\ell = e^{2i\delta_\ell} - 2iT'_\ell\, e^{2i\delta_\ell} = e^{2i(\delta_\ell + \delta')} \tag{4.84}$$

is the ℓ-wave S-matrix element of the system with the phase δ'_ℓ defined by

$$e^{2i\delta'_\ell} = 1 - 2iT'_\ell. \tag{4.85}$$

The total T-matrix element $\bar{T}_\ell$ of the system is given by

$$\bar{T}_\ell = \frac{1}{2i}(1 - S_\ell) = T_\ell + T'_\ell\, e^{2i\delta_\ell}. \tag{4.86}$$

Substituting equations (4.82) and (4.58) into the above equation, we obtain

$$\bar{T}_\ell = T_\ell^0 + \frac{1}{\varepsilon + i}e^{2i\delta_\ell^0} + \frac{(\varepsilon + q)^2\Gamma'/2}{(\varepsilon^2 + 1)[E - E' - \widetilde{E}_{\text{shift}} + i\widetilde{\Gamma}/2]}e^{2i(\delta_\ell^0 + \delta_r)}, \tag{4.87}$$

where $T_\ell^0 = -e^{i\delta_\ell^0}\sin\delta_\ell^0$ is the background T-matrix element, with δ_ℓ^0 being the background phase shift. Recall that $\tilde{\Gamma} = (q + \varepsilon)^2/(\varepsilon^2 + 1)\Gamma'$. Clearly, $\bar{T}_\ell$ is a coherent superposition of the three terms on the rhs of the above equation. The scattering cross section $|\bar{T}|^2/k^2$ depends on the interference between these terms. It is interesting to note that for $q = -\varepsilon$, the second term vanishes; therefore, the cross section as a function of energy reduces to the usual resonant Lorentzian. On the other hand, for $q + \varepsilon \neq 0$, there is strong asymmetry in the energy profile of the resonance structure due to quantum interference. We will discuss the effects of quantum interference in resonances and spectra in more detail in the next two chapters.

4.6.3 Two coupled bound states, only one of which is coupled to a continuum

In the above derivation, we considered that both bound states were coupled to a common continuum. We now consider a case where $|b\rangle$ is coupled to $|E\rangle$ but $|b'\rangle$ is not coupled to $|E\rangle$, meaning that $\Lambda'_E = 0$ and so $\tilde{\Lambda}_E = gb_E$. We can now readily write down the T-matrix element for this case by taking the limit $\Lambda' \to 0$ in equation (4.87). The result is

$$\bar{T}_\ell = \frac{1}{\varepsilon + i} + \frac{\hbar\kappa_\ell}{E - E' - \mathscr{E}_\ell + i\hbar\kappa_\ell/2} e^{2i\delta_\ell}, \tag{4.88}$$

where

$$\kappa_\ell = \frac{2|g|^2}{\Gamma_\ell(\varepsilon^2 + 1)} \tag{4.89}$$

and

$$\mathscr{E}_\ell = \frac{4|g|^2}{\Gamma_\ell^2(\varepsilon^2 + 1)} E_{\text{shift}} \tag{4.90}$$

The above equation (4.88) can be rewritten in a different form by replacing

$$e^{2i\delta_\ell} = 1 - 2iT_\ell = 1 - \frac{2i}{\varepsilon + i}. \tag{4.91}$$

Thus, we have

$$\begin{aligned} \bar{T}_\ell &= \frac{1}{\varepsilon + i} \frac{E - \widetilde{E}' - i\hbar\kappa_\ell/2}{E - \widetilde{E}' + i\hbar\kappa_\ell/2} + \frac{\hbar\kappa_\ell/2}{E - \widetilde{E}' + i\hbar\kappa_\ell/2} \\ &= \frac{E - \widetilde{E}' - i\hbar\kappa_\ell/2 + \hbar\kappa(\varepsilon + i)/2}{(\varepsilon + i)[E - \widetilde{E}' + i\hbar\kappa_\ell/2]}, \end{aligned} \tag{4.92}$$

where $\tilde{E}_{b'} = E - E_{b'} - \mathscr{E}$. Therefore, the S-matrix can also be expressed in the following form:

$$\begin{aligned} S_\ell = 1 - 2iT_\ell &= \frac{(E - \widetilde{E}')(\varepsilon - i) - \hbar\kappa_\ell - i\hbar\kappa(\varepsilon + i)/2}{(\varepsilon + i)[E - \widetilde{E}' + i\hbar\kappa_\ell/2]} \\ &= \frac{(\varepsilon - i)(E - \widetilde{E}' - i\hbar\kappa_\ell/2)}{(\varepsilon + i)[E - \widetilde{E}' + i\hbar\kappa_\ell/2]}, \end{aligned} \tag{4.93}$$

which is manifestly unitary. The unitarity of the S-matrix is lost if the bound states decay irreversibly. A decaying bound state can be modeled in the present context by making the bound-state energy complex with a negative imaginary part. The analyticity of the S-matrix implies that a resonance appears at complex energies that have negative imaginary parts. The imaginary part of the bound-state energy corresponds to half the width of the spectrum of decay. We will come back to this topic in chapter 6, where bound-state decay will be discussed from the perspective of scattering theory.

4.7 Bound state in a continuum

A bound state in a continuum (BIC) is a discrete eigenstate with a positive-energy eigenvalue, implying that the bound state lies above the threshold of the lowest channel or in the continuum. The concept of a BIC was first proposed by von Neumann and Wigner in 1929 [4]. This is a counterintuitive concept, since a bound state is normally a negative-energy eigenstate. The amplitude of a BIC wave function falls off in space, making the wave function square-integrable like a normal bound state. Von Neumann and Wigner theoretically derived the BIC by an inverse method. Assuming the existence of a positive-energy square-integrable wave function, they sought an appropriate potential that could support such states. From the quantum scattering point of view, a BIC is formed due to the destructive interference of the outgoing waves scattered by the potential, creating an 'unusual' trap [5] for an electron [4]. Since its original formulation, the concept of the BIC has been extended to wider physical contexts over the years [6, 7], with applications to diverse areas [8–10]. Hsu *et al* [11] observed trapped light in a continuum as a realization of the BIC of radiation modes caused by the destructive interference of outgoing radiation amplitudes.

Four decades after its discovery [4], Stillinger and Herrick [7] considered a BIC in two-body interactions in atomic and molecular physics. In the case of two-body scattering, a BIC can be identified as a scattering resonance with zero width. As we discussed earlier, a resonance arises due to the existence of a quasi-bound state. In the absence of any dissipation, the finite width of a resonance is caused by the coupling of a bound or a quasi-bound state with the continuum of scattering states. This means that the width of a resonance vanishes if one creates an external condition by which the resonantly coupled bound or quasi-bound state can be decoupled from the continuum. This means that a resonance state with zero width becomes a BIC [6, 12]. The creation of a BIC in cold atomic collisions by manipulating photoassociative optical FRs with a pair of lasers tuned to two excited molecular bound states in the presence of MFR has been theoretically predicted [13]. Optical and magnetic FRs and the possibility of realizing a BIC in cold atoms will be discussed in the next chapter.

4.8 Resonant mean-field interaction

In chapter 2, we introduced Fermi's pseudo-potential or contact potential expressed in terms of of a_s only. This serves as the low-energy mean-field interaction potential for a dilute gas of particles. At resonances, a_s diverges, but the mean-field potential cannot diverge. The s-wave scattering amplitude given by

$$f_k = \frac{1}{k[\cot \delta_0(k) - i]} \tag{4.94}$$

can never diverge. The mean-field potential should be proportional to f_k under all conditions and not simply a_s. It is proportional to a_s only if $k\,|a_s| \ll 1$. In the limit $k \to 0$, when a_s diverges, f_k reduces to

$$f_k \simeq \frac{i}{k}, \tag{4.95}$$

which can easily be verified by replacing $\cot \delta_0(k)$ with its effective range expansion as discussed in chapter 2 and then taking the limit $k \rightarrow 0$. So, in this limit, the S-matrix element $S(k) = e^{2i\delta_0} = 1 + 2ikf_k = -1$, that is, $\delta_0 = (n + 1/2)\pi$, where n is an integer. Thus, at resonance, the scattering amplitude is limited by unitarity, and the energy or finite-range dependence of the scattering amplitude cannot be ignored. To describe resonant phenomena at low energy, one can make use of a class of finite-range model potentials [14] which are derived by the inverse scattering method of Jost and Kohn [15, 16] within the framework of the effective range expansion.

Let us recall the effective range expansion of the phase shift given by

$$\cot \delta_0(k) = -\frac{1}{ka_s} + \frac{1}{2}r_0 k + \cdots, \tag{4.96}$$

which is valid for $kr_0 < 1$ and $r_0 \ll |a_s|$. Jost and Kohn [15, 16] showed that for $a_s < 0$, one can derive the model potential

$$V_{-}(r) = -\frac{4\hbar^2}{\mu r_0^2} \frac{\alpha\beta^2 \exp(-2\beta r/r_0)}{[\alpha + \exp(-2\beta r/r_0)]^2}, \tag{4.97}$$

where $\alpha = \sqrt{1 - 2r_0/a_s}$, $\beta = 1 + \alpha$, and μ is the reduced mass. Here, α is bounded by $1 \leqslant \alpha < 2$. This potential holds good for $|\delta_0(E)| < \pi/2$. In the limit $a_s \rightarrow \pm{-\infty}$, equation (4.97) reduces to

$$V_\infty = -\frac{4\hbar^2}{\mu r_0^2 \cosh^2(2r/r_0)}, \tag{4.98}$$

which was employed by Carlson *et al* [17] for the quantum Monte Carlo simulation of a homogeneous superfluid Fermi gas with infinite negative scattering length.

We know that a large and positive a_s implies the existence of a near-zero-energy bound state. Therefore, finite-range model potentials for positive scattering lengths depend on the binding energy of the bound state. Let the binding energy be $E_b = -\hbar^2\kappa_0^2/2\mu$, where $\kappa_0 > 0$. As shown by Jost and Kohn [15, 16], one can derive another model potential that depends, apart from a_s and r_0, on a third parameter Λ that is related to κ_0 by

$$\kappa_0 = \frac{1}{r_0}[1 + \sqrt{1 - 2r_0/a_s}]\frac{1 + \Lambda}{1 - \Lambda}, \tag{4.99}$$

where $-1 < \Lambda < 1$ and $a_s > 2r_0$ for $r_0 > 0$. If one sets $\Lambda = -\sqrt{1 - 2r_0/a_s}$, the positive-$a_s$ model potential is given by [14]

$$V_{+}(r) = -\frac{4\hbar^2}{\mu r_0^2} \frac{\alpha\beta^2 \exp(-2\beta r/r_0)}{[1 + \alpha \exp(-2\beta r/r_0)]^2}. \tag{4.100}$$

This choice of Λ corresponds to $E_b \simeq \hbar^2/(2\mu a_s^2)$ for $2r_0/a_s \ll 1$.

Exercise

- Discuss energy-dependence of phase shift near a resonance in a single-channel scattering.
- Discuss analyticity of the **S**-matrix element near a single channel resonance. Discuss how bound states and resonances are related.
- Show that, for the scattering of a particle of mass m by a rectangular potential well with width r_0, the minimum potential depth for the occurrence of a zero-energy resonance is $\hbar^2\pi^2/(4mr_0^2)$.
- Derive s-wave **T**-matrix element of Feshbach resonance using two-channel model.
- What is Fano resonance? Make a comparative study between Feshbach and Fano resonances.

References

[1] Feshbach H 1958 Unified theory of nuclear reactions *Ann. Phys.* **5** 357–90
[2] Fano U 1961 Effects of configuration interaction on intensities and phase shifts *Phys. Rev.* **124** 1866–78
[3] Nussenzveig H M 1959 The poles of the s-matrix of a rectangular potential well of barrier *Nucl. Phys.* **11** 499–521
[4] Von Neumann J and Wigner E 1929 On unusual discrete eigenvalues *Z. phys.* **30** 465–7
[5] Smart A G 2013 A photonic crystal sets a peculiar trap *Phys. Today* **66** 14
[6] Fonda L 1961 Resonance reactions and continuous channels *Ann. Phys.* **12** 476–84
[7] Stillinger F H and Herrick D R 1975 Bound states in the continuum *Phys. Rev.* A **11** 446–4
[8] Bulgakov E N, Pichugin K N, Sadreev A F and Rotter I 2006 Bound states in the continuum in open Aharonov-Bohm rings *JETP Lett.* **84** 430–5
[9] Sadreev A F, Bulgakov E N and Rotter I 2006 Bound states in the continuum in open quantum billiards with a variable shape *Phys. Rev.* B **73** 235342
[10] Moiseyev N 2009 Suppression of Feshbach resonance widths in two-dimensional waveguides and quantum dots: a lower bound for the number of bound states in the continuum *Phys. Rev. Lett.* **102** 167404
[11] Hsu C W, Zhen B, Lee J, Chua S-L, Johnson S G, Joannopoulos J D and Soljačić M 2013 Observation of trapped light within the radiation continuum *Nature* **499** 188–91
[12] Friedrich H and Wintgen D 1985 Interfering resonances and bound states in the continuum *Phys. Rev.* A **32** 3231–42
[13] Deb B and Agarwal G S 2014 Creation and manipulation of bound states in the continuum with lasers: applications to cold atoms and molecules *Phys. Rev.* A **90** 063417
[14] Mal S, Adhikary K, Sardar D, Saha A K and Deb B 2019 Modeling atom–atom interactions at low energy by Jost–Kohn potentials *J. Phys. B: At. Mol. Opt. Phys* **52** 165004
[15] Kohn W 1962 The validity of Born expansions *Phys. Rev.* **87** 539
[16] Kohn W 1963 Variational methods for periodic lattices *Phys. Rev.* **87** 472
[17] Carlson J, Chang S-Y, Pandharipande V R and Schmidt K E 2003 Superfluid Fermi gases with large scattering length *Phys. Rev. Lett.* **91** 050401

Chapter 5

Cold atomic collisions

In the preceding four chapters, we have mainly described the fundamentals of formal scattering theory at low energy. In this chapter, we apply this theory to cold atomic collisions. We first discuss the applications to atomic scattering in the absence and then in the presence of an external field.

In the millikelvin or submillikelvin temperature regime, atomic hyperfine (HF) interactions—that is, interactions between the nuclear and electronic spins of an atom—become important in atomic collisions, even in the absence of any external field, because the energy scale associated with the HF interaction is typically on the order of 100 mK. Therefore, to understand low-energy atomic collisions, it is instructive to begin with a general introduction to atomic HF interactions. It is also essential to have a basic understanding of the origin and the method of constructing diatomic molecular potentials to understand low-energy atomic scattering from first principles. A detailed discussion of these topics, which would require some knowledge of atomic and molecular structure, is beyond the scope of this book.

5.1 Atomic hyperfine interactions

The effects of nuclear spin and the finiteness of the nuclear volume on the atomic (electronic) energy levels are called HF effects [1, 2]. These effects lead to splits and shifts of the energy levels. The effects that cause splitting are called HF structural effects, and the shifts occur due to isotopic effects [1]. The splits in the energy levels range from 1 MHz to 10 GHz. HF structure in an atom primarily arises due to the magnetic dipole and electric quadrupole moments of the nucleus. For an atom in the electronic ground state, HF structure is caused by the interaction between the magnetic moment of a valence electron of the atom and the magnetic dipolar field produced by the nuclear spin magnetic moment at the electron's position, or, conversely, the interaction of the nuclear spin magnetic moment with the dipolar field produced by the magnetic moment of the electrons at the nuclear position. For an electronically excited atom, HF structure originates from two sources:

doi:10.1088/978-0-7503-3959-9ch5

© IOP Publishing Ltd 2025. All rights, including for text and data mining (TDM), artificial intelligence (AI) training, and similar technologies, are reserved.

(1) the magnetic dipolar interaction between the nuclear spin magnetic moment and the magnetic moment of the electron, and (2) the interaction between the electric quadrupole moment of the protons in the nucleus and the gradient of the electric field produced by the electron at the position of the proton.

Let us consider the magnetic dipole HF interaction. Let $\mathbf{M}_n$ be the magnetic moment due to the nuclear spin. The corresponding dipolar field at a position given by the vector $\mathbf{r}$ (measured from the center of the nucleus) is:

$$\mathbf{B}_n = -\frac{\mathbf{M}_n}{r^3} + \frac{3(\mathbf{M}_n \cdot \mathbf{r})\mathbf{r}}{r^5}. \tag{5.1}$$

Let $\mathbf{M}_e$ be the net magnetic moment of the valence electrons of an atom. The Hamiltonian $\hat{H}_{hf} = -\mathbf{M}_e \cdot \mathbf{B}_n$ describing the HF interaction is given by

$$\hat{H}_{hf} = \frac{\mathbf{M}_n \cdot \mathbf{M}_e}{r^3} - \frac{3(\mathbf{M}_n \cdot \mathbf{r})\mathbf{M}_e \cdot \mathbf{r}}{r^5}. \tag{5.2}$$

Obviously, this is an anisotropic interaction, and its matrix elements depend on the coupled angular momentum states of the nuclear angular momentum and the total angular momentum (orbital plus spin angular momentum) of the electrons of the atom. In the absence of any external magnetic field, the total HF angular momentum of an atom a is $\mathbf{f}_a = \mathbf{i}_a + \mathbf{j}_b$ is conserved, where $\mathbf{i}_a$ and $\mathbf{j}_a$ denote the nuclear spin and the total electronic angular momentum, respectively. The expectation value of the operator $\mathbf{f}^2$ is $f(f+1)\hbar^2$, where f is called the HF quantum number. $\mathbf{M}_n = \mu_n g_n \mathbf{i}/\hbar$ and $\mathbf{M}_e = \mu_B g_j \mathbf{j}/\hbar$, where μ_n and g_n denote the nuclear Bohr magneton and the nuclear Landé g factor, respectively, and μ_B and g_j represent the electronic Bohr magneton and Landé g factor, respectively. $\hat{H}_{hf}$ can be expressed in the form $\hat{H}_{hf} = h_{hf}\mathbf{i} \cdot \mathbf{j}$, where h_{hf} represents the strength of the interaction that depends on the electronic wave function. For the ground-state atoms, $\mathbf{j} = \mathbf{s}$ is the total spin of the valence electrons, and the HF interaction reduces to the interaction between the nuclear and electronic spins.

The HF quantum number can be either an integer or an odd integral multiple of one-half. In fact, the HF quantum number determines the quantum statistical properties of the neutral atoms in an ensemble of atoms. Neutral atoms are classified into bosonic and fermionic atoms according to their HF quantum numbers. Atoms possessing integer HF quantum numbers are bosonic atoms, while those having odd integral multiples of one-half HF quantum numbers are fermions. Since the exchange symmetry of two indistinguishable neutral atoms is dictated by their HF quantum numbers, a low-energy quantum collision between two indistinguishable atoms is influenced by their HF quantum numbers.

In two-atom collisions, the central and noncentral interactions between the atoms at short separations lead to the coupling of the HF angular momenta of the two atoms. Coulombic interactions between the constituent nuclei and the electrons of a diatomic molecule, and those between the nuclei, lead to the central molecular potentials. The noncentral potentials result from the spin–spin or spin-dipolar interactions between the constituent particles of the molecule. At long separations, the joint wave function of the two atoms is product separable, and scattering channels can be represented by an uncoupled angular momentum basis. However, this is no longer true at short separations (typically below 20 a_0), where noncentral interactions cause spin–spin interactions. Therefore, it is necessary to express the short-range coupled basis in terms of the channel or uncoupled basis, or vice versa, as and when required. In the absence of any external magnetic field, the total HF angular momentum $\mathbf{F} = \mathbf{f}_a + \mathbf{f}_b$ of two atoms a and b is conserved. This means that the group of channels belonging to the HF number F does not mix with any channel of different F. However, this is no longer true for collisions in the presence of an external magnetic field, as the rotational symmetry of the system breaks down. As a consequence, the channels in the presence of a magnetic field depend on the magnetic quantum numbers (projections of various angular momenta along the magnetic field direction). In this case, F is no longer a good quantum number, but its projection M_F in the magnetic field direction remains a good quantum number. Low-energy collisions in the presence of an external magnetic field are complex phenomena that we will discuss in some detail in this chapter. First, let us understand how molecular or atom–atom potentials depend on the HF interactions in the absence of an external magnetic field.

5.2 Atom–atom potentials

The general features of single-channel low-energy atomic scattering are primarily determined by the long-range parts of the diatomic or atom–atom potentials, because the short-range potentials do not significantly influence the scattering wave function. The short-range parts of the potentials are generally much greater than the collision energy. As a result, for the s-wave, the wave function oscillates mainly with a short local wavelength and a small amplitude at short range. For higher partial waves, the wave amplitude diminishes at short separations, except for the shape resonances. This means that the probability for the two atoms to come together at short range is very small at low energy; in other words, two colliding atoms mostly experience the long-range part of the potentials. For multichannel collisions, the situation is different in the case of resonances, where the two atoms can temporarily bind together at short separations. The short-range parts of the potentials of a diatomic molecule form as the net result of several interactions: the Coulomb repulsion between the two positively charged nuclei, Coulomb attractions of the electrons toward the nuclei, exchange interactions between the overlapping electron clouds of the two atoms, spin–orbit interactions, etc. which are beyond the scope of discussion in this book. We refer the interested reader to [1, 3]. Here, we discuss some basic features of the long-range parts of the potentials that are relevant for cold atomic collisions.

5.2.1 Born–Oppenheimer approximation

The calculations of molecular potentials rely heavily on the Born–Oppenheimer (BO) approximation [4], which states that, to a first approximation, the electronic, vibrational, and rotational motions of a molecule are uncoupled, and therefore the total molecular wave function is product separable into electronic, vibrational, and rotational wave functions. In other words, this implies that when calculating electronic motion, one can disregard vibrational and rotational motions; the variables that describe the vibrational and rotational motions then appear as parameters for the electronic motion. This approximation is based on the smallness of the ratio of the mass m_e of the atomic electron to the mass m_n of the nucleus. This ratio is typically on the order of 10^{-4}. The vibrational energy is smaller than the electronic energy by a factor of $\sqrt{m_e/m_n}$, and the rotational energy is smaller again than the vibrational one by the same factor. For instance, the electronic energy is typically on the order of 10^{14} Hz, the vibrational energy is on the order of 10^{12} Hz, and the rotational energy is on the order of 10^{10} Hz. This means that both vibrational and rotational motions are much smaller than the electronic motion. This clear separation of the degrees of freedom (DOFs) of these three types of motion is the main basis of the BO approximation. The variation of these three types of motional energy with $\sqrt{m_e/m_n}$ can be understood from the following simple heuristic arguments. The long-range potentials between two atoms can also be obtained by performing a multipole expansion of the Coulomb potentials between the nuclei and the electrons in the center of mass (COM) frame of the two atoms and then averaging over the electronic states of the individual atoms.

As the two atoms approach each other to form a diatomic molecule, the valence electrons of the two atoms play the main role in forming the molecular bonding through their attractive Coulombic interactions with the nuclei of both atoms, while the inner-shell electrons remain practically unaffected. This means that the typical distance of the valence electrons from either of the nuclei is on the order of the atomic size, which is typically on the order of the Bohr radius, a_0. Therefore, the typical electronic energy is

$$E_e \sim \frac{\hbar^2}{m_e a_0^2}. \tag{5.3}$$

Considering that the motion of electrons and the vibration of the relative motion of the two nuclei are of a simple harmonic type (due to the same electrostatic force with a spring constant of K), it follows that the frequency of the electron's motion is $\omega_e = \sqrt{K/m_e}$, while that of the nuclear motion is $\omega_n = \sqrt{K/m_n}$. Thus, $E_e \simeq \hbar\omega_e$, and the energy of quantized vibrational motion is

$$E_v \sim \hbar\omega_n = \hbar\sqrt{\frac{K}{m_e}}\sqrt{\frac{m_e}{m_n}} = \sqrt{\frac{m_e}{m_n}}E_e. \tag{5.4}$$

Therefore, the ratio of vibrational to electronic energy follows $\sqrt{\frac{m_e}{m_n}}$. The rotational energy associated with the relative motion of the two nuclei is given by

$$E_r = \frac{\mathbf{L}^2}{2I} \sim \frac{\hbar^2}{m_n a_0^2} = \frac{m_e}{m_n}E_e = \sqrt{\frac{m_e}{m_n}}E_v, \tag{5.5}$$

where $\mathbf{L}$ is the angular momentum, which is $\sim\hbar$, and $I = m_n a_0^2/2$ stands for the moment of inertia of the configuration about the axis of rotation, which is perpendicular to the internuclear axis. Thus, E_r is smaller than E_v by $\sqrt{\frac{m_e}{m_n}}$.

The formation of molecular bonds is primarily due to the attractive interactions between the electrons and the nuclei. At short separations, the electron clouds of the two atoms overlap. For a diatomic molecule composed of two one-electron atoms, one can imagine two competing situations: (1) both electrons are more strongly attracted to one of the two nuclei, leading to ionic bonding; (2) both electrons are more or less equally attracted toward both nuclei, leading to covalent bonding. While covalent bonding is dominant at large separations, at short separations, both types of bonding are equally probable. Under the BO approximation, the first electronic wave functions are evaluated by solving the electronic eigenvalue equations, assuming that the motions of the nuclei are frozen (i.e. neglecting the kinetic energy terms associated with nuclear motion). These electronic eigenvalues depend on the nuclear coordinates parametrically. The vibrational and rotational wave functions are then calculated by solving the corresponding Schrödinger wave equations, where the electronic eigenvalues are treated as functions of the nuclear coordinates and hence as molecular potentials. These potentials are generally called adiabatic BO potentials. They depend on the internal states of the valence electrons of the atoms and the symmetry of the electronic states. The potentials of a particular symmetry do not cross, but potentials of different symmetries may cross. A perturbative approach to calculating the short-range part of the adiabatic BO potentials is to use molecular orbitals, which are linear combinations of single-electron atomic orbitals (single-electron bound wave functions of an atom) [1]. When two alkali-type single-electron atoms A and B form a molecule, one can use hydrogen-like atomic orbitals to construct molecular orbitals for a single electron. One then starts with the wave function of the valence electron of a diatomic molecular ion (for instance, a hydrogen molecular ion). Since this electron moves under the influence of the Coulomb forces produced by both nuclei, this is a two-center problem. Then, considering the COM of the system as the origin of the coordinate frame with the internuclear axis as the z-axis of this molecule- or body-fixed coordinate frame with fixed internuclear separation, one constructs the molecular orbital as a linear combination of two atomic wave functions—one wave function for atom A, for which the electron e is assumed to belong to the nucleus of A, and the other for which the same electron is assumed to belong to the nucleus of B. Obviously, these two atomic wave functions are non-orthogonal, since they describe the motion of the same electron with the same orbital symmetry. The normalization of this wave function depends on the overlap integral of the two wave functions. One thus calculates the first-order energy correction of the system, for which the unperturbed energy is simply the single-electron atomic energy of the specific orbital under consideration. This first-order energy correction is calculated for different fixed internuclear separations. When this energy is plotted as a function of separation, this function is the so-called adiabatic BO potential of the molecular ion. The molecular orbital of a neutral diatomic molecule formed from two single-electron atoms is constructed by taking the products of the atomic orbitals. We will not delve further into the calculations of the short-range part of the molecular potentials. For the details of molecular orbital theory, we refer the reader to [1].

The binding potentials, that is, the potentials with a minimum, can support a large number of vibrational states, which are characterized by the vibrational quantum number v. Usually, the ground (lowest-energy) vibrational level is labelled $v = 0$; accordingly, the higher-energy vibrational levels have ascending vibrational quantum numbers. However, for the purpose of describing low-energy resonances and near-threshold molecular bound states, we will designate the highest-energy vibrational level as $v_t = 0$, where the subscript 't' refers to the threshold. So, if a potential supports $N + 1$ vibrational levels, then the vibrational level with $v = N$ is assigned $v_t = 0$. In this way, one can count the vibrational levels starting from the threshold. Because diatomic molecular potentials are anharmonic, successive vibrational levels are unequally spaced; in this respect, they differ from harmonic potentials. Although a small portion of the potential in the vicinity of the equilibrium position appears harmonic, the anharmonicity increases as the separation between the atoms increases; in particular, the potential near the dissociation threshold is highly anharmonic. As a result, as the vibrational energy increases, the spacing between the levels decreases non-monotonically, and near the threshold, the vibrational level density becomes quite large. As the vibrational energy approaches the threshold limit, the levels form a quasi-continuum, and a continuum is reached when the molecule dissociates close to zero energy. Therefore, when two atoms collide with near-zero energy, there is a possibility of forming a bound state in the vicinity of the threshold, provided there exists an inelastic channel through which a third body can take the excess energy. This third body may be a photon or another atom.

A typical adiabatic diatomic molecular potential and the vibrational states supported by the negative-energy part of the potential are schematically shown in figure 5.1. The positive-energy states above the dissociation threshold belong to the

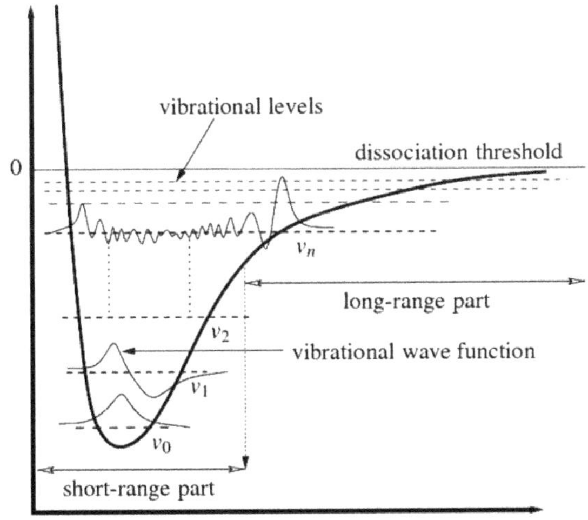

Figure 5.1. Schematic of a typical adiabatic diatomic molecular potential and a few representative vibrational wave functions.

class of scattering continuum states, while the negative-energy part below the threshold corresponds to the class of discrete states characterized by vibrational and rotational quantum numbers. In the short-range part, the electron clouds of the two atoms overlap, and the exchange interaction dominates. Typically, the short-range part is limited to below 20 a_0. The maximum amplitude of the ground vibrational wave function lies near the equilibrium position. As the vibrational quantum number increases, the amplitude of the vibrational wave functions near the equilibrium position decreases, and as the vibrational level goes upward toward the dissociation threshold, the amplitude of the wave functions starts to be increasingly concentrated near the two outer turning points. Also, as the vibrational quantum number increases, the energy spacing between the vibrational levels decreases, and close to the dissociation threshold, the vibrational levels form a quasi-continuum. In contrast, the spacing between the rotational levels increases with increasing rotational quantum number. The spacing between the rotational levels $J + 1$ and J is $2(J + 1)B_r$, where B_r is the rotational constant.

5.2.2 Long-range molecular potentials

For low-energy non-resonant collisions, the long-range part of the adiabatic potentials is the most important. Since a resonance at low energy is connected with the existence of a bound or quasi-bound state that primarily lies at a shorter range, the short-range part is also important for low-energy resonant collisions.

The long-range part originates from dispersive electromagnetic interactions between the two atoms at large separations, where the overlap between the charge clouds of the two atoms is negligible, but the separation is not so large that retardation effects are important. For two spherically symmetric atoms, the non-retarded long-range part of the potential can be expressed in the following form [5]:

$$V(R) = -\frac{C_6}{R^6} - \frac{C_8}{R^8} - \frac{C_{10}}{R^{10}} + \cdots, \tag{5.6}$$

where C_6, C_8, and C_{10} are generally called dispersion coefficients and R is the interatomic separation. Specifically, C_6, C_8, and C_{10} are called dipole–dipole (or van der Waals), dipole–octopole, and octopole–octopole coefficients, respectively. These coefficients for two ground-state homonuclear alkali metal atoms have been reported by Marinescu, Sadeghpour, and Dalgarno [6]. The origin of these terms may be understood in the following way. In general, the Coulomb potential of the charges of a diatomic molecule can be expanded in a multipolar form when the two nuclei are largely separated, considering the molecule as a two-center system. An electronic charge belonging to either center (nucleus) is coupled to the Coulomb potential created by another electronic charge belonging to the other center, leading to an electrostatic interaction Hamiltonian that can be treated as a perturbation in calculating electronic wave functions and energies under the BO approximation. For two S atoms (spherically symmetric atoms), the second-order perturbation calculations yield the above expression for the long-range potential. The detailed calculations are lengthy; we refer interested readers to the article by Dalgarno and

Davison [5]. For one atom in the S state and another in the P state, the long-range potential in the leading order varies as $1/R^3$, while for two P atoms, the long-range potential varies as $1/R^5$ in the leading order. Dispersion coefficients for excited-state alkali metal dimers have been calculated by Marinescu and Dalgarno [7].

Now, let us discuss how these analytical forms of the atom–atom long-range potentials provide us with valuable information about some general qualitative features of cold atomic collisions. In section 2.4, we derived the Wigner threshold laws, according to which, for a potential that becomes a van der Waals type of potential as $R \to \infty$, for $\ell \leqslant 3/2$, i.e. for $\ell = 0$ and $\ell = 1$, we have $f_\ell(k) \simeq -a_\ell k^{2\ell}$, while for all other $\ell \geqslant 1$, $f_\ell(k) \sim k^{n-3}$ as $k \to 0$. This means that for a potential that asymptotically varies as $1/R^6$, only the s-wave scattering amplitude in the limit $k \to 0$ becomes a constant, which is the negative of the s-wave scattering length. Scattering amplitudes for other higher partial waves go to zero as $k \to 0$. Similarly, for a potential that varies as $1/R^4$ as $R \to \infty$, only the s-wave amplitude is constant. The interaction between an ion and a neutral atom scales as C_4/R^4 at large separations, where C_4 is the coefficient. The origin of the C_4/R^4 interaction lies in the polarization of the atom by the Coulomb field of the ion—the ionic field induces the dipole moment $\mathbf{d}_a = \alpha \mathbf{E}_\mathbf{i}$ in the atom, where $\mathbf{E}_i = \hat{R} q_i/(4\pi\varepsilon_0 R^2)$ is the electric field produced by the ion of charge q_i at the position of the atom, with $\hat{R}$ being the unit radial vector. The interaction between $\mathbf{d}_a$ and the field itself results in the C_4/R^4 potential.

5.2.3 Fundamentals of symmetry concepts in diatomic molecules

We discussed the interplay between resonances and bound states in the previous chapter. In the case of cold binary atomic collisions, the bound states obviously refer to diatomic ro-vibrational molecular states. It is therefore instructive to discuss some basics of diatomic molecular physics, in particular, the general properties of molecular bound states near the dissociation threshold, not only for understanding scattering resonances but also for forming molecules close to the threshold through three-body collisions. Of course, pure two-body collisions cannot produce a real bound state (a negative-energy bound state, that is, a bound state below the threshold), because the formation of such a bound state by collisions requires the excess energy (the collision energy minus the bound-state energy) to be carried away by a third body. Here, the third body may be a third particle or a photon. Thus, molecule formation by collisions is essentially an inelastic or reactive process. Cold collisions in the presence of an external magnetic or laser field have enabled access to hitherto unexplored regimes of molecular physics and the creation of exotic long-range diatomic molecules by associating two cold atoms. As a result, a new field of spectroscopy, namely association spectroscopy, has emerged.

Since we are primarily interested in diatomic molecules for our purpose of binary atomic collisions, we confine ourselves to the symmetry properties of diatomic molecules. For details, we refer the reader to the book by G Herzberg on the structure and spectra of diatomic molecules [8]. The symmetry of a molecular potential is determined by the various molecular angular momenta (such as electronic, spin, rotational, and HF) and their mutual coupling, as well as the coupling of the angular momenta with the molecular symmetry axes. A diatomic molecule has cylindrical symmetry—the internuclear axis is an axis of symmetry. This axis is the molecule-fixed or body-fixed z-axis, with the origin of the body-fixed coordinate frame being the COM of the diatom. We let (x', y', z') denote the Cartesian coordinates of the body-fixed frame and let (x, y, z) denote those in the laboratory frame. We call the body-fixed frame the 'prime' frame. The projection $J_{z'}$ of the total angular momentum $\mathbf{J}$ along the z'-axis is a conserved quantity. $\mathbf{J}$ is composed of the total electronic orbital angular momentum $\mathbf{L} = \mathbf{l}_1 + \mathbf{l}_2$, the total spin angular momentum $\mathbf{S} = \mathbf{s}_1 + \mathbf{s}_2$, and the angular momentum $\mathbf{R}_n$ of rotation about the z'-axis due to the relative motion of the two nuclei. For two free atoms, if one ignores the mass of the atomic electrons compared to that of the nuclei, then $\mathbf{R}_n$ is equivalent to the angular momentum ℓ about the z'-axis of the relative motion of the two atoms. Thus, the eigenvalue $\hbar^2\ell(\ell + 1)$ of ℓ^2 is the continuum counterpart of the eigenvalue $\hbar^2 R_n(R_n + 1)$ of $\mathbf{R}_n$, implying that the quantum number ℓ is the continuum counterpart of R_n. The total electronic angular momentum of the molecule is $\mathbf{J}_e = \mathbf{L} + \mathbf{S}$. Note that since the nuclear spin is much more weakly coupled to the z'-axis compared to the couplings of $\mathbf{L}$ and $\mathbf{S}$ or $\mathbf{J}_e$, nuclear spins are neglected as a first approximation. The magnitude of the projection of $\mathbf{L}$ on the z'-axis, that is, $|L_{z'}|$, is denoted by Λ. For a nonzero Λ, $L_{z'}$ can assume two values: $L_{z'} = \pm\Lambda$. The vector $\vec{\Lambda}$ is defined such that Λ is its magnitude and its direction is aligned with the positive or negative z'-axis. The projection of $\mathbf{S}$ onto the z'-axis is denoted by Σ, which can be treated similarly to the vector $\vec{\Sigma}$. The projection $\mathbf{J}_e$ on the z'-axis is represented by Ω, which, along with its direction along the z'-axis, can be treated as the vector Ω. The projection of $\mathbf{R}_n$ on the z'-axis is zero, since $\mathbf{R}_n$ is perpendicular to the z'-axis.

In order to classify molecular electronic states, let us first consider different energy scales in molecular dynamics. There are three main energy scales, namely the electronic, spin–orbit, and rotational energy scales. The symmetry properties of electronic molecular states can be defined in certain limiting cases of these energy scales by considering the splitting in energy levels produced by molecular Coulomb, spin–orbit, and rotational interactions. Here, by rotational interaction, we mean the interaction of the electronic angular momentum with the internuclear axis rotation $\mathbf{R}_n$. The electronic energy levels depend on the internuclear separation r. The electronic energy scale of a particular electronic state may be defined in terms of the splitting between the BO potentials. Since BO potentials depend on the internuclear distance, one may define the electronic energy scale of a particular molecular electronic state using the BO potential energy gap ΔE_e at the equilibrium position of that state. The spin–orbit energy scale may be defined by an average splitting $\Delta E_{\rm so}$ due to the molecular spin–orbit interaction. Similarly, the rotational energy scale may be defined by a rotational splitting $\Delta E_{\rm rot}$. For diatomic molecules, the symmetry of the potentials can be understood by applying Hund's coupling schemes, also known as Hund's cases. Molecular electronic states are classified according to some limiting conditions on these energy scales. These classifications are known as Hund's cases:

Case (a):

$\Delta E_e \gg \Delta E_{so} \gg \Delta E_{rot}$. In this case, since electronic energy dominates over the spin–orbit interaction energy of an electron, the total orbital angular momentum $\mathbf{L}$ of electrons is coupled to the symmetry axis (in the case of a diatomic molecule, it is the internuclear axis) more strongly compared to the strength of the spin–orbit coupling. As a result, both orbital and spin angular momenta precess around the symmetry axis independently, making their projections Λ and Σ onto the z'-axis good quantum numbers. The molecular electronic states are characterized according to the values of Λ. The state with $\Lambda = 0$ is called a Σ state (not to be confused with the projection of $\mathbf{S}$ on the z'-axis), while the states with $\Lambda = 1$ and $\Lambda = 2$ are called Π and Δ states, respectively. The total angular momentum is $\mathbf{J} = \Omega + \mathbf{R}_n$. Note that $\Omega = \Lambda + \Sigma$.

Case (b):

$\Delta E_e \gg \Delta E_{rot} \gg \Delta E_{so}$. In this case, as in case (a), Λ is a good quantum number, but $\mathbf{R}_n$ is first coupled to Λ to form the resultant angular momentum $\mathbf{N} = \mathbf{R}_n + \Lambda$. The total angular momentum is given by $\mathbf{J} = \mathbf{N} + \mathbf{S}$.

Case (c):

$\Delta E_{so} \gg \Delta E_e \gg \Delta E_{rot}$. In this case, neither Λ nor Σ is a good quantum number, but Ω is a good quantum number, and $\mathbf{J} = \mathbf{J}_e + \mathbf{R}_n$.

Case (d):

$\Delta E_{rot} \gg \Delta E_e \gg \Delta E_{so}$. In this case, neither Λ nor Σ is a good quantum number, but $\mathbf{N} = \mathbf{L} + \mathbf{R}_n$ is a good quantum number and $\mathbf{J} = \mathbf{N} + \mathbf{S}$. For $\mathbf{L} = 0$, Hund's cases (b) and (d) are equivalent.

Case (e):

$\Delta E_{rot} \gg \Delta E_{so} \gg \Delta E_e$. In this case, the total angular momentum is $\mathbf{J} = \mathbf{J}_e + \mathbf{R}_n$.

We now discuss the different term symbols that specify the nature of the electronic states and symmetries. For a diatomic molecule, one of the symmetries is reflection symmetry, which states that under reflection of the electron positions upon any plane containing the internuclear axis, the electronic wave function remains the same (positive symmetry) or acquires a negative sign (negative symmetry). Homonuclear molecules have a COM inversion symmetry: if the positions of both electrons are reversed in the prime frame (COM frame), the wave function remains the same or changes sign. For Hund's case (a), the term symbol for a homonuclear molecule with $\Lambda = 0$ is denoted by ${}^{2S+1}\Sigma^{\pm}_{g(u)}$, where $(2S + 1)$ is spin multiplicity and $g(u)$ stands for 'gerade (ungerade),' which refers to the COM inversion symmetry; gerade implies positive (symmetric) COM inversion symmetry, and ungerade implies negative (antisymmetric) COM inversion symmetry. The superscript $\pm$ refers to the reflection symmetry: '+' for positive and '-' for negative symmetry. For a heteronuclear molecule, there is no COM inversion symmetry, and so the subscript 'g' or 'u' is absent in the term symbols. Reflection symmetry is applicable only for the Σ state because for $\Lambda \neq 0$, this symmetry is lifted by an effect called Λ-doubling, which is a perturbative phenomenon due to the interaction between the electronic orbital angular momentum and the rotation of the internuclear axis. This effect mixes the two angular states corresponding to $L_{z'} = +\Lambda$ and $L_{z'} = -\Lambda$. This can be understood in the following way: consider reflection in the $x - z$ plane in the prime frame. This reflection is equivalent to the transformation $y' \to -y'$, which in turn implies $L_{z'} = x'\frac{\partial}{\partial y'} - y'\frac{\partial}{\partial x'} \to -L_{z'}$. The eigenfunction of $L_{z'}$ is $(1/\sqrt{2}\pi)\exp[i\Lambda\phi']$. So, the reflection symmetry for $\Lambda \neq 0$ implies $(1/\sqrt{2}\pi)\exp[-i\Lambda\phi'] \to (1/\sqrt{2}\pi)\exp[-i\Lambda\phi']$. In the absence of a spin rotation (or orbital angular momentum rotation) interaction, the eigenfunctions are $\Phi_\pm = (1/\sqrt{2}\pi)[\exp(i\Lambda\phi') \pm \exp(-i\Lambda\phi')]$. Thus, under reflection, $\Phi_+ \to \Phi_+$ and $\Phi_- \to -\Phi_-$. In the presence of the interaction between electronic orbital angular momentum and axial rotation, these two states are no longer degenerate, and so the reflection symmetry is not a good symmetry for $\Lambda \neq 0$. In general, for Hund's cases (a) and (b), the term symbols of homonuclear diatomic molecules with $\Lambda \neq 0$ are $^{2S+1}X_{g(u)}$ for gerade (ungerade) symmetry, respectively, where X stands for Π ($\Lambda = 1$), Δ ($\Lambda = 2$), etc. Again, for heteronuclear molecules, the subscript $g(u)$ is absent. For Hund's case (c), the term symbol of a homonuclear molecule is $\Omega^{\pm}_{g(u)}$ for $\Omega = 0$ and $\Omega_{g(u)}$ for $\Omega \neq 0$, while that for heteronuclear molecules is $\Omega^{\pm}$ for $\Omega = 0$ and simply Ω for $\Omega \neq 0$. The Σ states for different values of L are distinguished by placing an uppercase letter before the $'\Sigma'$ symbol: X Σ for $L = 0$.

5.3 Collisions between cold atoms

The advancement of laser and evaporative cooling and trapping of dilute atomic gases in the 1980s and 1990s [9, 10] has opened a vast scope for research into atomic and molecular physics at ultralow temperatures. One of the landmark achievements enabled by the development of the technology of cooling and trapping is the demonstration of Bose–Einstein condensation [11–13] in dilute and weakly interacting atomic gases. The early developments in this endeavor focussed on understanding the collision phenomena of trapped atoms in laser light with the aim of minimizing heating, spin exchange, and the loss of atoms from a trap due to collisions. This fascinating story of scientific developments in the late 20th century has been chronicled in a number of reviews, monographs, and books. To name a few, we refer to the book by Arthur Ashkin [14], the book by Letokhov [15], and the Nobel lectures by Claude Cohen-Tannoudji [16], William Phillips [17], and Steven Chu [18], as well as the review of cold and ultracold atomic collisions by Weiner *et al* [19]. Here, our main purpose is to discuss the essential physics behind cold atomic collisions rather than to provide a detailed review of the field.

Since an ensemble of cold atoms is prepared in a trap, either in a magneto-optical trap (MOT) or in an optical trap, the ensemble is not uniform, and so the linear momentum is not a good quantum number. However, for a shallow or large-sized trap for which trapping frequencies are small, the atoms in the central part of the trap (where the atomic number density is much larger than that at the edges of the trap) are considered to be almost uniformly distributed. We consider collisions between the cold atoms in the central region of the trap, and so the methods of free-space (uniform system) collisions, as discussed in the preceding four chapters, are applicable. Therefore, in discussing collisions between cold atoms, particularly resonances, we do not explicitly consider the effects of trapping magnetic and optical fields on the collision phenomena. Thus, when we discuss collisions in an 'external field,' we are implying an applied field other than the trapping fields.

In this section, we focus on collisions in the absence of an external field. For two one-electron atoms (by 'one-electron,' we mean atoms having only one valence electron, such as alkali metal atoms), in the absence of any spin-dependent interaction and external fields, the adiabatic potentials are central potentials that depend only on the interatomic separation. These potentials can be categorized into two types: singlet and triplet potentials, depending on whether the joint state of the electronic spins of the two atoms is an antisymmetric (singlet) or symmetric combination (spin triplet) of the individual spin states. The Hamiltonian for the collision between two atoms is

$$\hat{H}_0 = \hat{H}_k + \hat{V}_c + \hat{H}_s + \hat{H}_{\mathrm{hf}}, \tag{5.7}$$

where $\hat{H}_k = -\frac{\hbar^2}{2\mu}\nabla^2$ is the kinetic part; $\hat{V}_c = V_0(R)P_0 + V_1(R)P_1$ is the central interaction part that consists of singlet and triplet potentials $V_0(R)$ and $V_1(R)$, respectively; and $P_0 = (1/4)|S = 0\rangle\langle S = 0|$ and $P_1 = (3/4)|S = 1\rangle\langle S = 1|$ are the projection operators onto the spin-singlet and spin-triplet manifolds, respectively. Here, S is the total spin quantum number such that $\mathbf{S} = \mathbf{s}_1 + \mathbf{s}_2$, and $\hat{H}_s$ represents the spin-dipolar and spin–spin interactions. The HF interaction part given by

$$H_{\mathrm{hf}} = \frac{a_{\mathrm{hf}}}{\hbar^2}[\mathbf{s}_1 \cdot \mathbf{i}_1 + \mathbf{s}_2 \cdot \mathbf{i}_2] \tag{5.8}$$

describes the HF interaction between the two atoms. This can be rewritten in the form $H_{\mathrm{hf}} = V_{\mathrm{hf}}^{+} + V_{\mathrm{hf}}^{-}$, where

$$V_{\mathrm{hf}}^{+} = =\frac{a_{\mathrm{hf}}}{2\hbar^2}[(\mathbf{s}_1 + \mathbf{s}_2) \cdot (\mathbf{i}_1 + \mathbf{i}_2)] \tag{5.9}$$

$$V_{\mathrm{hf}}^{-} = \frac{a_{\mathrm{hf}}}{2\hbar^2}[(\mathbf{s}_1 - \mathbf{s}_2) \cdot (\mathbf{i}_1 - \mathbf{i}_2)]. \tag{5.10}$$

The inclusion of noncentral interactions such as HF, spin–spin, and spin- dipolar interactions makes the scattering process essentially a multichannel problem. To calculate the multichannel scattering properties, the first step is to select the appropriate channel states, that is, the internal or angular momentum states of the two atoms. In the absence of an external field, these channel states are denoted by $|f_1, f_2; (F, \ell; J, M_J)\rangle$, where f_1 and f_2 are the HF quantum numbers of atoms 1 and 2, respectively; F is the total HF quantum number such that $\mathbf{F} = \mathbf{f}_1 + \mathbf{f}_2$. Here, $\mathbf{J} = \mathbf{F} + \mathbf{L}_r$ is the total angular momentum, $\mathbf{L}_r$ is the relative angular momentum in the COM frame with a corresponding quantum number of the partial wave ℓ, and M_J is the projection of $\mathbf{J}$ on the internuclear axis. This is a useful basis in the asymptotic limit, but at short separations, because of the overlap of atomic charges, one has to use the coupled or molecular basis $|S, M_S; IM_I(F, \ell, J, M_F)\rangle$, where I is the total nuclear spin quantum number with $\mathbf{I} = \mathbf{i}_1 + \mathbf{i}_2$. Since the projection m_ℓ of $\mathbf{L}_r$ on the internuclear axis is zero, we have $M_J = M_F$, where M_F is the projection of $\mathbf{F}$ along the internuclear axis. In the absence of an external magnetic field, both J and M_F are conserved. In the presence of a magnetic field, only M_F is conserved. Using the angular momentum addition theorem, one is then required to express the uncoupled basis as a superposition of the coupled basis. This procedure enables one to calculate the matrix elements of various terms of molecular interactions that are particularly relevant at short separations.

In the absence of an external magnetic field, both J and M_F are conserved. Usually, the coupling between the HF spin and the relative motion of the two atoms or the rotation of the internuclear axis is very weak or negligible. Therefore, both F and ℓ remain approximately good quantum numbers, and so the angular momentum state of the two atoms in the COM frame is separable into the total HF and rotational (axial rotation) states. The number of channels is then determined by examining how many different combinations of f_1, f_2, and ℓ are possible for a particular F and M_F. The threshold energy of a channel is the sum of the HF interaction energies of the two atoms. The time-independent multichannel scattering problem can then be formulated in matrix form, as described in chapter 3. The problem can be solved numerically by close-coupling or other computational methods, as discussed in chapter 8. Most of the computational techniques are based on the Numerov–Cooley algorithm [20], which seeks to calculate the wave function in matrix form at a third point $r = r_3$ given the wave function at two previous adjacent points r_1 and r_2. Thus, the Numerov–Cooley algorithm progressively calculates the wave function from short separations, where boundary conditions for $r \to 0$ are imposed, to asymptotically large separations. From the values of the wave function in the asymptotic regime, the scattering matrices are deduced by imposing asymptotic boundary conditions. More details of the numerical methods are given in chapter 8.

The discovery of a Bose–Einstein condensate (BEC) in trapped atomic gases in 1995 triggered a flurry of research activities in ultracold atomic collisions. In 1996, two groups of authors [21, 22] carried out detailed numerical calculations of spin-polarized Rb–Rb [21] and Na–Na [22] cold collisions in an MOT. Their main purpose was to understand the trap loss due to collisional relaxations at ultralow temperatures. Mies *et al* [21] presented five-channel scattering calculations showing the effects of spin–spin dipolar and second-order spin–orbit interactions on the collisional relaxation in ^{87}Rb atoms initially prepared in a spin-polarized HF state $|f = 2, m_f = 2\rangle$. Asymptotic channels are product separable in terms of the individual atomic angular momentum basis; these channels are called diabatic channels. At short separations, the channel states are not product separable due to coupling, and under the BO approximation, they are called adiabatic channels. These two sets of channels are connected by a unitary transformation according to angular momentum coupling schemes. The total angular momentum $\mathbf{J} = \mathbf{f}_a + \mathbf{f}_b + \ell = \mathbf{F} + \ell$, the parity $p = \pm 1$, and the total magnetic quantum number

M_J are good quantum numbers. In the five-channel calculations presented in [21], $J = 4$, $M_J = 4$, and since ^{87}Rb atoms are bosons, $P = +1$ and only even ℓ are allowed. The five diabatic channels indexed by the symbol 'γ' are $|\gamma;\, \ell,\, F, f_a, f_b\rangle \rightarrow \{|1;\, 2,\, 2,\, 1,\, 1\rangle,\, |2;\, 2,\, 2,\, 2,\, 1\rangle,\, |3;\, 2,\, 3,\, 2,\, 1\rangle,\, |4;\, 0,\, 4,\, 2,\, 2\rangle,\, |5;\, 2,\, 2,\, 2,\, 2\rangle\}$. In the absence of a magnetic field, the threshold energy is the sum of the single-atom HF interaction energies. Thus, channels $\gamma = 2$ and 3 have the same thresholds; similarly, channels 4 and 5 are asymptotically degenerate. Thus, there are only three thresholds. If one sets the zero of the energy scale to the highest threshold, that is, at the threshold of the $\gamma = 4,\ 5$ channels, then the channel energy of $\gamma = 2,\ 3$ is -328.011 mK and that of $\gamma = 1$ is -656.022 mK. On the other hand, if one considers the threshold of the lowest channel $\gamma = 1$ to be the zero of the energy scale, then the channel energy of $\gamma = 2,\ 3$ is 328.011 mK and that of $\gamma = 4,\ 5$ is 656.022 mK. The adiabatic channels indexed by the symbol α are characterized by Λ, the total electronic spin $S \equiv 0,\ 1$, the COM symmetry $\sigma \equiv g,\ u$, and the reflection symmetry $\rho \equiv \pm$. The five adiabatic channels are $|\alpha;\, \Lambda,\, S,\, (^{2S+1}\Lambda_\sigma^\rho)\rangle \rightarrow \{|1;\, 0,\, 0,\, (^1\Sigma_g^+)\rangle,\, |2;\, 0,\, 1,\, (^3\Sigma_u^+)\rangle,\, |3;\, 0,\, 1,\, (^3\Sigma_u^+)\rangle,\, |4;\, 0,\, 1,\, (^3\Sigma_u^+)\rangle,\, |5;\, 0,\, 1,\, (^3\Sigma_u^+)\rangle\}$. The close-coupled multichannel calculations performed by Mies *et al* [21] provided bounds of 0.43×10^{-15} and 2.4×10^{-15} cm^3 s^{-1} for the coefficient of loss of atoms from the $|f = 2,\ m_f = 2\rangle$ state. Physically, this coefficient means a rate of two atoms per collision event.

5.4 Collisions in a magnetic field

At low energy, resonances in atomic collisions can be created and manipulated by external electric, magnetic, and electromagnetic fields. From our study of resonances and bound states in the previous chapter, we understood that the depth of a potential well must be sufficiently large to support at least one bound state in order to obtain a resonance in the scattering. This means that resonances imply the existence of a strong potential or strong particle–particle interaction. From the scattering point of view, an effective binary interaction depends on the scattering amplitude or phase shift. In other words, a resonance means the maximization of scattering amplitude. Therefore, creating and manipulating resonances with an external field implies generating and controlling strong interactions. Cold atoms are particularly suited for exploring such external-field-controlled interactions. Such external-field-induced resonances allow us to explore strongly interacting quantum many-body physics. Therefore, understanding cold atomic collisions in external fields is important not only for the advancement of atomic physics and chemistry but also for exploring some aspects of condensed-matter physics.

5.4.1 Magnetic Feshsbach resonance

The application of an external magnetic field lifts the degeneracy of the magnetic quantum numbers of various HF channel potentials. The strength of the magnetic fields that are applied in laboratory experiments on cold-atom Feshbach resonances generally does not exceed 1000 Gauss (G) or 0.1 Tesla. The magnetic field primarily influences the asymptotic part of the potentials by lifting the degeneracy of the HF spin states of individual atoms. However, any small change in the asymptotic potentials has significantly large effects on cold atomic collisions due to the modification of threshold behavior. Thus, in the presence of a magnetic field, the channel angular momentum states change due to the inclusion of magnetic quantum numbers; therefore, the number of nondegenerate channels increases.

We have already discussed resonances, bound states, and their interrelations in chapter 4. A magnetic Feshbach resonance (MFR) is essentially a multichannel resonance that results from the modification of the HF channel potentials—particularly the lowest channel and one or more higher channel potentials. This can be treated using the formalism of Feshbach resonances discussed in section 4.5. The original treatment of Feshbach made use of the projection operator technique. In the context of low-energy atomic collisions, the early theoretical studies of magnetically induced Feshbach resonances were carried out in the 1970s and 1980s using spin-polarized hydrogen [23], deuterium [24], and lithium atoms [25]. In the context of trapped ultracold atoms, theoretical predictions of near-zero-energy MFR were made in the early 1990s [22, 26, 27, 28], even before the realization of a BEC. The first experimental demonstration of MFR in a BEC of Na atoms was reported by Inoue *et al* in 1998 [29].

Although calculations of Feshbach resonances require multichannel scattering problems to be solved numerically, a simplified two-channel model, as discussed in section 4.5.1, suffices to elucidate the essential physical effects of an s-wave Feshbach resonance. We recall the expression for the resonant part of the **T**-matrix element given by equation (4.39), which we rewrite here:

$$T_r(k) = \frac{1}{\varepsilon + i} e^{2i\delta_0}, \tag{5.11}$$

where ε is defined in section 4.5.1 as a dimensionless detuning parameter given by

$$\varepsilon = \frac{E - \bar{E}_c}{\hbar\Gamma/2}. \tag{5.12}$$

Here, $\bar{E}_c = E_c + E_{\text{shift}}$ is the shifted energy of the closed-channel bound state, with the shift E_{shift} being caused by the coupling between the open (continuum) and closed channels. The linewidth Γ appears due to the same coupling. Thus, the resonant scattering cross section

$$\sigma_r = |f_r|^2 = \frac{1}{k^2} \frac{(\hbar\Gamma/2)^2}{(E - \bar{E}_c)^2 + (\hbar\Gamma/2)^2} \tag{5.13}$$

which clearly shows that the resonant scattering cross section is a Lorentzian.

In an MFR, the s-wave scattering length a_s is basically tuned by a magnetic field. Under the resonant magnetic field B_r, a_s diverges. So, as the magnetic field is varied near B_r, a_s goes from small to large positive values on one side of the resonant point (B_r) and from small to large negative values on the other side, with a discontinuity at B_r. The detuning parameter $E - \bar{E}_c$ is a function of the magnetic field detuning $B - B_r$. This means that by varying B across B_0, one can tune the bound-state energy across the lowest open-channel threshold. Resonance occurs when the tuning of the field makes the bound-state energy coincide or nearly coincide with the collision energy. When B is varied close to B_r (so that $|E - \bar{E}_c| \ll \hbar\Gamma/2$), in many cases of MFR, $E - \bar{E}_c$ is a linear function of $B - B_0$, particularly near the threshold of the lowest channel. Let us now determine the relationship between the resonant part a_r of a_s and the detuning. For this, let us recall that $-\cot \delta_r(k) = -\varepsilon$. Therefore,

$$a_r = \lim_{k \to 0} \frac{\hbar\Gamma/2}{k(E - \bar{E}_c)} = -\lim_{k \to 0} \frac{\hbar\Gamma/2}{k\bar{E}_c}. \tag{5.14}$$

This is independent of k if the linewidth Γ is a linear function of k. It is positive if $\bar{E}_c < 0$, that is, if the bound state lies below the threshold of the lowest channel; in other words, the bound state is stable in the absence of any other perturbation. Therefore, it is said that a large positive scattering length reveals the existence of a bound state close to the threshold. On the other hand, a large negative scattering length indicates that the bound state lies above the threshold; that is, it is a quasi-bound state embedded in the continuum. Thus, it is said that a large negative energy scattering length reveals the absence of a real bound state. The two-channel model of s-wave resonances, as discussed in section 4.5, shows that the total phase shift $\delta = \delta_0 + \delta_r$. Hence, $a_s = \lim_{k \to 0} \tan \delta/k \simeq a_{\rm bg} + a_r$, where $a_{\rm bg}$ is called the background scattering length. Clearly, a_s reduces to $a_{\rm bg}$ when the coupling between the open channel and the closed-channel bound state vanishes.

Let us now derive an expression for a_s as a function of the magnetic field tuned near B_r. The detailed calculations for this can be found in [26]. Consider again the two-channel model. In the absence of any magnetic field, the open-channel threshold energy is zero. An applied magnetic field shifts the thresholds of the channels and also the binding energy of a bound state due to Zeeman effects. Thus, for two ground-state atoms in the same HF spin state $|fm_f, fm_f\rangle$, the open-channel threshold experiences the shift $E_{th}^{\rm shift}(B) = -2\mu_s B$, where μ_s is the matrix element of the spin magnetic moment of the individual atom. Here, it is assumed that the field $\mathbf{B}$ is directed along the z-axis. At $B = B_r$, the bound-state energy $\tilde{E}_c$ coincides with the open-channel threshold. Since a_r is inversely proportional to $\tilde{E}_c$ and a_r diverges at $B = B_r$, we must set $\tilde{E}_c(B_r) = 0$. Let μ_c be the magnetic moment of the bound state. Then we have $\tilde{E}_c = -\mu_c(B - B_r)$. Similarly, the open-channel threshold value must be set to zero at $B = B_r$, implying that we have to set the threshold value at $E_{th}(B) = -2\mu_s(B - B_r)$. For the purpose of the scattering calculations, all energies are measured from the lowest threshold. Therefore, $\tilde{E}_c = (2\mu_s - \mu_c)(B - B_r)$, where $2\mu_s > \mu_c$. Let $W(R)$ be the coupling potential between the closed and open channels. Then, the Feshbach width Γ is given by

$$\Gamma = \frac{2\pi}{\hbar} \left| \int dR \phi_c(R) W(R) \psi_E(R) \right|^2, \tag{5.15}$$

where $\phi_c(R)$ is the closed-channel bound-state wave function and $\psi_E(R)$ is the energy-normalized s-wave scattering wave function in the open channel with collision energy $E = \hbar^2 k^2/2\mu$. In the limit $k \to 0$, $\psi_E(R)$ may be replaced by its asymptotic form $\psi_E(R) \simeq \sqrt{2\pi\mu/(\hbar^2 k)}\sin(kr + \delta_0)$, where δ_0 is the s-wave phase shift which may be equated to $-ka_{\rm bg}$ at low energy. Substituting $\sin(kr + \delta_0) \simeq k(r - a_{\rm bg})$, one can write $\Gamma = ka_{\rm bg}\Gamma_0$, where

$$\Gamma_0 = \frac{4\pi^2\mu}{\hbar^3 a_{\rm bg}} |\int dR\phi_c(R)W(R)(r - a_{\rm bg})|^2. \tag{5.16}$$

Note that since $a_{\rm bg}$ can be positive or negative, Γ_0, which is independent of k, can also be positive or negative accordingly. Substituting $\tilde{E}_c = (2\mu_s - \mu_c)(B - B_r)$ and $\Gamma = ka_{\rm bg}\Gamma_0$ into equation (5.17), we obtain

$$a_r = -\frac{a_{\rm bg}\hbar\Gamma_0/2}{(\mu_c - 2\mu_s)(B - B_r)}, \tag{5.17}$$

where $\Delta_B = \frac{\hbar\Gamma_0/2}{\mu_c - 2\mu_s}$. Thus, $a_s = a_{\rm bg} + a_r$ can be expressed in the form

$$a_s(B) = a_{\rm bg}\left(1 - \frac{\Delta_B}{B - B_r}\right). \tag{5.18}$$

For $B = \Delta_B + B_r$, $a_s(B)$ is zero. Therefore, this value of B, denoted by B_0, is called the zero crossing. Thus, experimental measurement of B_0 allows us, in turn, to determine the experimental value of Γ. In general, Γ and the resonant phase shift can depend on k nonlinearly. However, for a broad Feshbach resonance ($\hbar\Gamma \gg \tilde{E}_c$), one can ignore the nonlinear dependence on k. In contrast, for a narrow Feshbach resonance, one cannot ignore the nonlinear dependence or the effective-range expansion of the phase shift.

So far, we have discussed MFR with a two-channel model. We now briefly discuss how the MFR is calculated numerically by a multichannel scattering method as discussed in chapter 3. The Hamiltonian is $H = H_k + H_I$ with $H_I = H_C + H_s + H_{\rm hf} + H_B$, where H_k is the kinetic part; $H_C = V_0(R)P_0 + V_1(R)P_1$ is the Coulombic interaction part that consists of the singlet and triplet potentials $V_0(R)$ and $V_1(R)$, respectively; and $P_0 = (1/4)|S = 0\rangle\langle S = 0|$ and $P_1 = (3/4)|S = 1\rangle\langle S = 1|$ are the operators used for projection onto the spin-singlet and spin-triplet manifolds, respectively. Here, H_s represents the spin-dipolar and spin–spin interactions, and

$$H_B = \sum_{j=1,2}\left[\gamma_e s_z^{(j)} - \gamma_n i_z^{(j)}\right]B \tag{5.19}$$

is the Zeeman interaction part, where γ_e and γ_n denote the electronic and nuclear Landé g factors, respectively, and $s_z^{(j)}$ and $i_z^{(j)}$ are the z-components of the electronic and nuclear spins of the jth atom, respectively. In the presence of a magnetic field, the appropriate channel states are constructed by diagonalizing the Zeeman Hamiltonian H_B in the basis of $|f_1, m_{f1}; f_2, m_{f2}(F, \ell; J, M_F)\rangle$, where m_{f1} and m_{f2} are the projections of f_1 and f_2 onto the magnetic field, respectively. The asymptotic channels are the various eigenstates obtained by diagonalizing H_B. Let the jth eigenstate be denoted by $|u_j\rangle$, where u stands for the uncoupled basis, since $|f_1, m_{f1}; f_2, m_{f2}(F, M_F)\rangle$ is written as the product of the individual atomic angular momentum states. This is a useful basis in the asymptotic limit, but at short separations, due to molecular interactions, one has to use the coupled basis $|S, M_S; IM_I(F, \ell, J, M_F)\rangle$, where S is the total spin quantum number with $\mathbf{S} = \mathbf{s}_1 + \mathbf{s}_2$, and I is the total nuclear spin quantum number with $\mathbf{I} = \mathbf{i}_1 + \mathbf{i}_2$. Using the angular momentum addition theorem, one is then required to express the uncoupled basis as a superposition of coupled bases. This procedure enables one to calculate the matrix elements of various terms of molecular interactions which are particularly relevant at short separations.

Let the total number of channels be N. The interaction Hamiltonian H_I is written in an $N \times N$ matrix form in the $|u_j\rangle$ basis. Let this matrix be denoted by $\mathbf{W}(R)$, whose diagonal elements are the channel potentials and whose off-diagonal elements represent interchannel couplings. The jth channel threshold is given by $E_{\text{th}}^{(j)} = W_{jj}(R \to \infty)$. We label the channel with the lowest threshold channel number 1 and accordingly enumerate the channel numbers in ascending order with increasing thresholds. The channel energies are measured with respect to the threshold of channel 1, implying that the asymptotic channel energy of the jth channel is $E_j = E_{\text{th}}^{(j)} - E_{\text{th}}^{(1)}$. The jth channel wave number is given by $k_j = \sqrt{2\mu(E - E_j)}/\hbar$, where E is the collision energy. If jth channel is open, then k_j is real and positive or zero; otherwise, it is a pure imaginary number. The potential matrix is given by $bfV(R) = \mathbf{W}(R) - \mathbf{E_{th}}$, where $\mathbf{E_{th}}$ is an $N \times N$ diagonal matrix whose jth diagonal element is $E_{\text{th}}^{(j)}$. The ith channel wave function can be written as a superposition of all the channel states in the following form:

$$\psi_i(R) = \sum_j \phi_{ji}(R) u_j(\tau), \tag{5.20}$$

where $u_j(\tau)$ denotes the jth channel's angular momentum wave function and τ are the angular momentum (spin) variables. The channel wave functions $u_j(\tau)$ form a complete orthonormal basis with $\sum_\tau u_i^*(\tau) u_j(\tau) = \delta_{ij}$. Here, $\phi_{ji}(R)$ is the jth component of the ith wave function. Thus, in this basis, $\psi_i(R)$ can also be represented in the form of a vector whose jth component is $\phi_{ji}(R)$. The total wave function then can be expressed in matrix form, with its ith column being the ith wave function in vector form. Let this matrix be denoted by $\mathbf{F}(R)$. The Schrödinger equation in matrix form reads:

$$\left[\left\{-\frac{d^2}{dR^2} + \frac{\ell(\ell+1)}{R^2}\right\}\mathbf{1}_N + \mathbf{U}(R)\right]\mathbf{F}(R) = \mathbf{K}^2\mathbf{F}(R), \tag{5.21}$$

where $\mathbf{U}(R) = 2\mu\mathbf{V}(R)/\hbar^2$, $\mathbf{1}_N$ is an $N \times N$ identity matrix, and $\mathbf{K}^2$ is an $N \times N$ diagonal matrix whose jth element is k_j^2. To solve the above equation numerically, $\mathbf{F}(R)$ is evaluated by solving the coupled Schrödinger equations numerically using mainly the Numerov–Cooley algorithm as described in chapter 8. If the values of the wave function at two positions R_1 and R_2 with infinitesimally small separation are known, then its value at the third adjacent position R_3, which is again infinitesimally close to R_2, can be evaluated using this algorithm. In this way, the wave function in matrix form is propagated from a small separation to asymptotically large separations where scattering boundary conditions are imposed on the wave function. To start the propagation at a small separation, the values of the wave functions at two points must be given based on the physically relevant boundary conditions at $R = 0$.

As discussed in section 3.2.2, one must apply a transformation to the wave function matrix $\mathbf{F}(R)$ to obtain an $N \times N_o$ physical wave matrix $\tilde{\mathbf{F}}(R)$, where N_o is the number of open channels and $N_c = N - N_o$ is the number of closed channels. The $N_o \times N_o$ submatrix of $\tilde{\mathbf{F}}(R)$ involving all the open channels provides all the information about the multichannel scattering properties when the asymptotic multichannel scattering boundary conditions discussed in section 3.1.1 are imposed. The rest of the $N_c \times N_o$ submatrix provides the wave function components of the closed channels—these components must vanish as $R \to \infty$.

Depending on the width, MFRs are categorized into two types: narrow or closed-channel-dominated ones, and broad or open-channel-dominated ones. The width of a Feshbach resonance is quantified by the dimensionless parameter $s_{res} = \frac{\delta\mu\Delta}{\bar{E}}\frac{a_{bg}}{\bar{a}}$, where $\delta\mu$ is the difference between the magnetic moments of the bound state and the two separated atoms, a_{bg} is the background scattering length, and Δ is the width of the resonance in units of magnetic field strength. For two ground-state atoms, the diatomic potential at long range reduces to the van der Waals potential of the form C_6/r^6 in the leading order. Accordingly, one can define a long-range length scale $R_{\rm vdW} = (2\mu C_6/\hbar^2)^{1/4}/2$. For ground-state potentials, Gribakin and Flambaum [30] defined a length scale called the mean scattering length $\bar{a} = \frac{2\pi}{\Gamma(1/4)^2}\left(\frac{2\mu C_6}{\hbar^2}\right)^{1/4}$ and the corresponding energy scale $\bar{E} = \frac{\hbar^2}{2\mu\bar{a}^2}$. Narrow resonances are characterized by $s_{res} < 1$, while for broad or entrance-channel-dominated resonances, $s_{res} > 1$. Gao [31] and Flambaum *et al* [32] defined an effective range r_e for the ground-state atom–atom potential, namely $r_e \approx \left(\frac{\Gamma(\frac{1}{4})^4}{6\pi^2}\right)\bar{a}\left[1 - 2\left(\frac{\bar{a}}{a_S}\right) + 2\left(\frac{\bar{a}}{a_S}\right)^2\right]$. It has been shown [33, 34, 35] that r_0 near a narrow Feshbach resonance may become quite large, negative, and magnetic field dependent. Experimental observations [34] have shown that near the narrow Feshbach resonance ($B_0 = 58.9$ G) of ^{39}K, r_0 sharply varies with the magnetic field. Similar results have been found for ^{6}Li MFR near 543.3 G [35]. In contrast, r_0 near a broad Feshbach resonance is positive, small, and close to r_e. For $|a_s| \gg R_{\rm vdW}$, r_0 is about three times $R_{\rm vdW}$ [31, 32], which is typically of the order of ten Bohr radii.

5.4.2 Outline of the advancement in MFR

The first experimental observation of MFR was reported by Inouye *et al* [29] from MIT in 1998. They observed it in an optically trapped BEC of Na atoms. Figure 5.2 depicts their experimental results, which show a resonance under a magnetic field of $B = B_r = 907$ G. Although there was a huge loss of atoms near B_r due to three-body

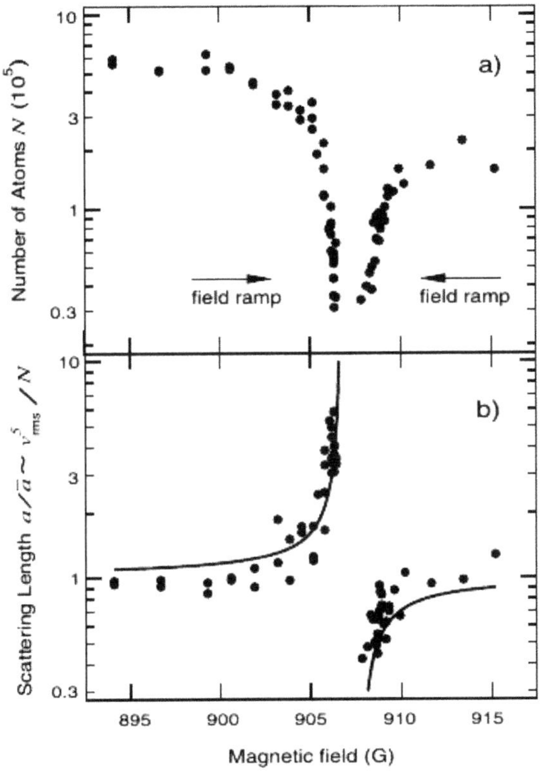

Figure 5.2. Observation of MFR in an optically trapped BEC of Na atoms. The upper panel shows a loss of atoms when the magnetic field B is tuned close to the resonance. The lower panel shows the scattering length normalized by the background scattering length a_{bg} as a function of B (in Gauss) near the resonance. Reproduced from [29], with permission from Springer Nature.

recombination, this early experiment clearly demonstrated that it is indeed possible to tune the scattering length using a magnetic-field-induced resonance. Here, we briefly discuss how they carried out the experiment.

First, they produced a BEC in the ground HF spin state $|F = 1; m_F = -1\rangle$ by laser cooling and evaporative cooling in a magnetic trap. Subsequently, the BEC was transferred into an optical dipole trap created at the focus of an infrared laser beam. An adiabatic radio-frequency (RF) field sweep was applied to flip the atomic spin to $|F = 1; m_F = 1\rangle$. Atoms were confined in a cigar-shaped optical dipole trap. In the presence of a variable-bias magnetic field, the density of the trapped atoms was measured by a rapid nondestructive *in situ* phase-contrast imaging technique. As figure 5.3 shows, the atoms in $|F = 1; m_F = 1\rangle$ were found to be completely lost from the trap at $B = 907$ G, while the atoms in $|F = 1; m_F = -1\rangle$ remained trapped under a wide range of magnetic field strengths around 907 G. This implies that only the atoms in $|F = 1; m_F = 1\rangle$ underwent resonant collisions at $B = 907$ G, as explained by Feshbach resonance theory. The resonance arises due to the magnetic coupling, which couples the scattering state to a quasi-bound dimer state of the second highest vibrational level $v = 14$ of the triplet ground-state potential. The loss

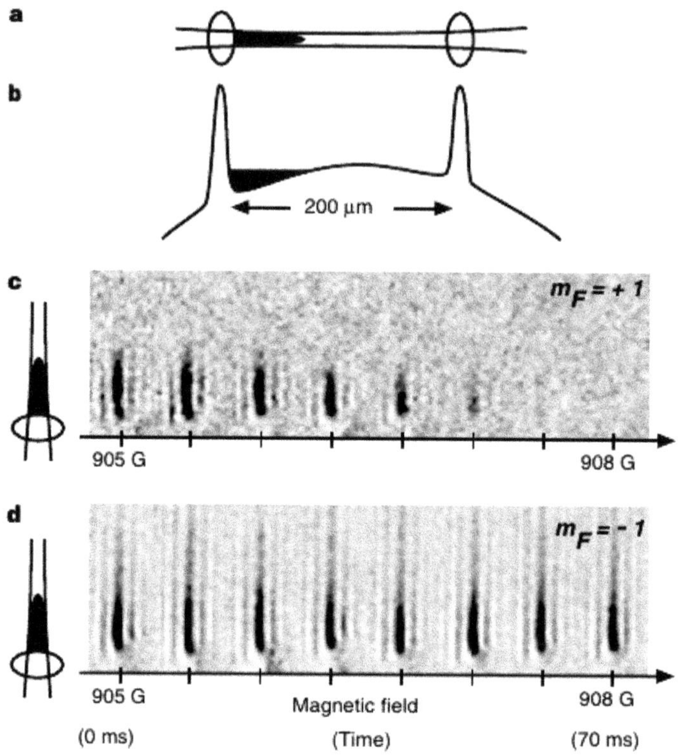

Figure 5.3. (a), (b): A BEC of Na atoms in a cigar-shaped optical dipole trap. (c), (d) *In situ* phase-contrast images of the trapped cloud under different magnetic fields for condensate atoms initially prepared in the HF spin state $|F = 1; m_F = 1\rangle$ (c) and the HF spin state $|F = 1; m_F = -1\rangle$ (d). Reproduced from [29], with permission from Springer Nature.

of atoms is quite prominent because the three-body inelastic process is significant in this particular resonance. Another MFR in a sodium BEC at 1195 G was reported in 1999 [36] by the same group. This resonance was also found to be accompanied by a strong inelastic loss of atoms.

Following the first observation of MFR in a Na BEC, several groups observed MFRs in other ultracold alkali atomic gases within a span of ten years: *s*-wave MFR in a two-component Fermi gas of ^{6}Li atoms [37, 38], *p*-wave MFR in spin-polarized ^{6}Li atoms [39], *s*-wave MFR in bosonic ^{39}K [40], *s*-wave MFR in fermionic ^{40}K [41, 42], and *p*-wave MFR in ^{40}K [41].

5.5 Collisions in a laser field

Like a magnetic field, a laser field can influence low-energy ground-state collision properties in a significant way. A laser can couple the low-energy scattering state (also called the free state) to a ro-vibrational molecular bound state by the process of photoassociation (PA) [43, 44]. In the dispersive regime, this free–bound coupling leads to a scattering resonance known as optical Feshbach resonance (OFR), as mentioned earlier.

5.5.1 Optical Feshbach resonances

OFR refers to the resonances that occur due to the coupling of an open-channel continuum state with a molecular bound state in a closed channel by a laser. It is thus based on free–bound or continuum–bound PA spectroscopy, in which two colliding atoms associate to form a molecular bound state in an electronically excited molecular potential. This phenomenon resembles that of the Fano state, as discussed in the previous chapter. The only difference is that in the present case, the molecular bound state undergoes spontaneous radiative decay due to electric dipole transitions. In contrast, the bound state involved in MFR is in an electronic ground-state channel, and so electric-dipole-facilitated decay is absent, though other decay processes may occur. Mitigating the loss due to spontaneous decay of the bound state is an important issue in OFR. To minimize this type of loss in OFR, the PA laser field is dispersively tuned away from the free–bound resonant transition frequency, because if the laser is tuned on resonance or near resonance, OFR can lead to PA of two atoms into an excited molecule that eventually disintegrates, disappearing from the trap that confines the atoms. The transition processes in one- and two-color OFR are schematically shown in figures 5.4 and 5.5(a), respectively. Figure 5.5(b) shows a scheme for Raman PA processes that form deeply bound

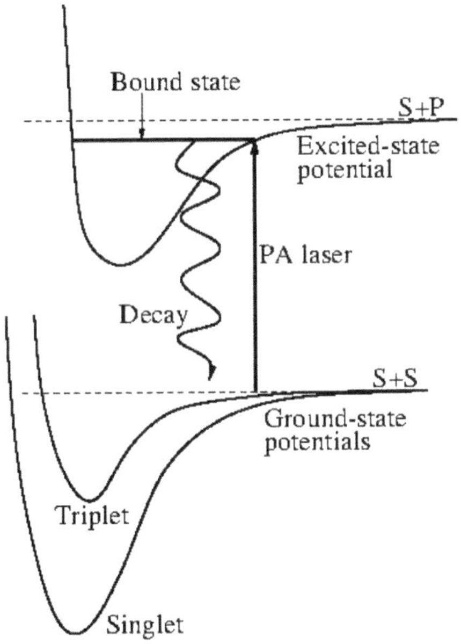

Figure 5.4. Schematic of one-color PA between two homonuclear atoms: a colliding pair of cold atoms absorbs a single photon from the PA laser and thereby forms an excited molecular bound state that eventually decays to two free atoms or a loosely bound ground-state molecule. The decay products generally have higher kinetic energy than the trap depth and so are ejected from the trap. The same schematic can be applied in the case of two heteronuclear atoms, but the excited-state potential has different long-range behavior, and the equilibrium positions of the ground- and excited-state potentials lie at nearly the same separation, in contrast to the homonuclear case (see text).

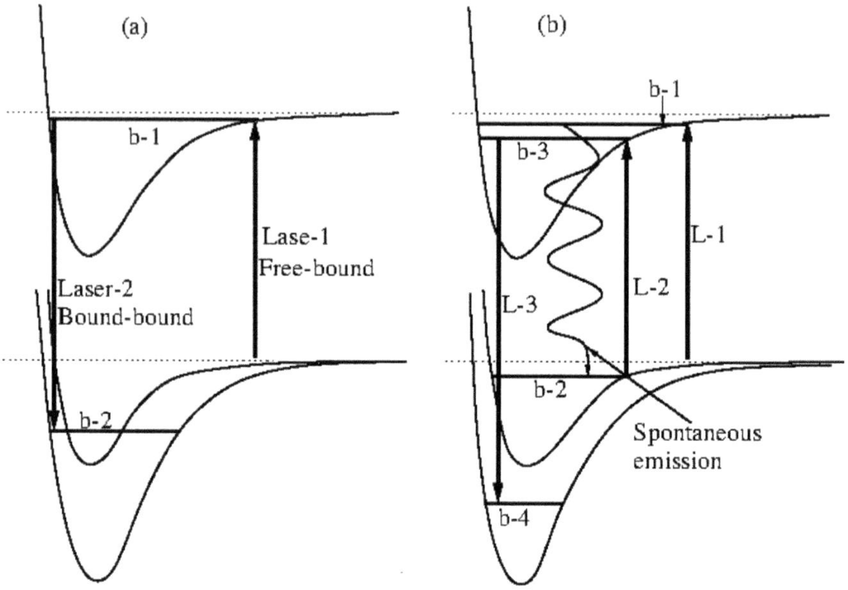

Figure 5.5. Schematics showing two-color PA (a) and multicolor Raman PA (b). L-1, L-2, and L-3 denote lasers; b-1, b-2, b-3, and b-4 denote bound states.

ground-state molecular states using multiple lasers. The formation of ground-state molecules by Raman PA will be discussed in the next chapter.

A comprehensive review of the early studies of OFR can be found in [44]. The OFR method was first proposed by Fedichev *et al* [45] in 1996, and early experiments demonstrating the implementation of the method were carried out a few years later [46–48]. To date, several experimental groups have successfully observed OFR. The first experimental realization of OFR was reported by Fatemi *et al* [46] in 2000 using cold Na atoms in an MOT at submillikelvin temperatures. They inferred the alteration of the collisional wave functions due to OFR through the detection of molecular ions that were formed by a two-color photoassociative method. Theis *et al* [48] demonstrated the tuning of the scattering length by OFR in a BEC of ^{87}Rb atoms by controlling the frequency, power, and duration of a PA laser tuned across a photoassociative resonance, as shown in figure 5.6. They detected the change in the scattering length by measuring the shifts in the mean-field interaction energy of the BEC through two-photon Bragg spectroscopy. Bragg spectroscopy makes use of a moving optical lattice created by two counterpropagating laser pulses with slightly different frequencies. The atoms of the BEC are diffracted from this lattice. The Bragg resonant frequency is shifted due to the mean-field interaction, which is proportional to the scattering length. By measuring the density of the Bragg-diffracted beams using absorption imaging, one can infer the change of scattering length due to OFR. Thus, Theis *et al* were able to alter the scattering length by one order of magnitude from 0.5 to 10 nm. Thalhammer *et al* [49] observed OFR in an

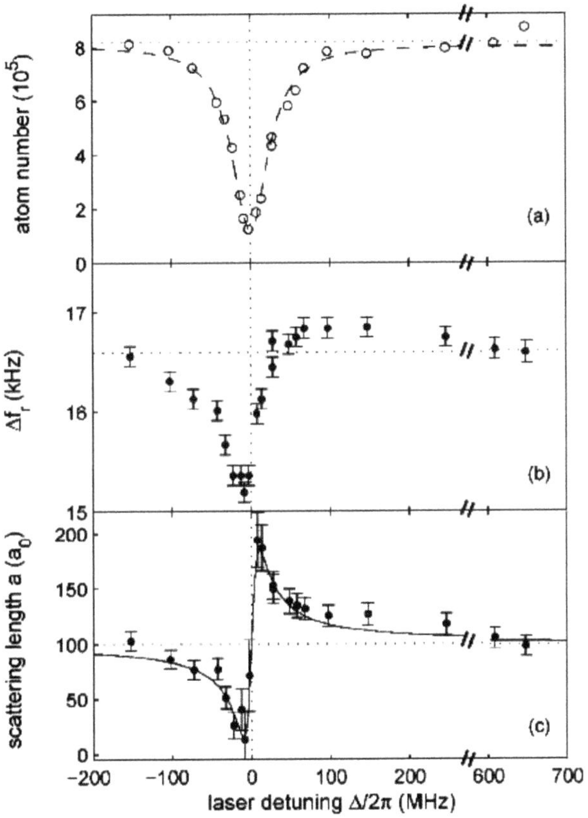

Figure 5.6. Observations of OFR. The number of atoms remaining in the trap (a), Bragg resonant frequency (b), and the scattering length (c) as a function of laser detuning. Figure reprinted with permission from [48], Copyright (2004) by the American Physical Society.

^{87}Rb BEC using two-color Raman photoassociative transitions. Their results are shown in figure 5.7. As in [48], the OFR was measured through Bragg spectroscopy. In two-color OFR, laser detuning from the bound–bound resonance can be used as an additional parameter that controls the tunability of the scattering length. Further details of two-color PA will be discussed in the next chapter.

5.5.2 Theory of OFR

As in the case of MFR, the essential process of OFR may be understood by considering a two-channel OFR model. The only difference from the two-channel MFR model is that the diatomic bound state in OFR is supported by an electronically excited molecular potential, so this bound state can decay spontaneously. Since OFR is based on the PA method, OFR can be understood by considering the scattering theory behind PA [51, 52], as we discuss in more detail in the next chapter. Here, we discuss the essential ideas from the point of view of the Fano state due to continuum–bound laser coupling. Let the excited bound molecular state be

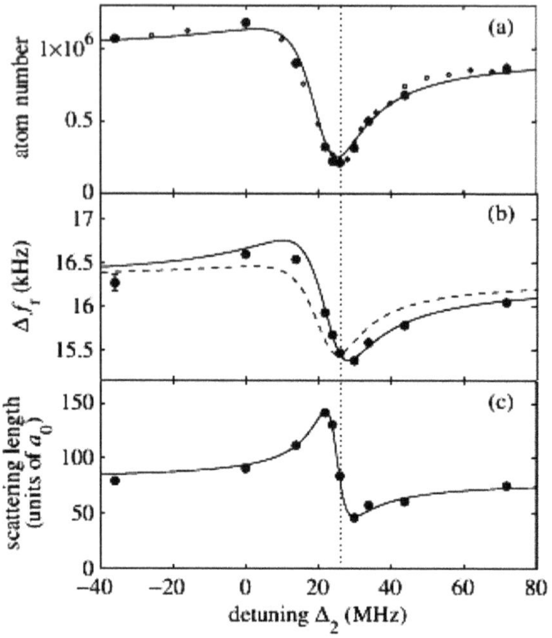

Figure 5.7. Observations of OFR produced using two-color Raman PA. (a) The number of atoms remaining after a 100 μs Raman pulse as a function of two-photon detuning Δ_2. (b) The Bragg resonant frequency Δf_r. (c) The scattering length. Figure reprinted with permission from [49], Copyright (2005) by the American Physical Society.

characterized by the vibrational quantum number v and the rotational quantum number J. Let $|E\rangle$ represent the energy-normalized free basis state (scattering state) of two ground-state atoms in a particular open channel and let $|b\rangle$ represent the excited molecular bound state in a closed channel in the absence of any external field. In order to write the Hamiltonian for the system, we first need to calculate the free–bound electric dipole transition matrix elements using the dipole approximation and the rotating-wave approximation (RWA). Let us consider that an electromagnetic field tuned closely to the photoassociative transition $|E\rangle \rightarrow |b\rangle$ is shone on two colliding alkali-type atoms A and B. The atom–field interaction Hamiltonian is

$$H_I = -\mathbf{d}_A. \cdot \mathbf{E}_L(\mathbf{r}_{eA}, t) - \mathbf{d}_B. \cdot \mathbf{E}_L(\mathbf{r}_{eB}, t), \tag{5.22}$$

where $\mathbf{E}_L(\mathbf{r}, t)$ denotes the electric field of the laser, while $\mathbf{r}_{eA}$ and $\mathbf{r}_{eB}$ represent the position vectors of the valence electrons of atoms A and B, respectively. The dipole approximation implies that we can replace the electronic position vectors with the respective atomic COM or nuclear positions, i.e. $\mathbf{r}_{eA} \rightarrow \mathbf{r}_A$ and $\mathbf{r}_{eB} \rightarrow \mathbf{r}_B$, where $\mathbf{r}_A$ and $\mathbf{r}_B$ are the respective atomic COM position vectors. Let the field be of a plane-wave type given by

$$\mathbf{E}_L(\mathbf{r}, t) = \frac{1}{2}[\hat{\varepsilon} E_L \exp[-i(\omega_L - \mathbf{r} \cdot \mathbf{k}_L)] + \text{C.c}], \tag{5.23}$$

where C.c stands for complex conjugate; $\hat{\varepsilon}$, ω_L, and $\mathbf{k}_L$ represent the polarization unit vector, angular frequency, and propagation vector of the laser, respectively:

$$H_I = \frac{e^{i\omega_L t}}{2} E_L[-\mathbf{d}_A \cdot \hat{\varepsilon} \exp[-i\mathbf{k}_L \cdot \mathbf{r}_A] - \mathbf{d}_B \cdot \hat{\varepsilon} \exp[-i\mathbf{k}_L \cdot \mathbf{r}_B]] + \text{h. c.}\,, \tag{5.24}$$

where h.c. denotes the Hermitian conjugate. Now, writing $\mathbf{r}_A = \mathbf{R} + m_B\mathbf{r}/(m_A + m_B)$ and $\mathbf{r}_B = \mathbf{R} - m_A\mathbf{r}/(m_A + m_B)$, where $\mathbf{R}$ and $\mathbf{r}$ are the COM and relative coordinates of the two atoms, respectively, and $m_{A(B)}$ is the mass of atom A(B), for a pair of homonuclear ($m_A = m_B$) atoms, one can express H_I in the following form:

$$H_I = e^{i\omega_L t} e^{-\mathbf{k}_L \cdot \mathbf{R}} E_L[-\mathbf{D} \cdot \hat{\varepsilon} \cos(\mathbf{k}_L \cdot \mathbf{r}/2) + i\mathbf{D}_{-} \cdot \hat{\varepsilon} \sin(\mathbf{k}_L \cdot \mathbf{r}/2)] + \text{h. c.}\,, \tag{5.25}$$

where $\mathbf{D} = \mathbf{d}_A + \mathbf{d}_B$ and $\mathbf{D}_{-} = \mathbf{d}_A - \mathbf{d}_B$. Since $|\mathbf{k}_L \cdot \mathbf{r}/2| \ll 1$, one can ignore the second term within the third bracket. Moreover, for a homonuclear molecule, the matrix element of $\mathbf{D}_{-}$ vanishes. Here, $\mathbf{D}$ stands for the molecular dipole moment.

We now cast H_I into a form suitable for computation by making use of the RWA. In the present context, the identity operator can be written as $\mathbf{1} = \int dE|E\rangle\langle E| + |b\rangle\langle b|$. Thus, the interaction Hamiltonian in the COM frame can be expressed in the form

$$H_I = \mathbf{1}H_I\mathbf{1} = \int dE e^{i\omega_L t}[\Lambda_{bE}|b\rangle\langle E| + h.\ c.] + \int dE e^{-i\omega_L t}[\Lambda_{bE}|b\rangle\langle E| + \text{h. c.}], \tag{5.26}$$

where

$$\Lambda_{bE} = \frac{E_L}{2}\langle b|\mathbf{D} \cdot \hat{\varepsilon} \cos(\mathbf{k}_L \cdot \mathbf{r}/2)|E\rangle \tag{5.27}$$

is the free–bound or continuum–bound dipole matrix element. Note that $\langle E|H_I|E\rangle = 0$ and $\langle b|H_I|b\rangle = 0$ by the parity selection rule, since the channel states have definite parity. The RWA implies that we can drop the counter-rotating terms that rotate with the sum of the laser and atomic or molecular transition frequencies and retain only the corotating terms that oscillate at the difference between the laser and the transition frequencies. In other words, the RWA means that the atom–molecule lowering operator $|E\rangle\langle b|$ has to oscillate as $\exp[i\omega_L t]$, and the raising operator $|b\rangle\langle E|$ has to oscillate as $\exp[-i\omega_L t]$. Thus, under the RWA and the dipole approximation, the interaction Hamiltonian is given by

$$H_I = e^{-i\omega_L t}\int dE \Lambda_{bE}|b\rangle\langle E| + \text{h. c.} \tag{5.28}$$

In order to calculate Λ_{bE}, we need to specify the angular and radial wave functions of the continuum and the bound states. The angular basis state $|J, M\rangle$ in the space-fixed (laboratory) frame is transformed into the state

$$|J, M\rangle' = \mathscr{D}(\hat{R})|J, M\rangle = \sum_{M'} \mathscr{D}^{(J)}_{M'M}(\hat{r})|JM'\rangle \tag{5.29}$$

under the action of the rotational operator $\mathscr{D}(\hat{R})$ known as the Wigner **D**-matrix, which describes transformation under the rotation $\hat{R}$ of the frame from the space-fixed to the body-fixed (COM) frame. The frame rotation is characterized by the three Euler angles $(\alpha, \beta, \gamma) \equiv \hat{R}$. Explicitly, $\mathscr{D}(\hat{R}) = \exp[iJ_z\alpha]\exp[iJ_y\beta]\exp[iJ_z\gamma]$. Here, $|J, M\rangle'$ is an angular state that is described in terms of the space-fixed coordinates, where the prime superscript is used to indicate that this is obtained via a frame transformation from the basis $|J, M\rangle$. Here, $|J, M'\rangle$ is a basis described in terms of the body-fixed coordinates. For a diatomic molecule, $|J, M'\rangle$ is characterized by the total angular momentum quantum number J and its projection M' along the internuclear axis, which is a symmetry axis. Since $|M'|$ is a conserved quantity, different values of M' refer to the different energy eigenstates of a molecule. Thus, a molecular state is characterized by a definite value of $|M'|$. The function $\mathscr{D}^{(J)}_{M'M}(\hat{R})$, when normalized to unity, can be regarded as the angular wave function of a diatomic molecule. Explicitly, this is given by $\psi_{JM'M} = i^J\sqrt{\frac{2J+1}{8\pi^2}}\mathscr{D}^{(J)}_{M'M}(\hat{r})$. For $M' = 0$, $\mathscr{D}^{(J)}_{M'M}(\hat{R})$ reduces to the spherical harmonic $Y_{JM}(\beta, \gamma)$. Thus, a molecular angular state can be written as

$$|J, M'M\rangle = \psi_{JM'M}|JM'\rangle. \tag{5.30}$$

Next, one has to apply tensor calculus to calculate the matrix element of $\mathbf{D} \cdot \hat{\varepsilon}$, which is a product of two vectors. Note that this is a scalar product of two tensors of rank 1, since a vector is a tensor of rank 1. For this, one has to expand $\mathbf{D} \cdot \hat{\varepsilon}$ in the rank 1 tensor basis $\hat{e}_{\pm 1} = \mp(\hat{x} \pm i\hat{y})/\sqrt{2}$ and $\hat{e}_0 = \hat{z}$. Thus, we have $\mathbf{D} \cdot \hat{\varepsilon} = \sum_{\mu=\pm 1,0} D_\mu \varepsilon^*_\mu$, where $X_\mu = \mathbf{X} \cdot \hat{e}_\pm$ and $\mathbf{X}$ stands for either $\mathbf{D}$ or $\hat{\varepsilon}$. The component X_μ can be expressed in the tensor form $X_\mu = X\sqrt{4\pi/3}\,T^{(1)}_\mu$, where $T^{(1)}_\mu$ is the μth component of a tensor of rank 1. In the absence of any external magnetic field, it is convenient to consider the propagation direction of the electromagnetic field as the space-fixed z-axis. Then, $\varepsilon_0 = 0$ and the field polarization unit vector $\hat{\varepsilon}$ is decomposed into two components ε_+ and ε_-, which refer to the left (anticlockwise) and right (clockwise) circularly polarized field components, respectively, when viewed facing the incoming photon; that is, in the direction opposite to the propagation direction. Now, since the state $|J, M', M\rangle$ refers to both space- and body-fixed coordinate frames, we need to express $T^{(1)}_\mu$ in terms of the tensor components in the body-fixed frame. Explicitly, this is given by

$$T^{(1)}_\mu = \sum_{\mu'} \mathscr{D}^{(1)}_{\mu',\mu} T^{(1)}_{\mu'}. \tag{5.31}$$

The task at hand is to calculate the matrix element of $T^{(1)}_\mu$ between the states $|J_1, M', M_1\rangle$ and $|J_2, M', M_2\rangle$. This is done by applying the Wigner–Eckart theorem. Thus, we have

$$\begin{aligned}
&\sum_{\mu'}\langle J_2, M', M_2|T^{(1)}_\mu|J_1, M', M_1\rangle \\
&= i^{J_1 - J_1}\frac{\sqrt{(J_2+1)(2J_1+1)}}{8\pi^2}\sum_{\mu'}\int\!\!\int\!\!\int \sin\beta d\beta d\alpha d\gamma \\
&\times \mathscr{D}^{(J_2)*}_{M', M_2}\mathscr{D}^{(J_1)}_{M', M_1}\mathscr{D}^{(1)}_{\mu', \mu}\langle J_2, M'|T^{(1)}_{\mu'}|J_1, M'\rangle.
\end{aligned} \tag{5.32}$$

5.5.3 PA Hamiltonian

Let us consider two colliding alkali atoms undergoing PA transitions. Ignoring the HF interactions for the sake of simplicity, the Hamiltonian can be written as $H = H_0 + H_I$, where

$$H_0 = H_{\text{elec}}(\mathbf{r}_1, \mathbf{r}_2; \mathbf{r}_\mathbf{A}, \mathbf{r}_\mathbf{B}) - \frac{\hbar^2}{2\mu}\nabla_r^2 - \frac{\hbar^2}{2M}\nabla_R^2 + H_I \tag{5.33}$$

describes the motion of the two atoms, including the motion of the atomic electrons described by H_{elec}, with $\mathbf{r}_{1(2)}$ being the position vector of the valence electron of atom A (B). Here, $\mathbf{r}_\mathbf{A}$ and $\mathbf{r}_\mathbf{B}$ denote the position vectors of the nuclei of atoms A and B, respectively; ∇_r and ∇_R are the Laplacian operators corresponding to the relative coordinate $\mathbf{r} = \mathbf{r}_\mathbf{A} - \mathbf{r}_\mathbf{B}$ and the COM coordinate $\mathbf{R} = (\mathbf{r}_\mathbf{A} + \mathbf{r}_\mathbf{B})/2$, respectively. Ignoring the COM motion, under the BO approximation, the wave function $\Phi = \chi(\mathbf{r}_1, \mathbf{r}_2; \mathbf{r})\phi(\mathbf{r})$ is separable into the electronic wave function χ and the nuclear wave function ϕ. The Hamiltonian H_{elec} satisfies the eigenvalue equation

$$H_{\text{elec}}\chi = V\chi, \tag{5.34}$$

where V denotes electronic energy.

We now consider two molecular electronic states—one ground state, denoted by χ_g, in which both valence electrons are in S orbitals, and an excited state, denoted by χ_e, where one of the valence electrons is in a P orbital and the other is in an S orbital—coupled by the PA laser. Let the corresponding electronic energies be denoted by V_g and V_e. Under this two-state approximation and the RWA, the Hamiltonian can then be expressed as

$$H = \sum_{i=g,e} V_i|\chi_i\rangle\langle\chi_i| - \frac{\hbar^2}{2\mu}\nabla_r^2 + [e^{-i\omega_L t}\Lambda_{eg}|\chi_e\rangle\langle\chi_g| + \text{h.c.}], \tag{5.35}$$

where $\Lambda_{eg} = (1/2)\langle\chi_e|-\mathbf{D}\cdot\varepsilon E_L|\chi_g\rangle$ is the laser coupling between the two electronic states. Now, making a unitary transformation $U^\dagger HU$, where $U = \exp[i\omega_L t|\chi_e\rangle\langle\chi_e|]$, we can work with the following time-independent Hamiltonian:

$$\widetilde{H} = V_g|\chi_g\rangle\langle\chi_g| + (V_e - \hbar\omega_L)|\chi_e\rangle\langle\chi_e| - \frac{\hbar^2}{2\mu}\nabla_r^2 + [\Lambda_{eg}|\chi_e\rangle\langle\chi_g| + \text{h.c.}]. \tag{5.36}$$

5.5.4 T-matrix element

Let the eigenstate $|\Psi\rangle$ of H be expanded in the two-electron basis in the following form:

$$|\Psi\rangle = \phi|\chi_e\rangle + \psi|\chi_g\rangle. \tag{5.37}$$

We now write the wave function $\Psi(\mathbf{r}, \mathbf{r}_1, \mathbf{r}_2) = \langle\mathbf{r}, \mathbf{r}_1, \mathbf{r}_2|\Psi\rangle$ in the following form:

$$\Psi(\mathbf{r}, \mathbf{r}_1, \mathbf{r}_2) = \frac{1}{r}\left[\sum_{J,M_J}\phi_{vJM}(r)Y_{J,M_J}(\hat{r})\chi_e(\mathbf{r}, \mathbf{r}_1, \mathbf{r}_2) + \sum_{\ell,m_\ell}\psi_{E\ell m_\ell}(r)Y_{\ell,m_\ell}(\hat{r})\chi_g(\mathbf{r}, \mathbf{r}_1, \mathbf{r}_2)\right], \tag{5.38}$$

where $\phi_{vJM}(r)Y_{J,M_J}(\hat{r})$ represents the excited ro-vibrational state and $\psi_{E\ell m_\ell}(r)Y_{\ell,m_\ell}(\hat{r})$ represents the ground continuum state. Here, we have assumed that the rotational states of a single vibrational level v can be coupled to the continuum of scattering states by the PA laser. The above wave function is similar to the Fano wave function $\Psi_F(\mathbf{r}, \mathbf{r}_1, \mathbf{r}_2) = \langle \mathbf{r}, \mathbf{r}_1, \mathbf{r}_2 | E\rangle_F$, where $|E\rangle_F$ can be identified with the Fano state of equation (4.43) if one makes the expansion $|E\rangle = \sum_{\ell m_\ell} |E, \ell, m_\ell\rangle |\chi_g\rangle, |b\rangle \equiv \sum_{J,M_J} |v, J, M_J\rangle |\chi_e\rangle$ and further identifies $\phi_{vJM}(r)Y_{J,M_J}(\hat{r}) = b_E\langle \mathbf{r}|v, J, M_J\rangle$ and $\psi_{E\ell m_\ell}(r)Y_{\ell,m_\ell}(\hat{r}) = \int dE' C_E(E')\langle \mathbf{r}|E', \ell m_\ell\rangle$.

From the Schrödinger equation $\widetilde{H}|\Psi\rangle = E|\Psi\rangle$, we obtain two coupled equations:

$$\left[-\frac{\hbar^2}{2\mu}\frac{d^2}{dr^2} + B_J(r) + V_e(r) - \hbar\delta - E - i\hbar\frac{\gamma}{2}\right]\phi_{vJM} = -\sum_{\ell m_\ell} \Lambda_{JM;\ell m_\ell}\psi_{E\ell m_\ell} \tag{5.39}$$

$$\left[-\frac{\hbar^2}{2\mu}\frac{d^2}{dr^2} + B_\ell(r) + V_g(r) - E\right]\psi_{E\ell m_\ell} = -\sum_{M} \Lambda_{\ell m_\ell;\, JM}\phi_{vJM}, \tag{5.40}$$

where $B_J(r) = \hbar^2/(2\mu r^2)[J(J+1) - |J_{z'}|^2]$ is the rotational term of the excited state and $B_\ell(r) = \hbar^2/(2\mu r^2)\ell(\ell+1)$ is the centrifugal term in the collision between two ground-state atoms. The free–bound coupling matrix element is

$$\Lambda_{JM;\ell m_\ell} = \langle JM\Omega|\Lambda_{eg}|\ell m_\ell 0\rangle, \tag{5.41}$$

and $\Lambda_{\ell m_\ell;\, JM} = \Lambda^*_{JM;\ell m_\ell}$. Note that here we have phenomenologically introduced the term $-i\hbar\gamma/2$ by substituting $V_e \to V_e - \hbar\gamma/2$, where $\gamma/2$ is the spontaneous emission linewidth of the excited molecular state. This is consistent with the analyticity of resonances, since a resonance occurs at a complex energy with a negative imaginary part. The potential $V_e(r)$ varies as $-1/r^3$ for $r \to \infty$, while the potential $V_g(r)$ behaves as the van der Waals potential $-1/r^6$ in the asymptotic regime. V_e supports several ro-vibrational bound states. Here, $\delta = \omega_L - \omega_a$ is the detuning of the laser frequency ω_L from the atomic frequency ω_a. Note that the bound-state energy is measured from the threshold of the excited-state potential, which is $\hbar\omega_a$ with respect to the threshold of V_g.

We next solve the two coupled equations (5.39) and (5.40) using a Green's function. Let ϕ^0_{vJ} be the bound-state solution of the homogeneous part ($\Lambda_{eg} = 0$) of (5.39) with ro-vibrational energy E_{vJ}. Note that we have removed the subscript M from the labeling of wave functions for simplicity. The Green's function for the bound state is given by

$$G_v(r, r') = -\frac{\phi^0_{vJ}(r)\phi^0_{vJ}(r')}{\Delta E_{vJ} + i\hbar\gamma/2}, \tag{5.42}$$

where $\Delta E_{vJ} = \hbar\delta + E - E_{vJ}$ and ϕ^0_{vJ} is the bound-state solution without PA coupling ($\Lambda = 0$). Thus, one can write

$$\phi_{vJM}(r) = \int_{E'} dE' \sum_{\ell m_\ell} A_{JM;\ell m_\ell}\phi^0_{vJ}(r), \tag{5.43}$$

where

$$A_{JM;\ell m_\ell} = \frac{\int dr' \Lambda_{JM;\ell m_\ell}(r')\phi^0_{vJ}(r')\psi_{E\ell m_\ell}(r')}{\Delta E_{vJ} + i\hbar\gamma/2}. \tag{5.44}$$

Let $\psi^{0,reg}_{E\ell}(r)$ and $\psi^{0,irr}_{E\ell}(r)$ denote the regular and irregular solutions, respectively, of the homogeneous part of equation (5.40). Note that $\psi^{0,reg}_{E\ell}(r)$ vanishes at $r = 0$, while $\psi^{0,irr}_{E\ell}(r)$ is defined by boundary condition at $r \to \infty$ only. The Green's function for the ground-state scattering can be written as

$$\mathscr{K}_\ell(r, r') = -\pi[\psi^{0,reg}_{E\ell}(r)\psi^{0,irr}_{E\ell}(r') + i\psi^{0,reg}_{E\ell}(r)\psi^{0,reg}_{E\ell}(r')] \qquad r' > r \tag{5.45}$$

$$\mathscr{K}_\ell(r, r') = -\pi\left[\psi^{0,\,reg}_{E\ell}(r')\psi^{0,\,irr}_{E\ell}(r) + i\psi^{0,\,reg}_{E\ell}(r)\psi^{0,\,reg}_{E\ell}(r')\right]r' < r. \tag{5.46}$$

Substituting equation (5.43) into equation (5.40) and using $\mathscr{K}_\ell(r, r')$, one obtains

$$\begin{aligned}\psi_{E\ell m_\ell}(r) = \exp(i\eta_\ell)\psi^0_{E\ell} + \sum_{\ell' m_{\ell'} M} A_{JM;\ell' m_{\ell'}}(E) \\ \times \int \mathscr{K}_\ell(r, r')\Lambda_{\ell m_\ell;\, JM}(r')\phi^0_{vJ}(r')dr',\end{aligned} \tag{5.47}$$

where $\psi^0_{E\ell} = \psi^{0,\,reg}_{E\ell}$ and η_ℓ is the phase shift in the absence of a laser field. Now, upon substituting equation (5.47) into equation (5.44) and doing some algebra, one finally obtains

$$A_{J,\,M;\ell,\,m_\ell} = \frac{f_{J,\,M;\ell m_\ell} + \left(E^{\text{shift}}_{J\ell} - i\hbar\Gamma_{J\ell}/2\right)\widetilde{A}_J}{\Delta E_{vJ} + i\hbar\gamma/2}, \tag{5.48}$$

where

$$E^{\text{shift}}_{J\ell} = \int\int dr'dr\phi^0_{vJ}(r')\Lambda_{JM;\ell m_\ell}(r')\text{Re}[\mathscr{K}_\ell(r', r)]\Lambda_{\ell m_\ell;\, JM}(r)\phi^0_{vJ}(r) \tag{5.49}$$

is the shift that depends on the radial correlation between continuum and bound states. The Wigner threshold laws dictate that in the limit $k \to 0$, the shift becomes independent of collision energy for all partial waves. For large k, the shift is small. Here,

$$f_{JM;\ell m_\ell} = \exp(i\eta_\ell)\int \phi^0_{vJ}(r')\Lambda_{JM;\ell m_\ell}(r')\psi^0_{E\ell}(r')dr', \tag{5.50}$$

$$\Gamma_\ell = \frac{2\pi}{\hbar}\left|f_{JM;\ell m_\ell}\right|^2 \tag{5.51}$$

and

$$\widetilde{A}_J = \sum_{\ell, m_\ell, M} \frac{f_{JM;\ell m_\ell}}{\Delta E_{vJ} + i\hbar\gamma/2 - E^{\text{shift}}_J + i\hbar\Gamma/2}, \tag{5.52}$$

where $E^{\text{shift}}_J = \sum_{\ell, m_\ell, M} E^{\text{shift}}_{J\ell}$ and $\Gamma = \sum_{\ell, m_\ell, M}\Gamma_\ell$.

Using the asymptotic boundary conditions of regular and irregular scattering wave functions, the T-matrix element in the presence of the PA laser can be expressed as

$$T_\ell = T_\ell^0 + T_\ell^L = \frac{1}{2i}(1 - e^{2i\delta_\ell}), \tag{5.53}$$

where $\delta_\ell = \eta_\ell + \eta_\ell^L$; the laser-coupling-dependent complex phase shift η_ℓ^L is given by

$$e^{2i\eta_\ell^L} = 1 - 2iT_\ell^L, \tag{5.54}$$

with

$$T_\ell^L = \sum_{\ell' M} \frac{\pi f_{\ell m_\ell;\, JM} f_{JM;\ell' m_{\ell'}}}{\hbar\delta + E - \tilde{E}_{vJ} + i\hbar\gamma/2 + i\hbar\Gamma/2} \tag{5.55}$$

being the T-matrix element that explicitly depends on the laser coupling. Here, $\tilde{E}_{vJ} = E_{vJ} + E_J^{\text{shift}}$ and

$$T_\ell^0 = \frac{1}{2i}(1 - e^{2i\eta_\ell}) = -e^{i\eta_\ell} \sin \eta_\ell \tag{5.56}$$

is the background T-matrix element that is independent of the laser coupling. The above equations reveal that the T-matrix element for a partial wave can be modified by second-order dispersive PA coupling.

At low energy, the free–bound laser coupling is finite for s-wave scattering and it is small for higher partial waves due to the Wigner threshold laws. Therefore, we need to primarily consider the tuning of the s-wave scattering length by OFR, as in the case of MFR. Let us recall the definition of the s-wave scattering length: $1/a_s = -\lim_{k\to 0} k \cot \delta_0$, which is related to the low-energy T-matrix element by $T_0 = 1/(-\cot \delta_0 + i) = ka_s/(1 + ika_s)$. Now, comparing this low-energy expression for the T-matrix element with $T_{\ell=0}^L$, as given by equation (5.55), and ignoring the PA coupling for all higher partial waves ($\ell \neq 0$), one can define an energy-dependent and laser-modified complex s-wave scattering length

$$a_s^L = \left(\frac{\hbar\Gamma_0}{2k}\right) \frac{1}{\hbar\delta + E - \widetilde{E}_{vJ} + i\hbar(\gamma + \Gamma)/2}, \tag{5.57}$$

where Γ_0 is the stimulated linewidth for the transition between the s-wave scattering state and the excited molecular bound state. Γ_0 is proportional to the square of the free–bound Franck–Condon overlap integral. In the limit $k \to 0$, the quantity $\frac{\hbar\Gamma_0}{k}$ becomes independent of energy, as can be understood from the Franck–Condon overlap integral between a bound state and the energy-normalized scattering state

$$\psi_{\ell=0} = \sqrt{\frac{\pi k}{E}} \sin\left[kr + \eta_0(k)\right]. \tag{5.58}$$

According to the Wigner threshold laws, in the limit $k \to 0$, the s-wave phase shift $\eta_0(k) \sim k$. Clearly, in the limit $k \to 0$, the quantity $|\psi_{\ell=0}|^2/k$ is independent of k, and so it is proven that $\frac{\hbar\Gamma_{J0}}{k}$ is independent. Therefore, at low energy, a_s^L is a constant given by

$$a_s^L = \frac{G}{\hbar\delta - \widetilde{E}_{vJ} + i\hbar(\gamma + \Gamma_J)/2}, \tag{5.59}$$

where $G = \lim_{k\to 0} \frac{\hbar\Gamma_{J0}}{2k}$; thus, the total scattering length is $a_s = a_s^0 + a_s^L$, with a_ℓ^0 being the background scattering length. So, in the limit $k \to 0$, both the real and imaginary parts of the light-modified complex scattering length are energy independent. When $\delta_{vJ} = \hbar\delta - E_{vJ} \geqslant 0$, the real part of a_ℓ^L is positive, since $E_J^{\text{shift}} < 0$. This implies that when the PA laser is tuned to the blue side of the resonance, the modified two-body interaction is repulsive. On the other hand, when $\delta_{vJ} < E_J^{\text{shift}}$, the real part of a_ℓ^L is negative. When the PA laser is tuned to the red side of the resonance by an amount exceeding $|E_J^{\text{shift}}|$, the modified interaction becomes attractive. The ratio of the imaginary part to the real part of a_s^L must be sufficiently small in order to observe a finite modification of the scattering length or the elastic scattering amplitude. This implies that $\gamma/2$ must be smaller than the magnitude of the detuning $|\delta - \tilde{E}_{vJ}/\hbar|$. On the other hand, for $|\delta - \tilde{E}_{vJ}/\hbar| \gg \gamma/2$, the real part of a_s^L is proportional to $G/|\delta - \tilde{E}_{vJ}/\hbar|$. So, making the detuning arbitrarily large diminishes the elastic scattering amplitude. This analysis shows that unlike MFR, OFR offers only a limited ability to tune the scattering length.

Let us now see how an energy-independent complex scattering length is related to elastic and inelastic scattering cross sections. One can introduce an s-wave complex phase shift $\delta_c = \delta_r + i\delta_i$, where both the real part δ_r and the imaginary part δ_i go to zero in the limit $k \to 0$. Therefore, the S-matrix element $S = e^{2i\delta_c} = e^{2i\delta_r}e^{-2\delta_i}$. Accordingly, the scattering length is

$$a_s = -\lim_{k\to 0}\frac{\tan\delta_c}{k} \simeq A + iB, \tag{5.60}$$

where $A = -\lim_{k\to 0}\frac{\delta_r}{k}$ and $B = -\lim_{k\to 0}\frac{\delta_i}{k}$. The elastic cross section is given by

$$\sigma_{\text{el}} = g_s\frac{\pi}{k^2}\left|1 - S\right|^2 \simeq 4\pi(A^2 + B^2) = 4\pi|a_s|^2 \tag{5.61}$$

where $g_s = 1(2)$ for two distinguishable (indistinguishable) particles and the inelastic scattering cross section is

$$\sigma_{\text{inel}} = g_s\frac{\pi}{k^2}\left(1 - |S|^2\right) \simeq 4\pi B^2 = 4\pi\left(\text{Im}[a_s]\right)^2 \tag{5.62}$$

Therefore, in the low-energy limit, the light-modified elastic scattering cross section is given by

$$\sigma_{\rm el} = 4\pi g_s \frac{|G|^2}{(\hbar\delta - \tilde{E}_{vJ})^2 + (\hbar(\gamma + \Gamma)/2)^2} \tag{5.63}$$

and the inelastic cross section is given by

$$\sigma_{\rm inel} = \pi g_s \left[\frac{\hbar |G| \gamma}{(\hbar\delta - \widetilde{E}_{vJ})^2 + (\hbar(\gamma + \Gamma)/2)^2} \right]^2 . \tag{5.64}$$

The above formula for the laser-modified s-wave scattering length applies to single-color OFR. The S-matrix element for two-color OFR is derived in the next chapter.

5.6 Collisions in both magnetic and laser fields

A resonance can be manipulated simultaneously by both magnetic and laser fields. If the magnetic field is tuned to an MFR and the optical field is tuned to an OFR, both resonances can interfere, which can result in Fano resonances [53].

5.6.1 Magneto-optical Fano–Feshbach resonance

We briefly introduced single and doubly dressed Fano states in the previous chapter. The resonances in external fields can be described in terms of these dressed states.

Here, we consider the scattering of two cold atoms in the presence of both an optical (laser) field and a magnetic field. The magnetic field is tuned near an MFR and the laser is tuned near an OFR. The magnetic field creates a singly dressed Fano state $|E\rangle_F$ involving the continuum of scattering states and the closed-channel quasi-bound state in the electronic ground-state manifold. When a PA laser is applied that couples $|E\rangle_{\rm F}$ to an excited molecular bound state, a doubly dressed Fano state $|E\rangle_{\rm FF}$ results, causing Fano resonances. The detailed mathematical formalism of Fano resonances has already been given in the previous chapter. Here, we apply this method to describe such dual field-induced resonances, which we term magneto-optical Fano–Feshbach resonances (MOFRs). A distinctive feature of MOFRs is their asymmetric and split resonance profile [54].

Early experiments [55, 56, 57] explored the manipulation of PA spectra with a tunable MFR. The effects of MFR on PA were studied theoretically in several papers [58–62, 53]. An experiment by Junker *et al* [55] on PA in a thermal gas of ultracold ^{7}Li atoms in the vicinity of a tunable MFR at temperatures ranging between 9 and 18 μK clearly demonstrated an asymmetric spectral line shape reminiscent of a Fano profile. Their experimental results are displayed in figure 5.9. This observation led to the theoretical development of the magnetically induced Fano effect in PA [53] and MOFR [54]. The PA rate as a function of laser detuning (or the magnetic field for fixed laser detuning in the case of MOFR) constitutes the PA spectral line shape. The method used to calculate PA spectra will be presented in the next chapter. Here, we mention that strong PA coupling in the presence of an MFR can give rise to quantum interference between multiple transition pathways. As a result, the PA spectrum can exhibit a Fano-like shape. Figure 5.10 displays the

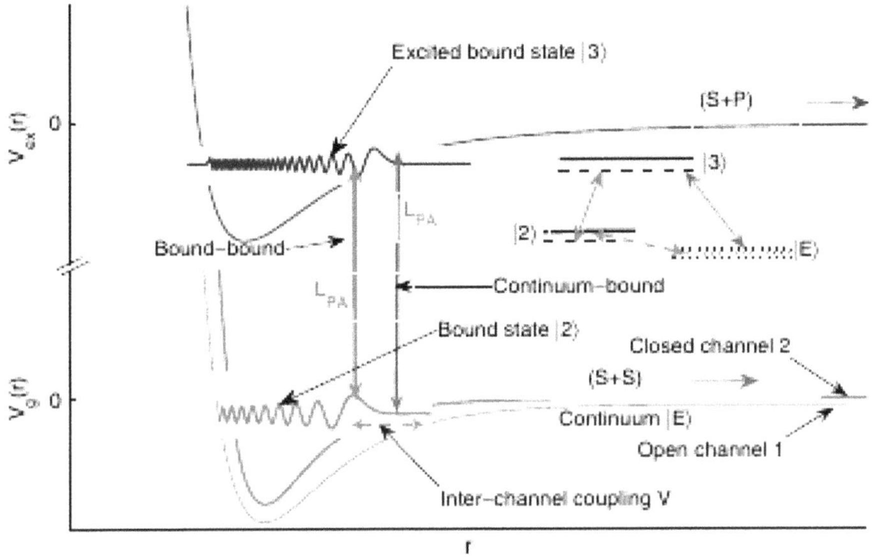

Figure 5.8. Schematic of magneto-optical Feshbach resonance, showing the couplings (vertical double arrows) of the bound state $|2\rangle$ and the continuum $|E\rangle$ with the excited bound state $|3\rangle$ produced using the same laser L_{PA}. The dashed double arrow indicates the coupling between the open channel and the closed channel. Reproduced from [50]. © IOP Publishing Ltd. All rights reserved.

Fano effect in the numerically calculated PA spectra of ultracold ^7Li atoms due to an MFR. The saturation effect in strong-coupling PA in the presence of MFR can also be interpreted in terms of Fano's theory [50] for a wide range of laser intensities.

The physical situation in the case of MOFR is similar to that in the case of Fano resonances. There are two competing transition pathways—one is the continuum–bound–continuum pathway, and the other is the bound–bound–continuum pathway. Fano resonances were originally studied in the context of autoionization, where an atom can transition to the electron–ion or ionization continuum via two competing optical transition pathways—one is the direct bound–continuum pathway, and the other is the bound–bound–continuum pathway. In the case of autoionizing Fano resonances, the first bound state belonging to a particular atomic configuration is embedded in the ionization continuum, while the second bound state belongs to another atomic configuration at a lower energy. In the case of autoionization, the bound–continuum transition is radiationless, and quantum interference occurs between two single-photon transition pathways. In contrast, in MOFR, both transition pathways involve two photons, as schematically shown in figure 5.8.

5.7 Higher partial-wave Feshbach resonances

So far, we have discussed the tuning of the s-wave scattering length a_s through MFR and OFR. However, in principle, it is possible to alter the higher partial-wave scattering amplitude using MFR and OFR. For instance, the partial-wave **T**-matrix element of equation (5.55) shows that not only the s-wave but also higher

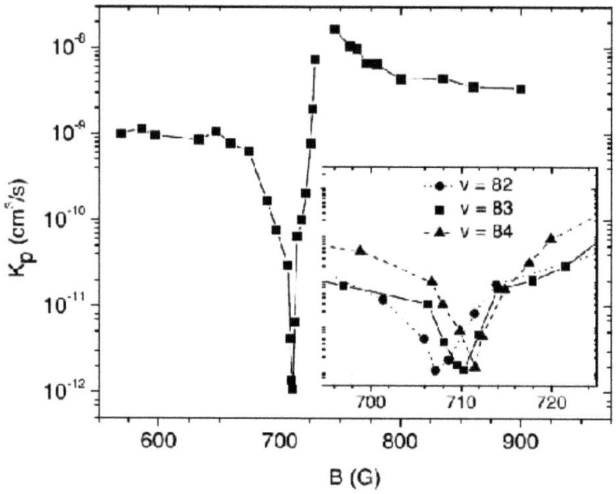

Figure 5.9. Experimentally obtained PA rate K_p for the transition to the vibrational level $v = 83$ of the $1^3\Sigma_g^+$ excited molecular state of $^7Li_2$ at a laser intensity of $I = 1.65$ W cm^{-2}. Figure reprinted with permission from [55], Copyright (2008) by the American Physical Society.

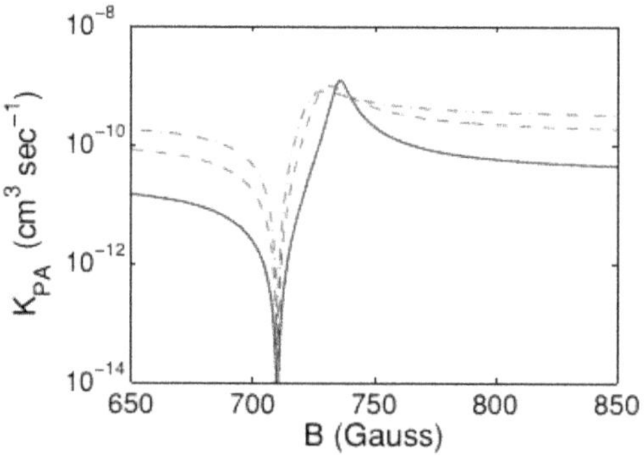

Figure 5.10. PA rate K_{PA} as a function of the magnetic field B in Gauss. Reproduced from [53]. © IOP Publishing Ltd. All rights reserved.

partial-wave scattering amplitudes can be influenced by second-order photoassociative coupling via the s-wave. The ability to control a higher partial-wave interaction of atoms using an external field would enable us to explore beyond s-wave quantum many-body physics with cold atoms. Successful demonstrations of s-wave superfluidity and other quantum many-body effects using bosonic and two-component fermionic atoms have been made possible by the availability of efficient field-based control of the s-wave interaction. As discussed in chapter 2, the mean-field interaction of dilute ultracold atomic gases can be expressed in terms of a_s; therefore, the ability to tune a_s using a Feshbach resonance allows us to control the mean-field interaction. Since the mean-field interaction primarily determines the phases of an interacting many-particle system, Feshbach resonances are routinely employed as a tool with which to control and explore the s-wave quantum many-body aspects of atomic gases.

To explore beyond s-wave many-body quantum physics with ultracold atomic gases, it is important to investigate ways to generate and alter the higher partial-wave scattering amplitudes of cold atoms using external-field-induced higher partial-wave resonances like the familiar s-wave MFR and OFR. Although, in principle, it is possible to alter the higher partial-wave scattering amplitude by MFR or OFR, the scope of this alteration is rather limited. This limitation primarily stems from the Wigner threshold laws, which dictate that at low temperatures, s-wave collisions are largely dominant over the higher partial-wave ones. Nevertheless, the use of MFR and OFR methods for manipulating p- and d-wave elastic scattering amplitudes of some specific atomic systems has been explored in the past. For indistinguishable fermionic atoms, s-wave scattering is forbidden, and the lowest partial wave that is allowed by exchange symmetry is the p-wave, which is particularly important for exploring the anisotropic superfluidity of spin-polarized fermionic atoms at low energy. Manipulating the d-wave interactions of cold atoms is important for exploring d-wave superfluidity, which may be related to high-temperature superconductivity. p-wave MFR in spin-polarized ^{40}K [63] and ^{6}Li [64] Fermi gases was experimentally observed in 2003 and 2004, respectively. Several optical methods [65, 66, 67] based on the use of PA coupling to specific excited molecular levels, such as purely long-range molecular states [68, 69], have been proposed for manipulating the p- and d-wave interactions of cold atoms. The experimental observation of p-wave OFR in fermionic ^{171}Yb atoms was reported in 2013. The ability to alter the higher partial-wave interactions of cold atoms using external fields under different physical and geometric conditions [70–72] is expected to enable us to explore beyond s-wave quantum effects in cold atomic gases.

5.8 Collisions in a static electric field

We now consider collisions between two ground-state atoms in the presence of a static (or DC) electric field. Marinescu and You [73] and Melezhik and Hu [74] suggested using a static electric field to control ultracold collisions in homonuclear atomic systems. The field distorts the spherical symmetry of a ground-state atom, leading to the polarization of the atom. As a result, the ground electronic state becomes hybridized with a small excitation to P (the electronic orbital angular momentum of the atom is $\ell_a = 1$), the electronic angular momentum in the direction of the applied electric field. Two polarized atoms then interact via the electric dipole–dipole potential, where the dipole moment of the atom is induced by the DC electric field and is therefore proportional to the static polarizability of the atom. This field-induced polarization of atoms primarily alters the long-range character of interatomic interaction potentials and thereby largely modifies the scattering properties in the limit of zero collision energy.

5.8.1 Scattering between a pair of homonuclear atoms

The DC electric-field-induced atom–atom potential between two atoms a and b, to the leading order in electric field amplitude $\mathscr{E}$, is given by [73]

$$\hat{V}_{\mathscr{E}}(R) = -\frac{C_{\mathscr{E}}}{R^3}P_2(\cos\theta), \tag{5.65}$$

where $C_{\mathscr{E}} = 2\mathscr{E}^2\alpha_a\alpha_b$ is the interaction coefficient, $\alpha_{a\,(b)}$ is the static polarizability of atom a (b), and $P_2(\cos\theta)$ is the Legendre polynomial of order two, with θ being the angle between the electric field and the internuclear axis.

The scattering between two atoms in the presence of a static electric field occurs at the potential

$$\hat{V}(\mathbf{R}) = \hat{V}_c(R) + \hat{V}_{hf} + \hat{V}_{\mathscr{E}}(R), \tag{5.66}$$

where $\hat{V}_c(R)$ is the usual adiabatic central potential and $\hat{V}_{hf}$ is the HF interaction. The potential $\hat{V}_{\mathscr{E}}(R)$ can be treated as a perturbation term for electric fields much smaller than the internal electric field of an atom. An electric field of 1000 kV cm^{-1} corresponds to about 10^{-4} a.u.; therefore, $V_{\mathscr{E}}(R)$ can be treated as a perturbation if the electric field is smaller than or of the order of 1000 kV cm^{-1}. Obviously, this potential is anisotropic because of the angular dependence. $P_2(\cos\theta)$ couples different partial waves with $\Delta\ell = \pm 2$, which means that only even partial waves couple to even ones and only odd partial waves couple to odd ones. Such couplings among the partial waves make the scattering phenomena essentially a multichannel problem. If $V_c(R)$ is a potential corresponding to a single set of internal electronic and spin DOFs, then different channels here correspond to different partial waves where all channels have the same threshold.

In the limit $R \to \infty$, the potential behaves as $1/R^3$, dictating the threshold behavior of the scattering. Since the potential is anisotropic, the usual partial-wave expansion for a central potential requires modification. The time-independent Schrödinger equation for different partial-wave components of the wave function becomes coupled subject to the conditions $\Delta\ell = \pm 2$ and $\Delta m_\ell = 0$. Since the electric field defines an axis of rotational symmetry (the z-axis in this case), the projection of the total angular momentum onto this axis is a constant of motion. Therefore, an electric field cannot couple two different ms together, and m_ℓ remains a good quantum number. For an electric field pointing in a particular direction, the Hamiltonian itself has odd parity and thus serves to change the parity of the atom. Therefore, an electric field can couple even and odd parity channels to each other but not to themselves.

The total wave function can be expanded in terms of the radial wave functions $\psi_{\ell,m_\ell}(\mathrm{R})$ and the rotational wave functions $Y_{\ell,m_\ell}(\hat{R})$, with $\hat{R}$ denoting the orientation of the vector $\vec{R}$ (e.g. (θ, ϕ)):

$$\Psi(\vec{R}) = \frac{1}{R}\sum_{\ell m_\ell}\psi_{\ell,m_\ell}(R)\,Y^*_{\ell m_\ell}(\hat{k})\,Y_{\ell m_\ell}(\hat{R}). \tag{5.67}$$

The asymptotic form of the wave function is given by

$$R\Psi(\vec{R}) \sim R\exp(i\vec{k}.\vec{R}) + f_{\vec{k},\vec{k}'}\exp(ikR), \tag{5.68}$$

where $\vec{k}$ and $\vec{k}'$ are the incident momentum and the scattered momentum, respectively. The on-shell elastic scattering is given by $f_{\vec{k},\vec{k}'}$, with $\vec{k}' = k\hat{R}$. Expanding the scattering amplitude $f_{\vec{k},\vec{k}'}$ into the complete basis $Y_{\ell m_\ell}$, one obtains

$$f_{\vec{k},\vec{k}'} = \frac{4\pi}{k}\sum_{\ell,m_\ell} T_{\ell m_\ell}\left(\vec{k}\right)Y_{\ell m_\ell}(\hat{R}), \tag{5.69}$$

where $T_{\ell m_\ell}$ is a T-matrix element which can also be expanded into the following form:

$$\frac{1}{k}T_{\ell m_\ell}(\vec{k}) = \sum_{\ell',m_{\ell'}} t_{\ell m_\ell}^{\ell' m_{\ell'}}(k)\,Y_{\ell' m_{\ell'}}(\hat{k}) \tag{5.70}$$

and

$$\exp(i\vec{k}.\vec{R}) = 4\pi\sum_{\ell,m_\ell} i^\ell j_\ell(kr)\,Y^*_{\ell m_\ell}(\hat{k})\,Y_{\ell m_\ell}(\hat{R}). \tag{5.71}$$

The incident plane wave can also be expanded in terms of spherical harmonics. Thus, we have

$$\begin{aligned} R\Psi\left(\vec{R}\right) \sim \frac{4\pi}{k}\sum_{\ell,m_\ell} i^\ell\Big[\,&Y^*_{\ell m_\ell}(\hat{k})\sin(kR - \ell\pi/2) + T_{\ell m_\ell}\left(\vec{k}\right) \\ &\times \exp(ikr - i\ell\pi/2)\Big]\,Y_{\ell m_\ell}(\hat{R}), \end{aligned} \tag{5.72}$$

where the asymptotic form $j_\ell(kR) \sim \sin(kR - \ell\pi/2)/(kR)$ has been used. The multichannel form of the scattering equation is given by

$$h_\ell\psi_{\ell m_\ell}(R) = \sum_{\ell',m_{\ell'}} i^{\ell'-\ell}\langle \ell m_\ell|\,V_\mathcal{E}(R)|\ell' m_{\ell'}\rangle\psi_{\ell' m_{\ell'}}, \tag{5.73}$$

where

$$h_\ell = -\frac{1}{2\mu}\frac{d^2}{dR^2} + \frac{1}{2\mu}\frac{\ell(\ell+1)}{R^2} + V - E. \tag{5.74}$$

Here, V is the single-channel central potential. The asymptotic boundary condition on $\psi_{\ell m_\ell}$ is

$$\psi_{\ell m_\ell} \sim \sin(kR - \ell\pi/2) + T_{\ell m_\ell}(k)\exp(ikR - i\ell\pi/2). \tag{5.75}$$

One can write the coupled equations in matrix form in the (ℓ) basis as follows:

$$\left[\left(-\frac{1}{2\mu}\frac{d^2}{dR^2}+V\right)\mathbf{I}+\frac{1}{2\mu R^2}\mathbf{L}^2+\mathbf{V}_{\mathcal{E}}\right]\Phi=E\Phi, \tag{5.76}$$

where $\mathbf{I}$ is the identity matrix and $\mathbf{L}^2\equiv$ diag[1(1+1), 2(2+1),........, $\ell(\ell+1)$.....] is a diagonal matrix. The wave function Φ is a matrix whose elements are given by $\phi_{\gamma'\gamma}$, where $\gamma\equiv\ell, m_\ell$ and $\gamma'\equiv\ell', m_{\ell'}$ are the incident and scattered angular states, respectively. Therefore, the overall wave function for an incident partial wave $\gamma(\ell, m_\ell)$ becomes

$$\psi_{\gamma=\ell,m_\ell}(E, R)=\sum_{\gamma'=\ell',m_{\ell'},}\phi_{\gamma'\gamma}(E, R)|\gamma'\rangle. \tag{5.77}$$

By imposing the boundary conditions on the partial waves, we can write the T-matrix elements $t_{\ell m_\ell}^{\ell' m_{\ell'}}$. The total elastic cross section is given by

$$\sigma=g_s 4\pi\sum_{\ell\ell'}\sum_{m_\ell m_{\ell\prime}}|t_{\ell m_\ell}^{\ell' m_{\ell'}}|^2. \tag{5.78}$$

where $g_s=2$ for two indistinguishable atoms; otherwise $g_s=1$.

The theoretical calculations of Marinescu and You [73] based on a model potential predicted several static electric-field-induced resonances in ultracold atomic collisions, as shown in figure 5.11. Deb and You [75] further interpreted these resonances in the limit $k\rightarrow 0$ as the shape resonances due to the electric-field-induced coupling between the s- and d-wave relative angular momenta. Figure 5.12

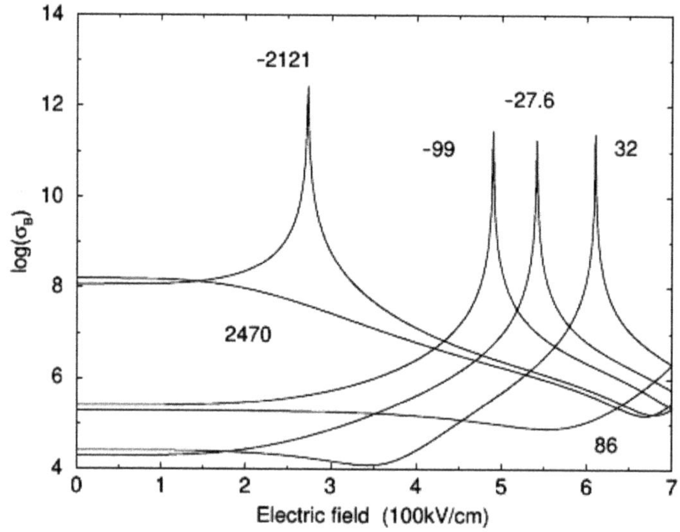

Figure 5.11. Resonances induced by a static electric field. The plot shows the logarithm of the total scattering cross section calculated using a model potential as a function of the electric field. The different curves are marked with their field-free scattering lengths, which are parameters of the model potential. Figure reprinted with permission from [73], Copyright (1998) by the American Physical Society.

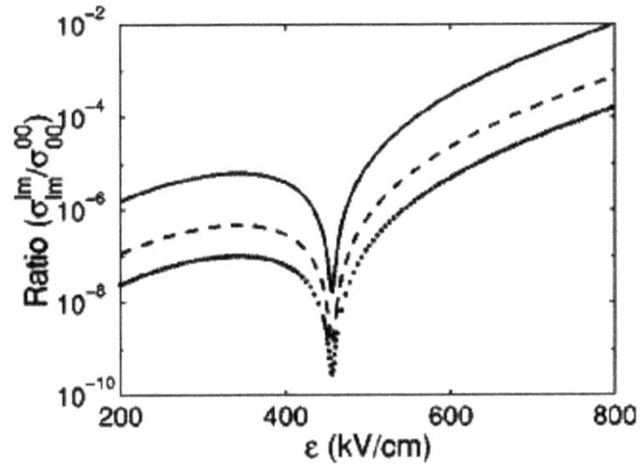

Figure 5.12. Scattering cross sections of a few low-lying partial waves compared to the s-wave cross section. The plot shows the ratios $\sigma_{20}^{20}/\sigma_{00}^{00}$ (solid), $\sigma_{30}^{30}/\sigma_{00}^{00}$ (dashed), and $\sigma_{40}^{40}/\sigma_{00}^{00}$ (dotted) at a collision energy of $E = 0.01$ nK. Figure reprinted with permission from [75], Copyright (2001) by the American Physical Society.

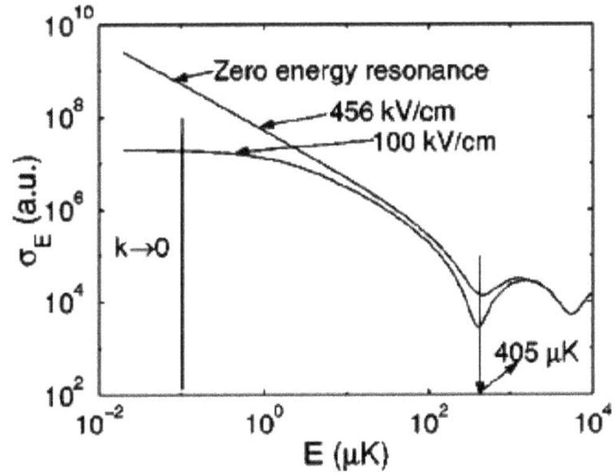

Figure 5.13. The total cross section σ_E as a function of collision energy E for a resonant electric field of $\mathcal{E} = 456\ kV$ cm^{-1} and an off-resonant field of $\mathcal{E} = 100$ kV cm^{-1}. Zero-energy resonance is clearly visible from the divergent behavior of the cross section as $k \rightarrow 0$, in which limit the cross section diverges as the inverse of the energy. Figure reprinted with permission from [75], Copyright (2001) by the American Physical Society.

shows the ratios of the $\ell = 2$, $\ell = 3$, and $\ell = 4$ wave cross sections to that of the $\ell = 0$ wave. Under a resonant electric field of $\mathcal{E} = 456$ kV cm^{-1}, the s-wave ($\ell = 0$) cross section dominates over the d-wave one by more than two orders of magnitude. This resonance is, in fact, a zero-energy resonance, as figure 5.13 shows; in the limit $k \rightarrow 0$, the resonant cross section diverges as the inverse of the energy, as per the unitarity limit, but under an off-resonant electric field of $\mathcal{E} = 100$ kV cm^{-1}, the cross section attains a constant value as $k \rightarrow 0$, in accordance with the Wigner threshold laws.

5.8.2 Scattering between a pair of heteronuclear atoms

In contrast to homonuclear atom pairs, a heteronuclear pair of atoms moving slowly near the dissociation threshold of the diatomic potential can possess an instantaneous electric dipole moment. This dipole moment can give rise to an electric dipole interaction due to an external electric field. Thus, the electric-field-induced interaction between a pair of atoms with different nuclei is directly proportional to the electric field strength. This interaction couples even and odd partial waves. In contrast, in the case of a homonuclear atom pair, the interaction is proportional to the square of the electric field strength, as it is a second-order process. A mixture of two different species of atoms in an electric field provides a platform that can manipulate not only low-energy scattering characteristics but also molecular properties near the ground-state dissociation threshold with electrostatic fields. It has been theoretically shown that an electric field can lead to an enhancement of the interaction between Li and Cs atoms in an ultracold collision [76, 77] due to the existence of an instantaneous dipole moment of the heteronuclear collision complex. The effect of a static electric field on the low-energy collision properties of a mixture of lithium and cesium atoms has been theoretically studied by Chakraborty and Deb [78], who solved the multichannel scattering problem using the Numerov–Cooley algorithm.

The electric-field-dependent potential is given by

$$\hat{V}_{\mathcal{E}} = -\mathcal{E}(\hat{e}_{\mathcal{E}}.\ \hat{e}_d)\sum_{S}\sum_{M_S}|SM_S\rangle d_S(R)\langle SM_S|, \tag{5.79}$$

where $\hat{e}_{\mathcal{E}}$ and $\hat{e}_d$ are the unit vectors along the electric field and the dipole moment, respectively. Here, $d_S(R)$ denotes the dipole moment functions of LiCs in the different spin states ($S = 0$ for singlet and $S = 1$ for triplet). Li and Krems [77] provided the analytical expression

$$d_S(R) = D \exp[-\alpha(R - R_e)^2], \tag{5.80}$$

using the numerical data computed by Aymar and Dulieu [79]. Here, the parameters are $r_e = 7.7\ a_0$, $\alpha = 0.1\ a_0^{-2}$, and $D = 6$ Debye for the singlet state and $D = 0.5$ Debye for the triplet state (where a_0 is the Bohr radius). The permanent dipole moment of the colliding pair of heteronuclear atoms leads to coupling between different angular momentum states and thus affects the ground-state scattering wave functions. The permanent dipole moment of the complex typically peaks around the equilibrium distance of the diatomic molecule in the ground state. Due to the strong spatial dependence of the permanent dipole moment at short separations, the interaction of a static electric field with the permanent dipole moment of the collision pair leads to new resonances with large modifications of the scattering wave function at short separations. Based on this short-range modification of the scattering wave function, a giant enhancement in the PA rate of the formation of polar molecules in a binary mixture of cold atoms has been predicted [78]. Figure 5.14 displays electric-field-induced resonances in a mixture of Li and Cs atoms, predicting a prominent resonance at $\mathcal{E} = 1298\ \mathrm{kV\ cm^{-1}}$.

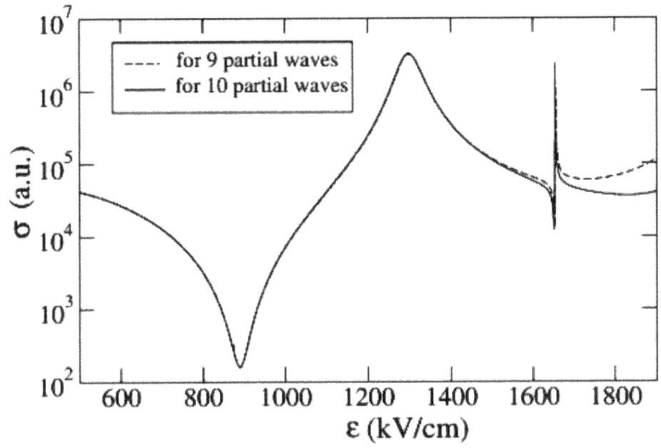

Figure 5.14. Total elastic cross section σ in a.u. as a function of the electric field $\mathcal{E}$ in kV cm^{-1} for two different total numbers of partial waves (chosen to show the convergence of the results). The first resonance appears at $\mathcal{E} = 1298$ kV cm^{-1}. Reproduced from [78]. © IOP Publishing Ltd. All rights reserved.

5.9 Ion–atom collisions at low energy

We mentioned earlier that the ion–atom potential at long separation varies as C_4/r^4. This means that, in the zero-energy limit, only the s-wave scattering cross section is finite and independent of energy; all other higher partial-wave cross sections vanish according to the Wigner threshold laws. Furthermore, the ion–atom potential has a much longer range compared to that between ground-state neutral atoms. As a result, even at cold temperatures, a larger number of partial waves make finite contributions to the total scattering cross section, unlike the case of neutral atoms. In order to reach the Wigner threshold regime in ion–atom collisions—that is, the regime of dominant s-wave collisions—one must reduce the relative kinetic energy between an ion and an atom to less than 1 μK. Experimentally, this is a challenging task. Although there are various techniques for cooling atoms to the microkelvin or even nanokelvin temperature regime, cooling ions in ion traps to such low temperatures using those techniques simply does not work, primarily due to the ion-trap-induced micromotion of the ions. Over the years, researchers have been developing hybrid ion–atom traps where ions and atoms can coexist in the same spatial zone. This has facilitated the study of ion–atom collisions through measurements of the collisional loss of atoms or ions from the hybrid trap. A comprehensive review of ion–atom collisions at low energy can be found in an article by Tomza *et al* [80].

The various transition processes that can happen in an ion–atom system are schematically illustrated in figure 5.15. One must first distinguish between homonuclear and heteronuclear collisions. In the homonuclear case, charge exchange symmetry leads to resonant charge exchange collisions, which are absent in the latter case. A detailed theoretical study of both cases [81] revealed several characteristic features of low-energy ion–atom collisions.

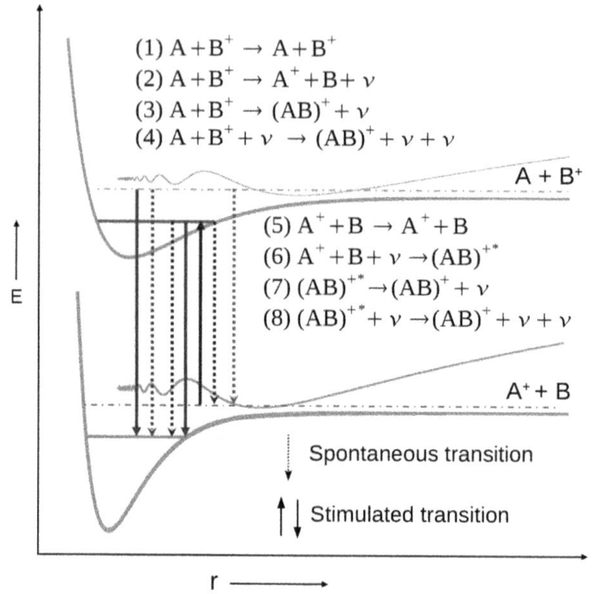

Figure 5.15. Possible transitions in ion–atom collisions in a laser field. Figure reprinted with permission from [81], Copyright (2011) by the American Physical Society.

Let us first consider heteronuclear collisions between a Be^+ ion and a Li atom. Both the ion and the atom have one electron in the outermost shell or subshell. The lowest molecular potential that asymptotically corresponds to separated $Be^+$and Li particles belongs to the electronic state $2^1\Sigma^+$. This has higher energy than the $1^1\Sigma^+$ potential, which asymptotically corresponds to the charge-exchanged system of Li^++Be. Since both Li^+ and Be have filled outermost S-shells, the $1^1\Sigma^+$ potential has lower energy than the $2^1\Sigma^+$ potential. The scattering cross section of the Be^++Li system as a function of energy is displayed in figure 5.16, which shows that as the energy is decreased below 1 μK, the cross section remains almost constant as $E \to 0$, marking the Wigner threshold regime where the total cross section is almost entirely determined by the s-wave. On the other hand, as the energy is increased above 10 mK, the total elastic cross section can be approximated by the analytical form [82]

$$\sigma_{\text{el}} \sim \pi\left(\frac{\mu C_4^2}{\hbar^2}\right)^{\frac{1}{3}}\left(1 + \frac{\pi^2}{16}\right)E^{-\frac{1}{3}}, \tag{5.81}$$

implying that at high energy, the elastic cross section decreases as the inverse of the one-third power of the energy. Figure 5.16 shows that the cross section attains its high-energy limit as the energy increases above 0.1 mK. The elastic cross sections for the two lowest potentials $^2\Sigma_g^+$ and $^2\Sigma_u^+$ of the homonuclear (Li^++Be) system are displayed in figure 5.17. This figure again reveals that the Wigner threshold regime can be achieved when the temperature is reduced below 1 μK.

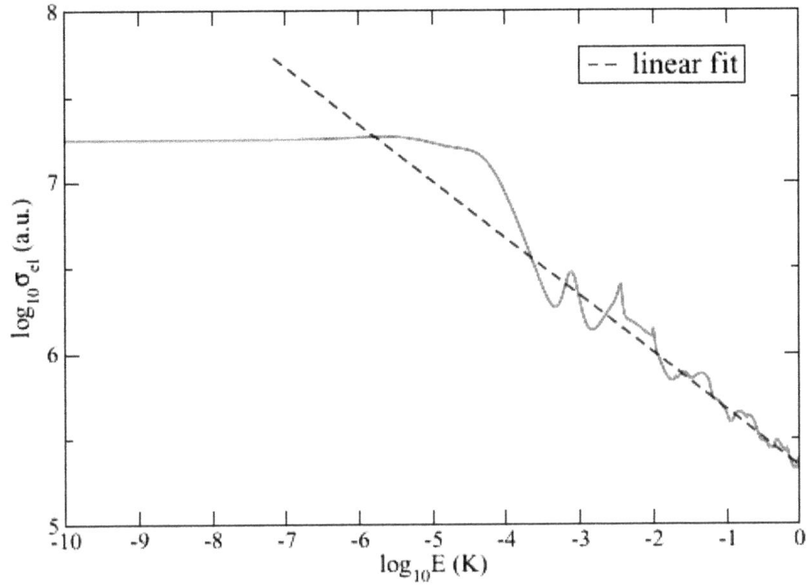

Figure 5.16. Logarithm of the elastic scattering cross section σ_{el} as a function of the logarithm of E in K for a $Li+Be^{+}$ ($2^{1}\Sigma^{+}$) collision. The dashed line is a linear fit at energies greater than 1 mK. Figure reprinted with permission from [81], Copyright (2011) by the American Physical Society.

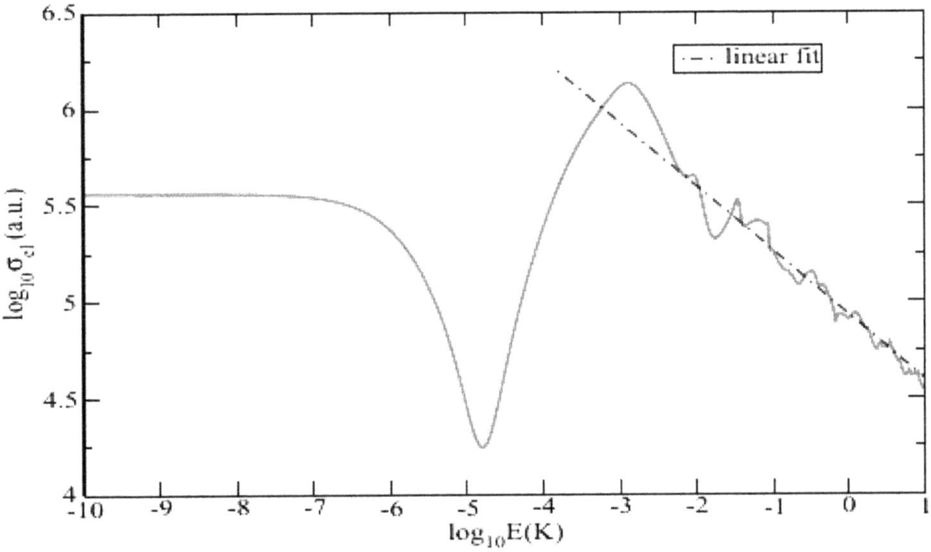

Figure 5.17. Logarithm of σ_{el} plotted against the logarithm of E in a.u. (1 a.u. $= 3.1577 \times 10^{5}$ K) for a Li^{+} +Li collision in the $^{2}\Sigma_{g}^{+}$ (dashed) and $^{2}\Sigma_{u}^{+}$ (solid) states. Figure reprinted with permission from [81], Copyright (2011) by the American Physical Society.

5.10 Bound state in a continuum of cold atoms

In the previous chapter, we introduced the concept of a bound state in a continuum (BIC). A BIC of two cold atoms can be formed due to quantum interference among multiple resonant transition pathways in the absence of any spontaneous emission.

In the context of cold atomic collisions, a BIC can be derived by analyzing the resolvent operator $\hat{G}(z) = (z - \hat{H})^{-1}$. Let us consider an MOFR model: a closed-channel bound state $|b_c\rangle$ is coupled to the continuum of scattering states $|E\rangle$ in the electronic ground-state configuration; both $|b_c\rangle$ and $|E\rangle$ are coupled to a pair of excited molecular bound states $|b_1\rangle$ and $|b_2\rangle$ by PA lasers L_1 and L_2, respectively. The binding energy of $|b_c\rangle$ and the threshold of the ground-state open channel can be tuned by a magnetic field to induce an MFR. The Hamiltonian is $\hat{H} = \hat{H}_0 + \hat{V}$, where

$$\hat{H}_0 = E_c|b_c\rangle\langle b_c| + \sum_{n=1,2}(E_n - \hbar\omega_{L_n})|b_n\rangle\langle b_n| + \int dEE|E\rangle\langle E|. \tag{5.82}$$

Here, ω_{L_n} is the frequency of the laser L_n and

$$\hat{V} = \sum_n \int dE\Lambda_n(E)|b_n\rangle\langle E| + \int dEV_E|b_c\rangle\langle E| + \sum_n \hbar\Omega_n|b_n\rangle\langle b_c| + \text{h.c.}, \tag{5.83}$$

where Ω_n is the Rabi frequency for the transition $|b_c\rangle \leftrightarrow |b_n\rangle$. One can introduce a projection operator $\mathbf{P}$ that projects the system to all three bound states

$$\mathbf{P} = |b_c\rangle\langle b_c| + \sum_{n=1,2}|b_n\rangle\langle b_n|, \tag{5.84}$$

and the projection operator $\mathbf{Q} = \mathbf{1} - \mathbf{P}$ given by

$$\mathbf{Q} = \int dE|E\rangle\langle E| \tag{5.85}$$

that projects the system into the continuum. These two projection operators have the properties

$$\mathbf{P}^2 = \mathbf{P}, \quad \mathbf{Q}^2 = \mathbf{Q}, \quad \mathbf{PQ} = \mathbf{QP} = 0, \quad \mathbf{P} + \mathbf{Q} = \mathbf{1}. \tag{5.86}$$

Thus, we have

$$\hat{G} = \hat{G}_0 + \hat{G}_0\hat{V}\hat{G} = \frac{1}{E - \hat{H}_0 + i\varepsilon} + \frac{1}{E - \hat{H}_0 + i\varepsilon}\hat{V}\hat{G}. \tag{5.87}$$

Next, we project out the continuum states to obtain an effective Hamiltonian in the following way:

$$\hat{V}\hat{G} = \hat{V}(\mathbf{P} + \mathbf{Q})\hat{G} = \hat{V}\mathbf{P}\hat{G} + \hat{V}\mathbf{Q}\hat{G} \tag{5.88}$$

$$\mathbf{Q}\hat{G} = \mathbf{Q}\hat{G}_0 + \frac{\mathbf{Q}}{E - \hat{H}_0 + i\varepsilon}(\hat{V}\mathbf{P}\mathbf{G} + \hat{V}\mathbf{Q}\hat{G}), \tag{5.89}$$

which leads to

$$\mathbf{Q}\hat{G} = \frac{1}{E - \hat{H}_0 - \mathbf{Q}\hat{V} + i\varepsilon}(\mathbf{Q} + \mathbf{Q}\hat{V}\mathbf{P}\hat{G}). \tag{5.90}$$

Upon substituting (5.90) into (5.88), after some algebra, one obtains

$$\mathbf{P}\hat{G}\mathbf{P} = \frac{1}{E - \hat{H}_0 - \mathbf{P}\hat{V}\mathbf{P} - \mathbf{P}\hat{V}\mathbf{Q}\frac{1}{E - \hat{H}_0 - \mathbf{Q}\hat{V}\mathbf{Q} + i\varepsilon}\mathbf{Q}\hat{V}\mathbf{P}}. \tag{5.91}$$

Thus, we can write the effective Hamiltonian

$$\hat{H}_{\text{eff}} = \hat{H}_0 + \mathbf{PRP}, \tag{5.92}$$

where

$$\mathbf{R} = \hat{V} + \hat{V}\mathbf{Q}\frac{1}{E - \mathbf{H}_0 - \mathbf{Q}\hat{V}\mathbf{Q} + i\varepsilon}\mathbf{Q}\hat{V}. \tag{5.93}$$

We next diagonalize H_{eff} in the subspace of the three bound states, leading to

$$\mathbf{Q}\hat{V}\mathbf{Q} = 0 \tag{5.94}$$

and

$$\begin{aligned}
&\hat{V}\mathbf{Q}\frac{1}{E - \hat{H}_0 - \mathbf{Q}\hat{V}\mathbf{Q} + i\varepsilon}\mathbf{Q}\hat{V} \\
&= \sum_{n,n'}\left[\Delta_{nn'}(E) - i\frac{\Gamma_{nn'}(E)}{2}\right]|b_n\rangle\langle b_{n'}| + \left[\Delta_f(E) - i\frac{\Gamma_f(E)}{2}\right]|b_c\rangle\langle b_c| \\
&+ \left[\sum_n\left\{\Delta_{nf}(E) - i\frac{\Gamma_{nf}(E)}{2}\right\}|n\rangle\langle b_c| + \text{h.c.}\right],
\end{aligned} \tag{5.95}$$

where

$$\Delta_{nn'}(E) = \mathcal{P}\int dE'\frac{\Lambda_n(E')\Lambda^*_{n'}(E')}{E - E'} \tag{5.96}$$

$$\hbar\Gamma_{nn'}(E) = 2\pi\Lambda_n(E)\Lambda^*_{n'}(E) \tag{5.97}$$

$$\Delta_f(E) = \mathcal{P}\int dE'\frac{|V_{E'}|^2}{E - E'} \tag{5.98}$$

$$\hbar\Gamma_f(E) = 2\pi|V_{E'}|^2 \tag{5.99}$$

$$\Delta_{nf}(E) = \mathcal{P}\int dE'\frac{\Lambda_n(E')V^*_{E'}}{E - E'} \tag{5.100}$$

$$\hbar\Gamma_{nf}(E) = 2\pi\Lambda_n(E)V_E^*. \tag{5.101}$$

Here, Γ_f is the resonance linewidth. The matrix elements of the effective Hamiltonian are

$$\begin{aligned}&\langle b_n|\hat{H}_{eff}|b_{n'}\rangle\\&= (E_n - \hbar\omega_{L_n})\delta_{nn'} + \Delta_{nn'}(E) - i\frac{\hbar\Gamma_{nn'}(E)}{2}\end{aligned} \tag{5.102}$$

$$\langle b_c|\hat{H}_{eff}|b_c\rangle = E_0 + \Delta_f(E) - i\frac{\hbar\Gamma_f(E)}{2} \tag{5.103}$$

$$\langle n|\hat{H}_{eff}|b_c\rangle = \Delta_{nf}(E) + \hbar\Omega_n - i\frac{\hbar\Gamma_{nf}(E)}{2}. \tag{5.104}$$

Explicitly, this Hamiltonian is given by

$$\begin{aligned}\hat{H}_{eff} = \hat{H}_0 + \sum_{n,n'=1,2}&\left[\left(\hbar\delta_{nn'} - i\frac{\hbar\Gamma_{nn'}(E)}{2}\right)|b_n\rangle\langle b_{n'}| + \left(\delta_c - i\frac{\hbar\Gamma_f(E)}{2}\right)|b_c\rangle\langle b_c|\right.\\&\left.+\sum_n\frac{\hbar\Gamma_{nf}(E)}{2}\{q_{nf} - i\}|n\rangle\langle b_c| + \text{h.c.}\right],\end{aligned} \tag{5.105}$$

where $\delta_{nn} = (E_n + \Delta_{nn})/\hbar - \omega_{L_n}$ is the detuning of the light-shifted nth bound state; $\delta_{nn'} = \hbar^{-1}\Delta_{nn'}$ $(n \neq n')$, where $\Delta_{nn'}$ is the real part of the quantity $\int dE'\Lambda_n^*(E')\Lambda_{n'}(E')/(E - E')$; and $\Gamma_{nn'}(E) = 2\pi\Lambda_n^*(E)\Lambda_{n'}(E)$. Here, $\delta_c = \hbar^{-1}[E_c(B) + \Delta_f - E_{\text{th}}]r$ is the detuning of the shifted closed-channel bound-state level from the the threshold E_{th} of the open channel. $\Delta_f = \mathscr{P}\int dE'|V_{E'}|^2/(E - E')$, where $\mathscr{P}$ stands for Cauchy's principal value, is the shift due to the coupling $V_{E'}$ between $|b_c\rangle$ and $|E'\rangle_{\text{bare}}$. q_{nf} is the Fano asymmetry parameter defined by

$$q_{nf} = \frac{\delta_{nf} + \Omega_n}{\Gamma_{nf}/2}, \tag{5.106}$$

and

$$\delta_{nf} = \hbar^{-1}\mathscr{P}\int dE'\Lambda_n^*(E')V_{E'}/(E - E') \tag{5.107}$$

is a shift of $|b_c\rangle \leftrightarrow |b_n\rangle$. Here, $\Gamma_{nf} = 2\pi\hbar^{-1}\Lambda_n^*(E)V_E$.

Let us define the dimensionless parameters $\tilde{\delta}_n = \delta_{nn}/(\Gamma_f/2)$ $g_n = \Gamma_{nn}/\Gamma_f$ and $g_{nn'} = \Gamma_{nn'}/(\Gamma_f/2)$ $(n \neq n')$. We assume that $\delta_{12} = \delta_{21} \simeq 0$. Furthermore, assuming that the two laser couplings Λ_{nE} are real quantities, we have $g_n = \Lambda_{nE}V_E/|V_E|^2 = \Lambda_{nE}/V_E$. Under these conditions, the effective Hamiltonian can be written in the matrix form

$$H_{eff} = \frac{\hbar\Gamma_f}{2}[\mathbf{A} + i\mathbf{B}], \tag{5.108}$$

where

$$\mathbf{A} = \begin{pmatrix} \tilde{\delta}_1 & 0 & q_{1f}\sqrt{g_1} \\ 0 & \tilde{\delta}_2 & q_{2f}\sqrt{g_2} \\ q_{1f}\sqrt{g_1} & q_{2f}\sqrt{g_2} & -(ka_s)^{-1} \end{pmatrix} \tag{5.109}$$

and

$$\mathbf{B} = \begin{pmatrix} -g_1 & -g_{12} & -\sqrt{g_1} \\ -g_{21} & -g_2 & -\sqrt{g_2} \\ -\sqrt{g_1} & -\sqrt{g_2} & -1 \end{pmatrix}. \tag{5.110}$$

The secular equation for the **B** matrix is $x^3 + (g_1 + g_2 + 1)x^2 = 0$; it has two roots that are exactly zero and a third that equals $-(g_1 + g_2 + 1)$. When the two eigenvectors of **B** with zero eigenvalues become simultaneous eigenvectors of **A** with real eigenvalues, we have two real roots of the effective Hamiltonian. In addition, when **A** and **B** commute, both these matrices are diagonalizable with simultaneous eigenstates. The commutative condition is

$$q_{1f} + \tilde{\delta}_1 = q_{2f} + \tilde{\delta}_2 = q_{1f}g_1 + q_{2f}g_2 - (ka_s)^{-1}. \tag{5.111}$$

The two eigenstates with two real eigenvalues are the coherent superpositions of the three bound states and represent two bound states in the continuum.

We have derived the BIC of cold atoms in an ideal situation, but we did not consider the effects of spontaneous emission. However, even a small amount of spontaneous emission can destroy the BIC [83]. A BIC of cold atoms manifests as an unusually sharp resonance. To realize such a resonance, the detection time should be much faster than the inverse of the spontaneous emission linewidth. A resonance width smaller than the spontaneous emission linewidth indicates the occurrence of a BIC. This may find important applications in association spectroscopy, as discussed in the next chapter.

Exercise

- Discuss physical origin of hyperfine interaction in atoms. Derive an expression for hyperfine interaction for an atom in ground state.
- What is Born-Oppenheimer approximation? Justify this approximation giving physical arguments.
- Discuss different angular momentum coupling schemes (Hund's cases) of diatomic molecules.
- Derive an expression for magnetic field dependence of s-wave scattering length in a magnetic Feshbach resonance applying two-channel model.

- What is optical Feshbach resonance? Derive an expression for s-wave scattering length in optical Feshabch resonance. Discuss its advantages and drawbacks as compared to a magnetic Feshbach resonance.
- Discuss how a static electric field can affect low energy atom-atom collisions.
- What is the physical origin of C_4/r^4 type long-range potential between an ion and a neutral atom? Discuss why it is difficult to achieve Wigner threshold law regime in ion-atom collisions as compared to that in atom-atom ones.

References

[1] Bransden B H and Joachain C J 2003 *Physics of Atoms and Molecules.* (India: Pearson Education)

[2] Arimondo E, Inguscio M and Violino P 1977 Experimental determinations of the hyperfine structure in the alkali atoms *Rev. Mod. Phys.* **49** 31–75

[3] Brown J M and Carrington A 2003 *Rotational Spectroscopy of Diatomic Molecules* (Cambridge University Press)

[4] Born M and Oppenheimer R 1927 Zur Quantentheorie der Molekeln *Ann. Phys.* **389** 457–84

[5] Dalgarno A and Davison W D 1966 The calculation of van der Waals interactions *Advances in Atomic and Molecular Physics* **vol 2** (Amsterdam: Elsevier) 1–32

[6] Marinescu M, Sadeghpour H R and Dalgarno A 1994 Dispersion coefficients for alkali-metal dimers *Phys. Rev.* A **49** 982–8

[7] Marinescu M and Dalgarno A 1995 Dispersion forces and long-range electronic transition dipole moments of alkali-metal dimer excited states *Phys. Rev.* A **52** 311–28

[8] Herzberg G 2008 *Molecular Spectra and Molecular Structure* **1** Read Books Ltd

[9] Cohen-Tannoudji C 1992 Atomic motion in laser light *Fundamental Systems in Quantum Optics, Les Houches, Session LIII* **1990** (Elsevier)

[10] Metcalf H J and van der Straten P 1999 *Laser Cooling and Trapping* vol 1 (New York, NY: Springer Science & Business Media)

[11] Cornell E A and Wieman C E 2002 Nobel lecture: Bose-Einstein condensation in a dilute gas, the first 70 years and some recent experiments *Rev. Mod. Phys.* **74** 875–93

[12] Anderson M H, Ensher J R, Matthews M R, Wieman C E and Cornell E A 1995 Observation of Bose-Einstein condensation in a dilute atomic vapor *Science* **269** 198–201

[13] Ketterle W 2002 Nobel lecture: when atoms behave as waves: Bose-Einstein condensation and the atom laser *Rev. Mod. Phys.* **74** 1131–51

[14] Ashkin A 2006 *Optical Trapping and Manipulation of Neutral Particles Using Lasers - A Reprint Volume with Commentaries* (World Scientific)

[15] Letokhov V S 2020 *Atom Optics with Laser Light* (London: CRC Press)

[16] Cohen-Tannoudji C N 1998 Nobel lecture: manipulating atoms with photons *Rev. Mod. Phys.* **70** 707–19

[17] Phillips W D 1998 Nobel lecture: laser cooling and trapping of neutral atoms *Rev. Mod. Phys.* **70** 721

[18] Chu S 1998 Nobel lecture: the manipulation of neutral particles *Rev. Mod. Phys.* **70** 685–706

[19] Weiner J, Bagnato V S, Zilio S and Julienne P S 1999 Experiments and theory in cold and ultracold collisions *Rev. Mod. Phys.* **71** 1

[20] Johnson B R 1977 New numerical methods applied to solving the one-dimensional eigenvalue problem *J. Chem. Phys.* **67** 4086–93

[21] Mies F H, Williams C J, Julienne P S and Krauss M 1996 Estimating bounds on collisional relaxation rates of spin-polarized ^{87}Rb atoms at ultracold temperatures *J. Res. Natl Inst. Stand. Technol.* **101** 521

[22] Tiesinga E, Williams C J, Julienne P S, Krauss M, Lett P D and Phillips W D 1996 A spectroscopic determination of scattering lengths for sodium atom collisions *J. Res. Natl Inst. Stand. Technol.* **101** 505

[23] Stwalley W C 1976 Stability of spin-aligned hydrogen at low temperatures and high magnetic fields: new field-dependent scattering resonances and predissociations *Phys. Rev. Lett.* **37** 1628–31

[24] Uang Y-H and Stwalley W C 1980 Close-coupling calculations of spin-polarized hydrogen-deuterium collisions *Phys. Rev. Lett.* **45** 627–30

[25] Uang Y-H, Ferrante R F and Stwalley W C 1981 Model calculations of magnetic-field-induced perturbations and predissociations in 6Li7Li near dissociation *J. Chem. Phys.* **74** 6267–70

[26] Moerdijk A J, Verhaar B J and Axelsson A 1995 Resonances in ultracold collisions of ^{6}Li, ^{7}Li, and ^{23}Na *Phys. Rev.* A **51** 4852

[27] Boesten H M J, Moerdijk A J and Verhaar B J 1996 Dipolar decay in two recent Bose-Einstein condensation experiments *Phys. Rev. A* **54** R29

[28] Vogels J M, Tsai C C, Freeland R S, Kokkelmans S J J M F, Verhaar B J and Verhaar D J 1997 Prediction of Feshbach resonances in collisions of ultracold rubidium atoms *Phys. Rev.* A **56** R1067

[29] Inouye S, Andrews M R, Stenger J, Miesner H-J, Stamper-Kurn D M and Ketterle W 1998 Observation of Feshbach resonances in a Bose-Einstein condensate *Nature* **392** 151

[30] Gribakin G F and Flambaum 1993 Calculation of the scattering length in atomic collisions using the semiclassical approximation *Phys. Rev. A* **48** 546–53

[31] Gao B 1998 Quantum-defect theory of atomic collisions and molecular vibration spectra *Phys. Rev. A* **58** 4222

[32] Flambaum V V, Gribakin G F and Harabati C 1999 Analytical calculation of cold-atom scattering *Phys. Rev. A* **59** 1998

[33] Blackley C L, Julienne P S and Hutson J M 2014 Effective-range approximations for resonant scattering of cold atoms *Phys. Rev. A* **89** 042701

[34] Roy S, Landini M, Trenkwalder A, Semeghini G, Spagnolli G, Simoni A, Fattori M, Inguscio M and Modugno G 2013 Test of the universality of the three-body Efimov parameter at narrow Feshbach resonances *Phys. Rev. Lett.* **111** 053202

[35] Dyke P, Pollack S E and Hulet R G 2013 Finite-range corrections near a Feshbach resonance and their role in the Efimov effect *Phys. Rev. A* **88** 023625

[36] Stenger J, Inouye S, Andrews M R, Miesner H-J, Stamper-Kurn D M and Ketterle W 1999 Strongly enhanced inelastic collisions in a Bose-Einstein condensate near Feshbach resonances *Phys. Rev. Lett.* **82** 2422

[37] Strecker K E, Partridge G B and Hulet R G 2003 Conversion of an atomic Fermi gas to a long-lived molecular Bose gas *Phys. Rev. Lett.* **91** 080406

[38] Bartenstein M *et al* 2005 Precise determination of ^{6}Li cold collision parameters by radio-frequency spectroscopy on weakly bound molecules *Phys. Rev. Lett.* **94** 103201

[39] Zhang J, Van Kempen E G M, Bourdel T, Khaykovich L, Cubizolles J, Chevy F *et al* 2004 P-wave Feshbach resonances of ultracold ^{6}Li *Phys. Rev.* A **70** 030702

[40] d'Errico C, Zaccanti M, Fattori M, Roati G, Inguscio M, Modugno G and Simoni A 2007 Feshbach resonances in ultracold 39k *New J. Phys.* **9** 223

[41] Regal C A, Ticknor C, Bohn J L and Jin D S 2003 Tuning p-wave interactions in an ultracold fermi gas of atoms *Phys. Rev. Lett.* **90** 053201

[42] Regal C A, Greiner M and Jin D S 2004 Observation of resonance condensation of fermionic atom pairs *Phys. Rev. Lett.* **92** 040403

[43] Weiner J, Bagnato V S, Zilio S and Julienne P S 1999 Experiments and theory in cold and ultracold collisions *Rev. Mod. Phys.* **71** 1–85

[44] Köhler T, Góral K and Julienne P S 2006 Production of cold molecules via magnetically tunable Feshbach resonances *Rev. Mod. Phys.* **78** 1311–61

[45] Fedichev P O, Kagan Y, Shlyapnikov G V and Walraven J T M 1996 Influence of nearly resonant light on the scattering length in low-temperature atomic gases *Phys. Rev. Lett.* **77** 2913

[46] Fatemi F K, Jones K M, Lett P D and Tiesinga E 2002 Ultracold ground-state molecule production in sodium *Phys. Rev.* A **66** 053401

[47] Enomoto K, Kasa K, Kitagawa M and Takahashi Y 2008 Optical Feshbach resonance using the intercombination transition *Phys. Rev. Lett.* **101** 203201

[48] Theis M, Thalhammer G, Winkler K, Hellwig M, Ruff G, Grimm R *et al* 2004 Tuning the scattering length with an optically induced Feshbach resonance *Phys. Rev. Lett.* **93** 123001

[49] Thalhammer G, Theis M, Winkler K, Grimm R and Denschlag J H 2005 Inducing an optical Feshbach resonance via stimulated Raman coupling *Phys. Rev.* A **71** 033403

[50] Deb B and Rakshit A 2009 Suppression of power broadening in strong-coupling photoassociation in the presence of a Feshbach resonance *J. Phys. B: At. Mol. Opt. Phys.* **42** 195202

[51] Bohn J L and Julienne P S 1996 Semianalytic treatment of two-color photoassociation spectroscopy and control of cold atoms *Phys. Rev.* A **54** R4637

[52] Bohn J L and Julienne P S 1999 Semianalytic theory of laser-assisted resonant cold collisions *Phys. Rev.* A **60** 414

[53] Deb B and Agarwal G S 2009 Feshbach resonance-induced Fano interference in photoassociation *J. Phys. B: At. Mol. Opt. Phys.* **42** 215203

[54] Deb B 2010 Magneto-optical Feshbach resonance: controlling cold collision with quantum interference *J. Phys. B: At. Mol. Opt. Phys.* **43** 085208

[55] Junker M, Dries D, Welford C, Hitchcock J, Chen Y P and Hulet R G 2008 Photoassociation of a Bose–Einstein condensate near a Feshbach resonance *Phys. Rev. Lett.* **101** 060406

[56] Winkler K *et al* 2007 Coherent optical transfer of Feshbach molecules to a lower vibrational state *Phys. Rev. Lett.* **98** 043201

[57] Ni K K *et al* 2008 A high phase-space-density gas of polar molecules *Science* **322** 231

[58] Mackie M, Fenty M, Savage D and Kesselman J 2008 Cross-molecular coupling in combined photoassociation and Feshbach resonances *Phys. Rev. Lett.* **101** 040401

[59] Pellegrini P and Côté R 2009 Probing the unitarity limit at low laser intensities *New J. Phys.* **11** 055047

[60] Pellegrini P, Gacesa M and Côté R 2008 Giant formation rates of ultracold molecules via Feshbach-optimized photoassociation *Phys. Rev. Lett.* **101** 053201

[61] Deb B and Rakshit A 2009 Suppression of power broadening in strong-coupling photoassociation in the presence of a Feshbach resonance *J. Phys. B: At. Mol. Opt. Phys.* **42** 195202

[62] Kuznetsova E *et al* 2009 Efficient formation of ground-state ultracold molecules via STIRAP from the continuum at a Feshbach resonance *New J. Phys.* **11** 055025

[63] Regal C A, Ticknor C, Bohn J L and Jin D S 2003 Tuning p-wave interactions in an ultracold Fermi gas of atoms *Phys. Rev. Lett.* **90** 053201

[64] Zhang J, Van Kempen E G M, Bourdel T, Khaykovich L, Cubizolles J, Chevy F, Teichmann M, Tarruell L, Kokkelmans S J J M F and Salomon C 2004 P-wave Feshbach resonances of ultracold ^{6}Li *Phys. Rev.* A **70** 030702

[65] Deb B and Hazra J 2009 Manipulating higher partial-wave atom–atom interactions by strong photoassociative coupling *Phys. Rev. Lett.* **103** 023201

[66] Goyal K, Reichenbach I and Deutsch I 2010 p-wave optical Feshbach resonances in ^{171}Yb *Phys. Rev.* A **82** 062704

[67] Deb B 2012 Resonant manipulation of d-wave interaction of cold atoms with two lasers and a magnetic field *Phys. Rev.* A **86** 063407

[68] Enomoto K, Kitagawa M, Tojo S and Takahashi Y 2008 Hyperfine-structure-induced purely long-range molecules *Phys. Rev. Lett.* **100** 123001

[69] Reichenbach I, Julienne P S and Deutsch I H 2009 Controlling nuclear spin exchange via optical Feshbach resonances in ^{171}Yb *Phys. Rev.* A **80** 020701

[70] Williams R A, LeBlanc L J, Jimenez-Garcia K, Beeler M C, Perry A R, Phillips W D and Spielman I B 2012 Synthetic partial waves in ultracold atomic collisions *Science* **335** 314–7

[71] Nishida Y and Tan S 2010 Confinement-induced p-wave resonances from s-wave interactions *Phys. Rev.* A **82** 062713

[72] Levinsen J, Tiecke T G, Walraven J T M and Petrov D S 2009 Atom-dimer scattering and long-lived trimers in fermionic mixtures *Phys. Rev. Lett.* **103** 153202

[73] Marinescu M and You L 1998 Controlling atom-atom interaction at ultralow temperatures by dc electric fields *Phys. Rev. Lett.* **81** 4596

[74] Melezhik V S and Hu C-Y 2003 Ultracold atom-atom collisions in a nonresonant laser field *Phys. Rev. Lett.* **90** 083202

[75] Deb B and You L 2001 Low-energy atomic collision with dipole interactions *Phys. Rev.* A **64** 022717

[76] Krems R V 2005 Molecules near absolute zero and external field control of atomic and molecular dynamics *Int. Rev. Phys. Chem.* **24** 99–118

[77] Krems R V 2008 Cold controlled chemistry *Phys. Chem. Chem. Phys.* **10** 4079–92

[78] Chakraborty D, Hazra J and Deb B 2011 Resonant enhancement of ultracold photoassociation rate by electric field induced anisotropic interaction *J. Phys. B: At. Mol. Opt. Phys.* **44** 095201

[79] Aymar M and Dulieu O 2005 Calculation of accurate permanent dipole moments of the lowest $\sigma + 1$, 3 states of heteronuclear alkali dimers using extended basis sets *J. Chem. Phys.* **122** 204302

[80] Tomza M, Jachymski K, Gerritsma R, Negretti A, Calarco T, Idziaszek Z and Julienne P S 2019 Cold hybrid ion–atom systems *Rev. Mod. Phys.* **91** 035001

[81] Rakshit A and Deb B 2011 Formation of cold molecular ions by radiative processes in cold ion-atom collisions *Phys. Rev.* A **83** 022703

[82] Côté R and Dalgarno A 2000 Ultracold atom-ion collisions *Phys. Rev.* A **62** 012709

[83] Naskar S, Sardar D, Deb B and Agarwal G S 2019 Suppressing deleterious effects of spontaneous emission in creating bound states in cold atom continuum *J. Phys. B: At. Mol. Opt. Phys.* **52** 245204

Chapter 6

Association spectroscopy

In contrast to atomic cooling achieved through laser techniques, molecular cooling generally involves not only cooling translational motion but also vibrational and rotational motion. Cooling molecules directly using the laser cooling method is an extremely difficult or even impossible-seeming task in most cases at present. An alternative approach is to associate two colliding cold atoms to form a molecule that is translationally and rotationally cold. The method by which two cold atoms bind together into a molecule in the presence of external fields is called 'association spectroscopy,' which is an inelastic collision process afforded by an external field. There are three kinds of association spectroscopy: (1) photoassociation (PA), (2) magnetoassociation, and (3) magneto-optical association. We briefly discussed PA in the previous chapter. This type of spectroscopy is a result of advancements in the field-assisted collision physics of cold atoms. In this chapter, we provide a description of these three types of association spectroscopy.

Cooling molecules to a very low temperature, namely the microkelvin regime, is of paramount interest in science today for several reasons. First, it will lead to unprecedented precision in molecular spectroscopy. Second, it will allow one to coherently control not only the internal degrees of freedom (DOFs) associated with electronic states but also external DOFs such as translational, vibrational, and rotational DOFs, making cold molecules a resource for emerging quantum technology. Third, it will enable researchers to explore the hitherto unexplored domain of ultracold chemistry. Fourth, cooling molecules to even lower temperatures, i.e. the quantum degeneracy regime, will enable scientists to produce quantum degenerate gases of molecules, namely, molecular Bose–Einstein condensates (BECs) and degenerate Fermi gases of molecules. Quantum degeneracy implies that the average occupancy of a unit phase-space volume h^3 (where h is Planck's constant) is equal to or greater than unity. For a uniform gas at thermal equilibrium, the thermal de Broglie wavelength is $\lambda_{th} = \hbar/\sqrt{2\pi m K_B T}$ [1]. Quantum degeneracy in a uniform gas or fluid of indistinguishable particles in thermal equilibrium implies

doi:10.1088/978-0-7503-3959-9ch6

© IOP Publishing Ltd 2025. All rights, including for text and data mining (TDM), artificial intelligence (AI) training, and similar technologies, are reserved.

$n\lambda_{th}^3 \simeq 1$, where n is the number density of the particles. In 2024, Bigagli *et al* experimentally produced a BEC of NaCs molecules using association spectroscopy [2]. Association spectroscopy is important not only as a new tool for extracting spectroscopic information at the interface of atomic and molecular domains but also for potential applications in the quantum simulation of interacting spin models [3] and quantum computation [4].

6.1 Photoassociation

In the previous chapter, we briefly discussed PA between cold atoms in the context of optical Feshbach resonance. Over the last three decades, PA spectroscopy [5–7] has been developed into an important tool for studying the spectroscopic properties of cold atoms and diatomic molecules. In PA, the scattering or continuum states of a pair of atoms are transformed into an electronically excited bound molecular state via the absorption of a single photon. The PA of cold atoms was predicted by Thorsheim *et al* [8] in the late 1980s. The development of cold-atom PA spectroscopy was preceded by observations of PA at ambient temperatures in the 1970s, 1980s, and early 1990s [9–11]. The progress of PA at ambient temperatures prior to the observation of cold-atom PA was reviewed by Tellinghuisen [12]. The major difference between PA at ambient and cold temperatures is that, at ambient temperatures, one cannot select any specific molecular bound state to be formed; however, at low temperatures, molecular states can be precisely selected. In cold-atom PA, it is possible to create ultralong-range or purely long-range molecules [13] for which the range can be on the order of or larger than 1 nm, but at ambient temperatures, the range of the molecules formed is typically much less than 0.5 nm [6]. The production of cold diatomic molecules by PA was first experimentally demonstrated by Fioretti *et al* in 1998 [14]. Subsequently, several experimental groups reported the photoassociative production of cold molecules [15–17].

Here, we first discuss the important salient features of cold-atom PA; we then provide a theoretical description of the PA processes and finally describe the applications of the theoretical methods. In terms of the number of photons involved in producing a single molecule by PA processes, PA can be classified into two categories: single-color (single-photon) and two-color (two-photon) PA. In single-color PA between atoms of the same nuclei, a colliding pair of ground-state atoms absorbs a single photon from the PA laser and thereby combines to form a ro-vibrational molecular bound state in an electronically excited molecular potential when the PA laser is tuned below the threshold of the potential and near the resonance with the free–bound transition. Both ground-state atoms have an equal probability of absorbing a single photon. Since a single photon can be absorbed in one PA event, if one of the atoms absorbs a photon, the other does not, and vice versa. The atom that absorbs the photon is excited to the P state, and the other remains in the ground S state. Because of the exchange symmetry between the two atoms, the long-range part of the excited potential results from the resonant dipole–dipole interaction and therefore varies as $-1/r^3$ as $r \rightarrow \infty$. Two-color PA involves two photons that cause two electric dipole transitions: one transition of the free–bound type and the other of the bound–bound type. Two-color PA can yield either a doubly excited or autoionizing molecule and a ground-state molecule. In the former case, the absorption of two photons can sequentially lead to the formation of a molecule composed of doubly excited (P) atoms; one photon first leads to the formation of a singly excited molecule, and the second photon causes the bound–bound transition to a doubly excited molecule, which may autoionize if the molecule is embedded in an ionization continuum. The resulting ionization process is called photoassociative ionization. In the latter two-color PA, also known as Raman PA, the second photon stimulates a bound–bound transition into a ground-state molecular state.

The important salient features of cold-atom PA can be understood by generalizing the basic principles of standard bound–bound molecular spectroscopy to free–bound transitions at low collision energies. One of the underlying principles in this context is the Franck–Condon (FC) principle, which states that, because of the clear separation of time scales in the vibrational, electronic, and rotational motions of a molecule, vibrational motion is frozen during the electronic transition from one state to the other. This implies that an electronic transition between two molecular states occurs at the same internuclear separation, or, in other words, the transition occurs vertically. The strength of the transition is proportional to the overlap integral between the initial and final vibrational states. Since, for a molecular vibrational state other than the ground vibrational state, a molecular vibrational wave function has a maximum amplitude near the two turning points, a bound–bound transition is most probable if the turning points of the initial and final bound states match or lie close to each other. In the case of a transition between two ground vibrational states belonging to two different molecular potentials, the transition is most probable if the equilibrium positions of the two potentials lie close to each other.

From the point of view of the free–bound FC principle and the properties of cold atomic collisions, it is easy to understand why single-photon PA can predominantly produce long-range excited-state molecules close to the dissociation threshold. From our description of the partial-wave analysis of single-channel scattering in chapter 2, we know that at ultralow energy, the amplitudes of scattering between two ground-state atoms are most significant for the s-wave and a few low-lying partial waves. The amplitudes of these waves are large at large separations, typically exceeding tens or hundreds of Bohr radii, while the amplitudes at short separations oscillate in space with high frequency and small amplitude. As a result, the free–bound FC integral becomes finite only at large separations, while it is negligible at short separations.

6.1.1 Single-photon photoassociation

A schematic of the basic physical processes in single-photon PA is shown in figure 5.6. When two cold atoms in the electronic ground (S) state collide under PA laser irradiation, a single-photon photoassociative transition can occur from the collisional state to a molecular bound state in an excited molecular potential. For homonuclear atoms, two atoms become asymptotically bound in the excited molecular potential due to a resonant dipole–dipole interaction in which one of the atoms in the electronic P-excited state exchanges its excitation with the other atom in the electronic S-state; thereby, the two atoms perpetually perform a throw-and-catch of a photon between themselves. This is possible for a homonuclear system of two atoms only. The PA laser must be tuned below the frequency of atomic transition $S \rightarrow P$. When the laser detuning $\delta = \omega_l - \omega_a$ (where ω_l denotes the laser frequency) coincides with the frequency offset $\delta_b = \omega_b - \omega_a$ of the molecular bound-state frequency ω_b from the atomic transition frequency ω_a, a PA resonance occurs. Usually, the formation of molecules by PA is detected by incoherent or loss processes—the excited molecule spontaneously decays to some atomic and molecular states that are not trapped and consequently become lost from the trapped system. After the PA process is complete, the atoms in the ground state are detected by an optical imaging method, and the reduction of the atomic number density due to loss is thereby inferred. Typically, the radiative lifetime of a P-excited state of an atom is on the order of 10 ns, meaning that the spontaneous emission linewidth is $\gamma = 2\pi \times 10$ MHz. In order to resolve the spectral lines, the spread in kinetic energy of the initial atoms must be much smaller than 10 MHz; equivalently, the temperature of the atomic gas must be much lower than $\hbar \times 10^7/K_B \sim 100\ \mu$K, where K_B is the Boltzmann constant.

From the scattering theory point of view, the PA rate is equivalent to the inelastic scattering rate due to the loss of atoms through the PA process. As discussed in chapter 3, the inelastic scattering cross section can be calculated by the multichannel scattering method when inelastic channels are included in the calculation. In the PA process, inelastic channels are basically the two-atom states that may be formed by the spontaneous emission of a photon from a photoassociated molecule. As a result, the number density n of the atoms decreases during the PA process. The rate of loss of atoms due to PA is given by

$$\dot{n} = -K_{\mathrm{PA}} n^2, \tag{6.1}$$

where K_{PA} is the PA rate. In writing the above equation, the atomic ensemble is assumed to be homogeneous. The rate of density depletion is proportional to the square of the density because the loss of atoms occurs due to the binary collisions of atoms that occur in the presence of the PA laser. It is obvious from the above equation that the dimension of the PA rate is $L^3\ s^{-1}$. For a thermal ensemble of atoms, the PA rate is given by

$$K_{\mathrm{PA}} = \langle v\ \sigma^{\mathrm{inel}} \rangle, \tag{6.2}$$

where $\langle \cdots \rangle$ implies averaging over the Maxwell–Boltzmann distribution of the relative velocity v between a pair of atoms and

$$\sigma^{\mathrm{inel}} = \frac{\pi}{k^2} \sum_{j \neq i} \sum_{\ell} (2\ell + 1) |S_{ji}^{(\ell)}(E, \omega_l)|^2, \tag{6.3}$$

where $S_{ji}^{(\ell)}(E, \omega_l)$ is the ℓ-wave S-matrix element for the transition from the incident or input channel i to the outgoing channel j and ω_l is the PA laser frequency. Here, j

refers to all inelastic channels. As discussed in chapter 3, an alternative formula for the inelastic scattering cross section is:

$$\sigma^{\rm inel} = \frac{\pi}{k^2}\sum_{\ell}(2\ell + 1)\left[1 - |S_{ii}^{(\ell)}(E, \omega)|^2\right]. \tag{6.4}$$

In PA, the input channel usually refers to the two colliding S (ground-state) atoms in a particular hyperfine state. Since, in the present context, the inelastic channels primarily include all product two-particle states that can be formed due to spontaneous emission from the excited molecule formed by PA, the inelastic channel includes the continuum of the electromagnetic vacuum. In principle, a rigorous treatment of PA should include the states of laser photons and the electromagnetic vacuum. Such a treatment becomes quite complicated. Instead, a practical approach is to incorporate spontaneous emission in a phenomenological way, as we discuss below.

If there were no inelastic channels, then the situation in single-photon PA would be analogous to that in magnetic Feshbach resonance (MFR). In the two-channel approximation of PA, the single open channel is the continuum state of two colliding ground-state atoms, and the closed channel belongs to an excited molecular potential that supports the bound state which is coupled to the continuum by the laser. We assume that this bound state is a decaying state with complex energy $E_c = E_b - i\hbar\gamma/2$, where E_b is the (real) bound-state energy and γ is its decay width. As we have seen in the previous chapter, for a two-channel model for which the closed channel supports a bound state, the scattering T-matrix element is given by

$$T_\ell(E) = \frac{\hbar\Gamma_\ell/2}{E - \widetilde{E}_b + i\hbar\Gamma_\ell/2}, \tag{6.5}$$

where $\tilde{E}_b = E_b + E_b^{\rm shift}$ is the shifted energy of the closed-channel bound state, with $E_b^{\rm shift}$ being the shift due to the coupling between the bound state and the open channel. The shift of a bound state due to its coupling with a continuum of states was defined in the previous chapter in the context of resonances. Here,

$$\Gamma_\ell = \frac{2\pi}{\hbar}|\langle E, \ell|V|b\rangle|^2 \tag{6.6}$$

is the free–bound linewidth of the bound state $|b\rangle$ due to its interaction V with the partial-wave energy-normalized scattering state $|E, \ell\rangle$ in the open channel. Here, we have assumed that only one partial-wave state can be coupled to the bound state. In the case of PA, $V = -\mathbf{d} \cdot \mathbf{E}_{\rm PA}/2$, with $\mathbf{d}$ and $\mathbf{E}_{\rm PA}$ being the dipole moment and the electric field amplitude of the PA laser, respectively. In the case of PA, the collision energy E must be replaced by $E \to E + \hbar\omega_l$, where ω_l is the laser frequency. Note that here, all energy quantities are measured from the open channel threshold; that is, the zero of energy is set at the open channel threshold. To express the denominator of equation (6.5) in terms of laser detuning from the closed-channel threshold, which is, in the present context, the frequency ω_a for the atomic transition from the electronic S state to the P state, we write $E + \hbar\omega_l - \tilde{E}_b = E + \hbar\delta - \hbar\Delta_b$, where $\delta = \omega_l - \omega_a$ and $\Delta_b = \tilde{E}_b/\hbar - \omega_a$ is the frequency offset of the shifted bound state from the atomic transition frequency. To derive a formula for the PA rate in a phenomenological way, we replace $\tilde{E}_b \rightarrow \tilde{E}_b - i\hbar\gamma/2$. So, the S-matrix element $1 - 2iT_\ell$ is given by

$$S^{(\ell)}(E, \omega_l) = \frac{E/\hbar + \delta - \Delta_b + i(\gamma - \Gamma_\ell)/2}{E/\hbar + \delta - \Delta_b + i(\gamma + \Gamma_\ell)/2} \tag{6.7}$$

which yields

$$|S^{(\ell)}(E, \omega_l)|^2 = 1 - \frac{\Gamma_\ell \gamma}{(E/\hbar + \delta - \Delta_b)^2 + \left(\gamma_t^2\right)/4}, \tag{6.8}$$

where $\gamma_t = \Gamma_\ell + \gamma$ is the total decay width of the bound state. Substituting the above equation into equation (6.4), one obtains

$$\sigma_{\text{inel}} = \frac{\pi}{k^2}\sum_\ell (2\ell + 1)\frac{\Gamma_\ell \gamma}{(\omega_E + \delta - \Delta_b)^2 + \left(\gamma_t^2\right)/4}, \tag{6.9}$$

where $\omega_E = E/\hbar$. Since Γ_ℓ is proportional to the laser intensity, we have expressed the inelastic scattering cross section as a function of laser frequency and intensity. The relaxation rate γ includes the spontaneous emission rate and any other loss rates, such as excited-state collision-induced decay. So, one has to use an experimentally determined or theoretically estimated γ in order to explain the PA spectrum by making use of the above formula. Alternatively, one can resort to semianalytic multichannel calculations by introducing 'artificial channels' that can take account of the bound-state decay, as shown by Bohn and Julienne [18].

In order to calculate the free–bound stimulated linewidth Γ_ℓ, one has to calculate the free–bound dipole coupling $\Lambda_{E,\ell} = \langle E, \ell | -\mathbf{d} \cdot \mathbf{E}_{\text{PA}} | b \rangle$. This can be done by applying the Wigner–Eckart theorem [19]. One has to specify the initial and final angular momentum states of the system. Since we are considering that both atoms are initially in the electronic ground state, the initial total electronic orbital angular momentum is zero. So, the incident channel is characterized by a ground-state electronic spin or hyperfine spin quantum numbers and rotation of the relative motion (ℓ).

So far, we have considered that the laser can couple only one ro-vibrational state $|b\rangle$. In fact, a tunable laser can excite multiple rotational states. The bound state $|b\rangle \equiv |v, J, M_J\rangle$ is characterized by a vibrational quantum number v and a molecular angular momentum $\mathbf{J} = \mathbf{R}_n + \mathbf{L} + \mathbf{S}$. The coupling schemes of various molecular angular momenta are determined by the Hund's case that defines the bound state. A general angular momentum state can be represented by $|J, M_J\rangle$, where J is the total angular momentum quantum number and its projection M_J along the quantization axis is known as the magnetic quantum number. Thus, the inelastic scattering cross section can be expressed in the form

$$\sigma_{\text{inel}} = \frac{\pi}{k^2} \sum_{\ell,J} (2\ell + 1) \frac{\Gamma_\ell \gamma}{(\omega_E + \delta - \Delta_b)^2 + (\gamma_t^2)/4}. \tag{6.10}$$

Now, in order to derive the thermally averaged PA rate constant as defined by equation (6.2), we have to integrate over the distribution of the relative speed between the two colliding atoms. In the center-of-mass (COM) frame, the total kinetic energy of the two atoms is the sum of the relative and COM kinetic energies. Since we have assumed that the potential depends only on the relative coordinates, the COM momentum is a constant of motion, and so is the COM kinetic energy. The Boltzmann factor $\exp[-(E_{\text{rel}} + E_{\text{COM}})/k_B T]$ is separable into two factors, $\exp[-(E_{\text{rel}})/k_B T]$ and $\exp[-E_{\text{COM}}/k_B T]$. Since the two Hamiltonian parts corresponding to the relative and COM motions commute, we can safely consider that the phase integral involving the same volume splits into two disjoint parts—one corresponding to relative motion and the other to COM motion. Since the observable σ_{inel} does not depend on the COM momentum, integration over the COM momentum gives a factor of unity; therefore, integration over the COM momentum distribution is of no interest. So, we need to evaluate an integral of the type $\int 4\pi v^2 dv\, P_{\text{MB}}(v) f(v)$, where v is the relative speed of the atoms, $f(v) = v\sigma_{\text{inel}}$ is a function of v, and

$$P_{\text{MB}}(v) = \left(\frac{\mu}{2\pi k_B T}\right)^{3/2} e^{-\mu v^2/2k_B T} \tag{6.11}$$

is the Maxwell–Boltzmann velocity distribution function, where k_B the denotes Boltzmann constant. The integral over the speed can be converted into an integral over energy by replacing $v = \sqrt{2E/\mu}$. The result is

$$K_{\text{PA}} = \frac{8\pi}{\mu^2} \left(\frac{\mu}{2\pi k_B T}\right)^{3/2} \int E e^{-E/k_B T} \sigma_{\text{inel}}(E) dE. \tag{6.12}$$

Now, substituting σ_{inel} as given by equation (6.10) into the above equation and using $k^2 = 2\mu E/\hbar^2$, we obtain

$$\begin{aligned} K_{\text{PA}} = & \frac{1}{hQ_T} \sum_{\ell,J} (2\ell + 1) \int e^{-E/k_B T} dE \\ & \times \frac{\Gamma_\ell \gamma}{(\omega_E + \delta - \Delta_b)^2 + (\gamma_t^2)/4}, \end{aligned} \tag{6.13}$$

where $Q_T = \left(\frac{2\pi\mu k_B T}{h^2}\right)^{3/2} = \lambda_{\text{th}}^{-3}$ and $\lambda_{\text{th}} = h/\sqrt{2\pi\mu k_B T}$ is the thermal de Broglie wavelength. Note that Γ_ℓ depends on J. The above expression (6.13) is the same as in equation (2) of the paper by Napolitano *et al* [20]. One can identify the quantity $\Gamma_\ell \gamma/[(\omega_E + \delta - \Delta_b)^2 + (\gamma_t^2)/4]$ with $|S_p|^2$ of [20].

6.1.2 Raman photoassociation: the creation of ground-state molecules

A schematic of two-color or Raman PA is given in figure 5.7. Two lasers are applied, with one laser tuned near a free–bound transition and the other near a bound–bound transition. One bound state belongs to the excited molecular potential and is coupled to the scattering states of two ground-state atoms by the first laser. This bound state is then coupled to another bound state at the ground-state molecular potential. It is thus a stimulated Raman process for creating molecules in the ground molecular potential from two colliding atoms.

The theoretical method for two-photon PA can be developed by generalizing the method for single-photon PA discussed in the previous section. In fact, a general theoretical method for treating continuum–bound coupled systems can be advanced along the lines of Fano–Feshbach resonances, as detailed in chapter 4. For the sake of simplicity, we consider that the PA laser couples only one excited ro-vibrational level. Let $|E\rangle$, $|b1\rangle$, and $|b2\rangle$ denote the energy-normalized scattering state with collision energy E, the excited bound state, and the bound state at the ground-state potential, respectively. As shown in the figure, laser L_1 with a frequency of ω_1 is tuned near the free–bound transition, and laser L_2 with a frequency of ω_2 is near-resonant with the bound–bound transition. The Hamiltonian under the dipole and rotating-wave approximations can be written as $H = H_1 + H_2$, where

$$H_1 = \int dE|E\rangle\langle E| + \hbar\omega_{b1}|b1\rangle\langle b1\rangle + \left[\int dE\Lambda_{E1}e^{-i\omega_1 t}|b1\rangle\langle E|+\text{h.c.}\right] \quad (6.14)$$

represents the scattering continuum, the excited bound state, and their laser coupling, with $\hbar\omega_1$ being the binding energy of $|b1\rangle$ and Λ_{E1} the free–bound dipole coupling between $|E\rangle$ and $|b1\rangle$. Here, h.c. denotes the Hermitian conjugate. Moreover,

$$H_2 = \hbar\omega_{b2}|b2\rangle\langle b2| + [ge^{-i\omega_2 t}|b1\rangle\langle b2| + \text{h.c.}] \quad (6.15)$$

describes the bound state in the ground-state potential and the bound–bound coupling, with $\hbar\omega_2$ being the binding energy of $|b2\rangle$ and g the bound–bound dipole coupling between $|b2\rangle$ and $|b1\rangle$. By making a unitary transformation defined by

$$U(t) = \exp[-i\omega_1 t|b1\rangle\langle b1| + i(\omega_1 - \omega_2)|b2\rangle\langle b2|], \quad (6.16)$$

we obtain an effective time-independent Hamiltonian given by

$$\widetilde{H} = U^\dagger H U - i\hbar U^\dagger \frac{dU}{dt} = \widetilde{H}_1 + \widetilde{H}_2. \quad (6.17)$$

The first term at the end of equations (6.17), namely

$$\widetilde{H}_1 = \int dE|E\rangle\langle E| - \hbar\delta_1|b1\rangle\langle b1\rangle + \left[\int dE\Lambda_{E1}|b1\rangle\langle E|+\text{h.c.}\right], \quad (6.18)$$

describes the scattering continuum, the excited bound state, and their laser coupling, while the second term

$$\widetilde{H}_2 = \hbar\delta_{b2}|b2\rangle\langle b2| + [\hbar g|b1\rangle\langle b2|. + \text{h.c.}]. \tag{6.19}$$

Here, $\delta_1 = \omega_1 - \omega_{b1}$, $\delta_{b2} = \omega_{b2} + \omega_1 - \omega_2$. Now, by comparing two-photon PA with the physical situation where only one bound state is coupled to a continuum and a second bound state is coupled to the former but not coupled to the continuum (as discussed in subsection 4.6.3), one can identify that the states $|b\rangle$ and $|b'\rangle$ of that section correspond to $|b1\rangle$ and $|b2\rangle$, respectively. Therefore, the energies E_b and $E_{b'}$ correspond to $-\hbar\delta_1$ and $\hbar\delta_{b2}$, respectively. So, the dressed states are similar in both cases, except that in two-photon PA, the excited bound state $|b1\rangle$ can decay spontaneously. Thus, as in the case of single-photon PA, we need to replace ω_{b1} by $\omega_{b1} - i\gamma/2$, where γ is the decay constant of the level $|b1\rangle$. Thus, we can explicitly calculate the S-matrix element and the PA rate for two-photon PA by following the Fano–Feshbach method. To do so, let us first recall the expression for the S-matrix element given by equation (4.93):

$$S_\ell = \frac{(\varepsilon - i)\left(E - \widetilde{E}_b' - i\hbar\kappa_\ell/2\right)}{(\varepsilon + i)\left[E - \widetilde{E}_b' + i\hbar\kappa_\ell/2\right]}, \tag{6.20}$$

where we need to replace

$$(\varepsilon + i) \rightarrow \frac{E/\hbar + \tilde{\delta}_1 + i(\Gamma_\ell + \gamma)/2}{\Gamma_\ell/2} \tag{6.21}$$

$$(\varepsilon - i) \rightarrow \frac{E/\hbar + \tilde{\delta}_1 + i(\gamma - \Gamma_\ell)/2}{\Gamma_\ell/2}. \tag{6.22}$$

Here, $\hbar\tilde{\delta}_1 = \hbar\omega_1 - (E_{b1} + E_{\text{shift}})$ and $\hbar\tilde{\delta}_2 = \hbar(\omega_1 - \omega_2) + (E_{b2}) + \mathcal{E}_\ell$, with E_{shift} and $\mathcal{E}_\ell$ being the shifts of levels $|b1\rangle$ and $|b2\rangle$, respectively, as defined by equations (4.78) and (4.90), respectively, and $\Gamma_\ell \equiv \Gamma$ is the stimulated linewidth given by equation (4.77). Substituting κ_ℓ as given by equation (4.88) and making the above substitutions, one can explicitly calculate the inelastic scattering cross section using formula (6.4), that is,

$$\sigma_{\text{inel}} = \frac{1}{k^2}[1 - |S_\ell|^2] \tag{6.23}$$

The effect of bound–bound Rabi coupling $2g$ leads to a dispersive shift and an increased linewidth in the PA spectrum. The terms associated with g are of the form $(\varepsilon \pm i)\mathcal{E}_\ell$ and $\pm i(\varepsilon\pm)\hbar\kappa_\ell/2$. Now, if we neglect all the shift terms due to free–bound coupling, assume $\gamma \ll \Gamma_\ell$, and approximate the terms $(\varepsilon \pm i)\mathcal{E}_\ell$ by their non-dispersive parts $(|\delta_1| \ll \Gamma)$, such that

$$\pm i(\varepsilon \pm i)\mathcal{E}_\ell/2 \simeq -4\hbar|g|^2/(\Gamma_\ell/2) \tag{6.24}$$

then the term $1 - |S_\ell|^2$ reduces to the following form:

$$1 - |S_\ell|^2 = \frac{(E - \hbar\delta_{b2})^2 \hbar^2 \gamma \Gamma}{(E - E_+)^2 (E - E_-)^2 + (\hbar\gamma_t/2)^2 (E - \hbar\delta_{b2})^2} \tag{6.25}$$

where $\gamma_t = \gamma + \Gamma_\ell$ and

$$E_\pm = \frac{\Delta_+}{2} \pm \frac{1}{2}\sqrt{(\Delta_-^2 + 4\hbar^2 |g|^2)} \tag{6.26}$$

with $\Delta_\pm = \hbar(\delta_{b2} \mp \delta_1)$. The above equation (6.25) coincides with formula (3) of [18], which was obtained using the multichannel method. Note that Δ_2 and Δ_1 in [18] correspond to $\hbar\delta_{b2}$ and $-\hbar\delta_1$, respectively.

The probability of forming a molecule in the electronic ground state by two-color PA can be calculated by applying the method of creating a doubly dressed Fano state, as discussed in section 4.6.2. The probability of forming the state $|b'\rangle$ is given by

$$P_{b'} = \int dE \hbar^2 |g|^2 |\tilde{b}_E|^2 dE \tag{6.27}$$

Consider that the bound state $|b'\rangle$ is the molecular bound state $|b2\rangle$ in the electronic ground state. So, P_{b2} is given by

$$P_{b2} = \int dE \frac{|g|^2 \Gamma_\ell}{2\pi\hbar\left[(E/\hbar + \tilde{\delta}_1)^2 + (\gamma_t/2)^2\right] |E/\hbar - \tilde{\delta}_2 + i\tilde{\kappa}_\ell|^2} \tag{6.28}$$

where $\tilde{\kappa}_\ell$ is of the same form as κ_ℓ given by equation (4.88) except that E_{b1} is replaced by $E_{b1} - i\hbar\gamma/2$.

An efficient way of transferring an atom or a molecule from one ground-state sublevel to another ground-state sublevel or a metastable state is stimulated Raman adiabatic passage (STIRAP) [21–23]. This works by suppressing the spontaneous emission loss as the passage from one state to the other occurs through a dark state (DS) in a Λ-type three-level configuration driven by two laser beams. A DS is an eigenstate of the system with zero eigenvalue. It is a coherent superposition of the two lower states of the three-state Λ system. The laser coupling of the DS with the excited state vanishes, and hence spontaneous emission is suppressed when stimulated Raman transition occurs adiabatically. As a result, the system remains almost in the DS during the passage. In a standard STIRAP, all three states are discrete. Now, the question arises of whether it is possible to employ a STIRAP-like state transfer involving a continuum of states, as in Raman PA, to produce ground-state cold molecules. The physical situation is similar to a Fano problem involving two bound states and one continuum of states, with one bound state being coupled to the other bound state and the continuum, as discussed in chapter 4. Here, the bound state with common coupling is the excited molecular state $|b_1\rangle$, the continuum is the scattering state $|E\rangle$ of the two ground-state atoms, and the other bound state is the ro-vibrational molecular state $|b_2\rangle$ in the molecular electronic ground state.

The problem is to transfer the initial state $|E\rangle$ to the target state $|b_2\rangle$ via the intermediate state $|b_1\rangle$ by a stimulated Raman process using appropriate laser pulses, so that the loss due to spontaneous emission from $|b_1\rangle$ is minimized.

In a STIRAP process, two slowly varying time-dependent laser pulses are applied in a counterintuitive way; that is, first, a laser pulse is applied between the target ($|b_1\rangle$) and the intermediate state ($|b_2\rangle$), which are initially empty; then, after a time delay, another laser pulse is applied between the initial and the intermediate state, subject to the fulfillment of an adiabatic condition. STIRAP becomes most efficient in discrete three-level Λ systems driven by two Raman lasers, due to the existence of a DS. An adiabatic evolution of the DS is achieved by slowly varying the intensity of the lasers in an appropriate way, resulting in almost lossless population transfer. For a Λ-like configuration involving a continuum, the existence of a DS is not obvious, and an exact DS is not possible, because the condition for the existence of an exact DS cannot be fulfilled in such a system. Nevertheless, under certain special conditions at low temperatures, typically below the millikelvin temperature regime, it is possible to create a quasi-dark state (quasi-DS) [24, 25] that can be used to achieve STIRAP in a continuum system with reasonably good efficiency.

6.1.3 A brief overview of photoassociation spectroscopy and its applications

In the previous subsections, we have presented the essential ideas and theoretical methods in PA spectroscopy from a quantum scattering perspective. Now, we briefly review its use as a probe of low-energy interacting atomic and molecular properties. Both theoretical and experimental developments in PA have been reviewed in several articles and monographs, as mentioned earlier. Here, we focus on the applications of the theoretical methods discussed above in some selected experimental scenarios.

First, let us consider the one-color PA rate $K_{\rm PA}$ given by the expression (6.12). In the context of cold atoms, Thorsheim *et al* [8] first theoretically discussed the PA of two cold Na atoms and derived an analytical expression for the PA rate. The major contribution to the energy integral in equation (6.13) comes from the resonant energy $E = \hbar(\delta - \Delta_b)$. Therefore, an approximate expression for $K_{\rm PA}$ can be written by replacing the Boltzmann factor $e^{E/k_BT} \simeq e^{-\hbar(\delta-\Delta_b)/k_BT}$ in the integrand. Thus, one obtains

$$K_{\rm PA} = \frac{1}{Q_T}\sum_{\ell J}(2\ell + 1)e^{-\hbar(\delta-\Delta_b)/k_BT}\frac{\Gamma_\ell \gamma}{\gamma_t}. \tag{6.29}$$

Let us consider that excited ro-vibrational states belong to Hund's case (b), namely the Σ molecular electronic state, which at long separation corresponds to separated $S + P$ atomic states. The ground $S + S$ scattering state also corresponds to Hund's case (b), i.e. Σ molecular potential. According to molecular symmetry as discussed in section 5.2.3, the angular momentum $\mathbf{N}$ is a conserved quantity in this case. The corresponding selection rules for molecular transitions dictate that $\Delta N = \pm 1$ and $\Delta S = 0$. So, there are two classes of transitions: (i) ${}^1\Sigma_g \rightarrow {}^1\Sigma_u$ and (ii) ${}^3\Sigma_u \rightarrow {}^3\Sigma_g$. If one ignores the mass of the electron in comparison to the nuclear mass, then $\ell \equiv N$.

Thus, we can write the ground state $|E, \ell\rangle \equiv |E, N\rangle$ and the excited bound state $|b\rangle \equiv |v, N'\rangle$.

Now, we explicitly calculate $\tilde{\Gamma}_\ell$ as defined by equation (6.6). The laser–atom interaction Hamiltonian $V = -\mathbf{d} \cdot \mathbf{E}_{\text{PA}}/2$. The matrix element $\langle E, N|V|vN'\rangle$ can be calculated using the Wigner–Eckart theorem. Let us write $\mathbf{E}_{\text{PA}} = \hat{\varepsilon}\mathscr{E}$ and $\mathbf{d} = \hat{\eta}d(r)$, where $\hat{\varepsilon}$ is the unit polarization vector of the field and $\hat{\eta}$ is the dipole moment unit vector, $\mathscr{E}$ is the electric field amplitude, and $d(r)$ is the magnitude of the molecular dipole moment, which may depend on the interatomic separation r. In spherical polar coordinates, we have

$$\hat{\varepsilon} \cdot \hat{d} = \sum_{\sigma=\pm,0} \varepsilon_\sigma^* \eta_\sigma, \tag{6.30}$$

where $x_\pm = \sin\theta_x e^{\pm i\phi_x}$ and $x_0 = \cos\theta_x$. Here, θ_x and ϕ_x refer to the polar and azimuthal angles of $\hat{x}$ ($x = \varepsilon, \eta$). For the particular case under consideration, only the $\varepsilon_0\eta_0$ part contributes to the transition amplitude. If we consider that the PA laser is unpolarized, then Γ_ℓ, as defined by equation (6.6), is given by

$$\begin{aligned} \Gamma_\ell &= \frac{\pi}{\hbar}|\mathscr{E}|^2 \left|\int d^3r \psi_\ell(r) d(r) \psi_v(r)\right|^2 |\langle N|\eta_0|N'\rangle|^2 \int d\theta_\varepsilon \sin\theta_\varepsilon \cos^2\theta_\varepsilon \int d\phi_\varepsilon \\ &= \frac{4\pi^2}{3\hbar}|\mathscr{E}|^2 |\langle d(r)\rangle|^2 h_{\ell N'} \end{aligned} \tag{6.31}$$

where $h_{\ell N'} = |\langle \ell|\eta_0|N'\rangle|^2$ is an angular factor known as the Hönl–London factor [8] and $\langle d(r)\rangle = \int d^3r \psi_\ell(r) d(r) \psi_v(r)$ is the radial part of the dipole matrix element, with $\psi_\ell(r)$ and $\psi_v(r)$ representing the ground scattering and excited bound molecular wave functions, respectively. Now, consider the relation between the photon flux ϕ and the field intensity $I = c|\mathscr{E}|^2$, where c is the speed of light, namely $\mathscr{E}^2 = \phi h\nu_l/c = h\phi/\lambda_l$, with $\nu_l = \omega_l/2\pi$ being the laser frequency. We can now express Γ_ℓ in the following form:

$$\Gamma_\ell = \frac{8\pi^3}{3\lambda_l}\left|\langle d(r)\rangle\right|^2 h_{\ell N'} \phi. \tag{6.32}$$

We derived the above expression for the stimulated linewidth by assuming that only one partial-wave state can be coupled to the bound state. Now, if multiple partial waves are coupled to a single bound state, then the stimulated linewidth is $\Gamma = \sum_\ell \Gamma_\ell$, which agrees with the expression given in [8] if one replaces ℓ with N. The spontaneous emission rate constant γ is the sum of the rates of spontaneous emission to all the possible continuum and bound states. These spontaneous emission rates are given by the formula for the Einstein A coefficient. The calculation of the cold-atom PA absorption coefficient for $\Sigma \to \Sigma$ was carried out by Thorsheim *et al* [8]. The results of their calculations are displayed in figure 6.1. Bohn and Julienne calculated the trap loss signal for the transition to $v = 64$ in the $^3\Sigma_g$ state of the Li_2 molecule for different laser intensities, as shown in figure 6.2.

Experimentally, rotationally resolved PA spectroscopy of ultracold atoms allows precise determination of the dissociation energy of a ground-state molecular

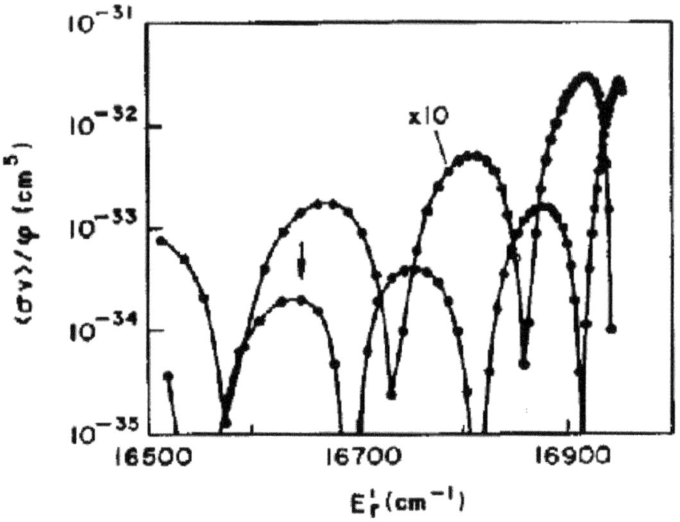

Figure 6.1. Calculated PA absorption coefficient per unit photon flux scaled by $1/N^2$ (N being the initial atom number density) as a function of the excited-state vibrational energy of Na_2 at a temperature $T = 10$ mK for $^1\Sigma_g \rightarrow ^1 \Sigma_u$ (solid circle) and $^3\Sigma_g \rightarrow ^3 \Sigma_u$ (open circle). The arrow marks the vibrational quantum number 107 of $^1\Sigma_u$. Figure reprinted with permission from [8], Copyright (1987) by the American Physical Society.

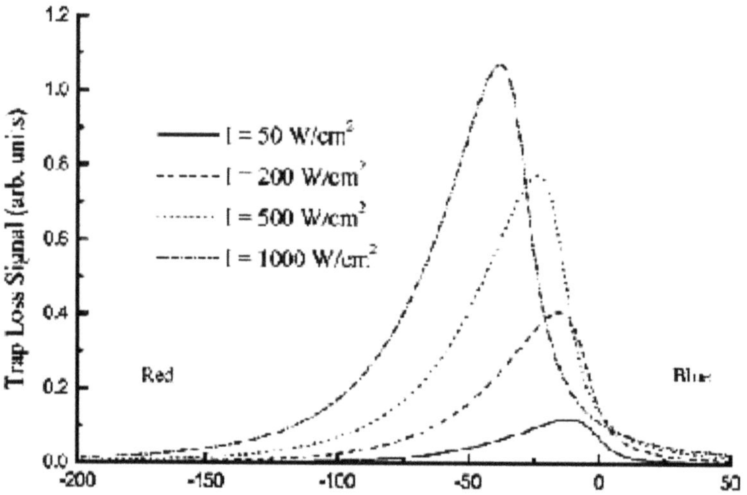

Figure 6.2. PA trap loss signal at 1 mK for vibrational level $v = 64$ in the excited state $^3\Sigma_g$ of Li_2. Figure reprinted with permission from [26], Copyright (1999) by the American Physical Society.

potential, unlike traditional bound–bound spectroscopy. A PA transition in the ultracold regime can connect the s-wave collisional state lying almost at the dissociation energy of the ground-state molecular potential to a rotational level $J = 1$ of a particular vibrational state in the excited molecule. When this information

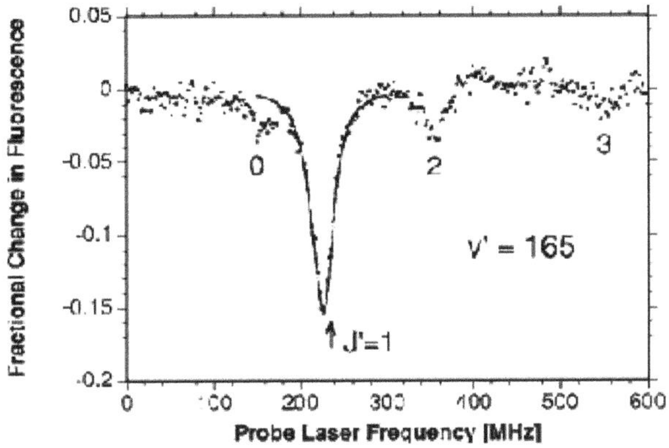

Figure 6.3. Trap-loss spectrum due to PA of Na. Clearly visible are the three lowest rotational spectra of the vibrational level 165 of the A $^1\Sigma_u^+$ state of Na_2. The solid line is a fit to the $J = 1$ rotational spectral shape. This spectrum allows precise determination of the dissociation energy of the ground-state potential, as discussed in the text. Reproduced from [27]. © IOP Publishing Ltd. All rights reserved.

about the energy of the ro-vibrational level relative to the threshold of the ground-state molecular potential is combined with bound–bound spectroscopic data that provides information about the relative energies of bound levels, one can readily determine the dissociation energy of the ground-state molecular potential. In this way, Jones *et al* experimentally determined the dissociation energy of the Na_2 molecule, as illustrated in figure 6.3.

Single-color PA produces excited-state transient molecules. For two atoms of the same nuclei, the excited molecule is formed with one atom in the electronic S state and the other in the P state. At long separations, the two atoms are bound into the molecule by the combined effects of resonant dipole–dipole and spin–orbit interactions. For two atoms of different nuclei, the two atoms remain bound by van der Waals interactions at large separations. The resonant dipole–dipole and spin–orbit interactions in an excited-state molecular configuration can lead to the formation of ultralong-range or purely long-range molecular states [13, 28], with their equilibrium positions lying at several tens of Bohr radii. Since such molecular states are fragile, these states are more likely to be accessible through the use of cold-atom PA. The excited-state molecular potentials are constructed using the Movre–Pichler method [29]. In this method, adiabatic central potentials are first constructed for the Σ and Π configurations. Then, the atomic spin–orbit terms are introduced as first-order perturbations. The molecular hyperfine interactions further modify these potentials. At ultracold temperatures, these interactions become important, since the thermal fluctuations may become smaller than the typical energy scales associated with these noncentral interactions. Thus, PA can probe minute molecular forces. The selection rules in molecular transitions are dictated by Hund's coupling cases, as discussed in the previous chapter. Hund's cases may change with the internuclear separation. In the case of PA transitions to ultralong-range molecular states, the usual bound–bound transition selection rules change due to the inclusion of spin–orbit and hyperfine interactions. These subtle molecular interactions are discussed at length in the review article by Jones *et al* [7].

Now, let us discuss the use of the Movre–Pichler method to construct the excited-state potential. We presented some general discussions about the calculations of Born–Oppenheimer (BO) molecular potentials. The standard procedure is to use molecular orbitals, which are a linear combination of atomic orbitals, and to apply the Heitler–London method. One can thus construct short-range molecular potentials that belong to either Hund's case (a) or (b). The long-range part of the potentials, which is calculated using dispersive electromagnetic interactions between the two atoms, is added to the short-range part by numerical extrapolation. These calculated potentials are corrected from time to time based on the available spectroscopic data that are experimentally obtained. The Movre–Pichler method seeks to include relativistic effects such as spin–orbit interactions in the potentials. It uses nonrelativistic molecular electronic states $|L\Lambda S\Sigma p\rangle$ as the zeroth-order states, where $p \equiv g$, u stands for the COM symmetry of the two homonuclear alkali atoms, and the other symbols have their usual meanings as introduced in the previous chapter. The atomic spin–orbit interaction term $A\sum_{i=a,b}\mathbf{l}_i \cdot \mathbf{s}_i$ is included as a first-order perturbation, where A is the atomic hyperfine constant and $\mathbf{l}_i$ and $\mathbf{s}_i$ denote the orbital and spin angular momenta of the valence electron of the ith atom. The long-range part of the corrected potentials for Hund's case (c), which stipulates that $\Omega_e = \Lambda + \Sigma$ is conserved, results from the diagonalization of the resulting perturbed Hamiltonian matrix written in the nonrelativistic molecular basis. For instance, figure 6.4 shows two ground-state and eight lowest excited-state nonrelativistic BO potentials of the Na_2 molecule for separations $R \leqslant 25a_0$, as calculated by Magnier *et al* [30]. For separations larger than $15a_0$, the excited potentials vary as $\pm C_3^{\Lambda}/R^3$,

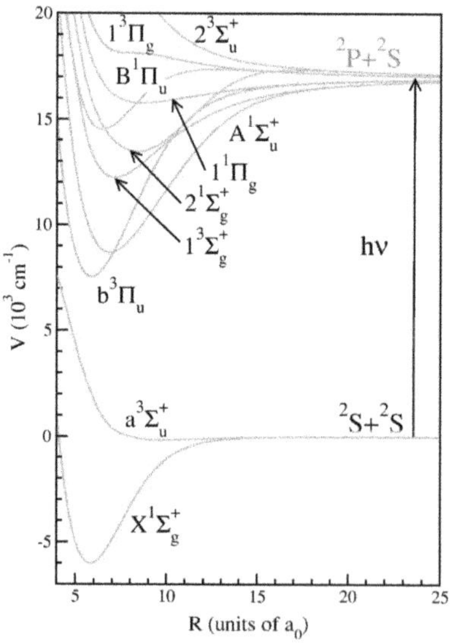

Figure 6.4. BO potentials of the sodium dimer as a function of internuclear separation R. Figure reprinted with permission from [7], Copyright (2006) by the American Physical Society.

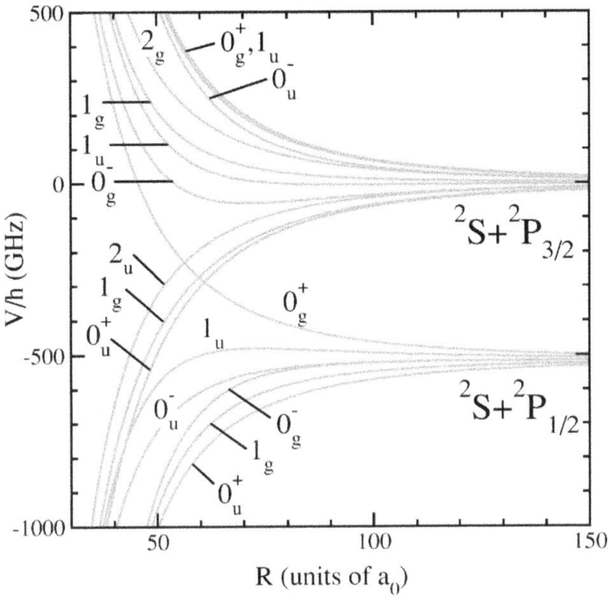

Figure 6.5. Relativistic adiabatic interaction potentials as a function of internuclear separation near the $2S + 2P$ dissociation limit of Na_2. Figure reprinted with permission from [7], Copyright (2006) by the American Physical Society.

while the ground-state potentials take the form of a $-C_6/R^6$ (van der Waals) potential, where C_3^Λ is the dispersion coefficient for Λ symmetry. When non-relativistic spin–orbit corrections are included, the long-range part of the excited potentials is characteristically modified, giving rise to Hund's case (c) potentials, as shown in figures 6.5 and 6.6. The magnified view of the 0_g potential in figure 6.6 clearly shows that it can support purely long-range molecular states with binding energies on the order of gigahertz. The PA signal strength is determined by the trap loss method. In this method, upon illuminating the atomic cloud with a PA laser, the number of atoms lost from the trap is measured by absorption imaging as a function of laser detuning and other system parameters of interest. Figure 6.7 shows relative PA signal strengths as a function of the vibrational level of the $A^1\Sigma_u^+$ state of the Li_2 molecule, as experimentally obtained by Abraham *et al* [31]. The signal exhibits oscillations due to the varying FC factor. These experimental results have been used to obtain singlet scattering lengths for ^{6}Li and ^{7}Li with error bars on the order of 10%. This is accomplished by comparing the numerically simulated signal strengths for different scattering lengths with experimental values, as this figure illustrates.

The processes of cold-molecule formation by one- and two-color PA are schematically illustrated in figure 5.6. One-color PA primarily forms excited-state molecules; however, a small fraction of weakly bound ground-state molecules may be produced by spontaneous emission from the excited molecular states. In contrast, two-color PA can lead to the formation of deeply bound ground-state molecules. The weakly bound ground-state molecules formed due to spontaneous emission can be transferred to

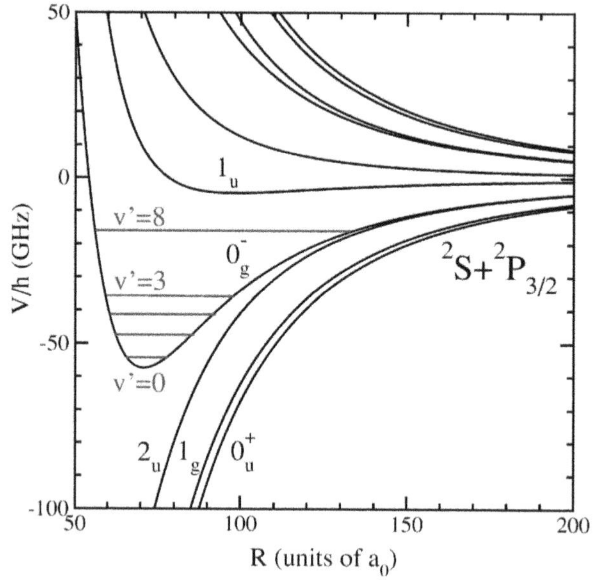

Figure 6.6. Magnified view of the relativistic adiabatic potential-energy curves of the previous figure. Figure reprinted with permission from [7], Copyright (2006) by the American Physical Society.

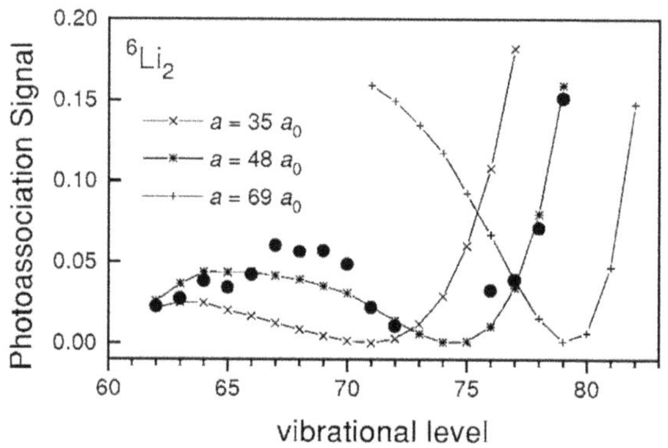

Figure 6.7. Relative trap loss PA signal strengths (dots) for transitions to different vibrational levels of the $A^1\Sigma_u^+$ state of the $^6Li_2$ molecule. Also shown are the calculated signal strengths for three different scattering lengths $a = 95a_0$ (solid with 'x'), $a = 48a_0$ (solid with '*') and $a = 69a_0$ (solid with '+'). Figure reprinted with permission from [7], Copyright (2006) by the American Physical Society.

deeply bound ro-vibrational levels in the ground-state potential by two-photon STIRAP, which is a coherent optical process. In the case of molecules composed of two homonuclear atoms, there is generally a mismatch between the equilibrium positions of the ground- and excited-state potentials due to the different asymptotic behaviors of the two potentials. Therefore, it is hard to prepare deeply bound homonuclear molecules using a single-step PA or STIRAP process. In contrast, heteronuclear molecules provide a more favorable bound–bound FC factor for preparing deeply bound molecules by one- or two-color PA. Heteronuclear molecules have permanent electric dipole moments. Ground-state polar molecules, such as KRb [32, 33], NaCs [34], and LiCs [35], were experimentally created using PA in the first decade of the twenty-first century. In 2005, Sage *et al* [36] produced RbCs polar molecules in the electronic, vibrational, and rotational ground state (absolute ground state) using a two-step process: in the first step, a PA laser created an excited molecule, which then decayed into a weakly bound ground-state molecule in the spin-triplet shallow potential. In the second step, a STIRAP pulse converted these weakly bound states into molecules in the absolute ground state. However, the yield of molecules from this two-step process was quite small, since the first step involved an incoherent process of spontaneous emission.

So far, we have discussed the PA of two neutral atoms into a cold molecule. However, it is possible to produce cold molecular ions using PA. One way to prepare cold molecular ions using PA is photoassociative ionization (PAI), whereby a photoassociative cold molecule is ionized by a laser [37, 38]. The creation of cold molecular ions by direct PA between colliding ions and atoms has been theoretically predicted [39]. Direct ion–atom PA is possible if the relative motion between an ion and an atom is cooled to the submillikelvin regime. The production of $NaCa^+$ molecular ions in the ground state using multiple optical transition pathways, including PA and spontaneous emission, has been theoretically described. It has been proposed that the STIRAP method with a quasi-dark state could be used to produce $(BeX)^+$ (where X stands for an alkali atom such as Na, K, or Rb) [40] and $(LiCs)^+$ [24] cold molecular ions.

In the preceding discussion in this chapter, we did not consider the effects of trap-induced confinement on PA spectroscopy. The trapping potential, particularly trap-induced tight confinement, can significantly impact the PA processes [41] of a pair of atoms in a trap. The trapping bound state of a pair of cold atoms interacting with a contact potential in an isotropic harmonic oscillator is exactly solvable [42]. Therefore, single-photon PA in such a trap means the formation of an excited molecular state through a dipole transition from the trapping bound state to the molecular bound state. A trap is generally anisotropic; if it is harmonic, then the relative and COM motions are separable. If the atom–atom interaction potential is central, then one is required to solve the relative motion of the two atoms only in the case of harmonic trapping. Trap confinement allows long interrogation times in PA and probing of the quasi two-dimensional regime of scattering (or the confinement-dominated three-dimensional regime), as experimentally demonstrated by Zelevinsky *et al* in narrow-line PA in a one-dimensional optical lattice [43]. One-dimensional and two-dimensional regimes of confinement are achieved by varying the ratio between axial and radial harmonic trapping frequencies of a trap with cylindrical symmetry. The effects of low dimensionality on PA and optical Feshbach resonance (OFR) have been theoretically analyzed by Naidon and Julienne [44]. Scattering theory in 1D and 2D will be discussed in the next chapter. Apart from the effects due to trap confinement or low dimensionality, another interesting effect originates from cavity quantum electrodynamics (CQED) when a pair of atoms is photoassociated inside a cavity [45, 46]. In essence, PA is an important tool for probing low-energy effects at the interface of atomic and molecular regimes [47] in free as well as confined space.

6.2 Magneto- and magneto-optical association: Feshbach molecules

The magnetoassociation of two cold atoms into a molecule makes use of an MFR. We have already elaborated on the physical origin of MFR in the previous chapter. An MFR occurs due to the magnetic field tuning of a near-zero-energy weakly bound state across the threshold of an open channel whose continuum states at low energy exhibit a Feshbach resonance. While this resonance is basically a result of multichannel elastic scattering, the formation of a true diatomic bound state (a negative-energy bound state or a bound state lying below the lowest channel threshold) from a pair of free atoms is essentially an inelastic process by which the excess energy of the two colliding atoms must be taken away by a third body (a photon, a particle, or a time-dependent field). The use of an MFR to produce weakly bound diatomic molecules was predicted in the late 1990s [48]. Here, we will briefly discuss how MFR facilitates the formation of exotic molecules known as Feshbach molecules (for a review of Feshbach molecules, see [49, 50]). The reason we call these molecules 'exotic' will be clear from the discussion below.

In chapter 4, we analyzed how scattering resonances and near-zero energy bound states are related. In the case of s-wave Feshbach resonances, the s-wave scattering length a_s diverges at the resonant magnetic field B_r at which the closed-channel bound state crosses the threshold. As the magnetic field B approaches B_r from the lower ($B < B_r$) or higher ($B > B_r$) side of the resonant point, a_s tends to $+\infty$ or $-\infty$ from a small positive or negative value. We know that a large positive a_s corresponds to the existence of a true bound state. Magnetoassociation is concerned with the question of how to populate this bound state. Such bound states can be formed only in the presence of an MFR, that is, when the magnetic field is tuned to the large positive side of the resonance. Since they are close to the threshold, they have very high vibrational quantum numbers, but they are rotationally and translationally cold, as they are formed from a pair of atoms undergoing an s-wave resonant collision. They are exotic because they can only be created in the laboratory by applying a magnetic field appropriately tuned to the large positive side of a Feshbach resonance. The wave function of such molecules is spatially extended far above the outer turning point, unlike that of the usual molecules.

There are two main methods for forming Feshbach molecules through tunable MFR in the absence of any optical field. One is via the relaxation of two colliding atoms into a molecule due to three-body inelastic collisions, while the other is via the appropriate sweeping of a time-dependent magnetic field. In the former method, as the magnetic field is tuned toward the large positive side of the resonance, the two resonantly interacting cold atoms can collide with a third atom that removes the excess energy, leaving the atom pair to relax into a negative-energy bound state. In the latter method, a time-dependent magnetic field is swept across the resonance, and a significant fraction of the atoms is coherently converted into molecules as the magnetic field is ramped into the positive side of the resonance. In this case, the excess energy is taken away from the atom pairs by the time-dependent field through the temporal Zeeman effect. Since both the resonant continuum and the molecular wave functions have huge spatial spreads, there is a significant free–bound FC overlap between the wave functions. Using both these methods, several experimental groups demonstrated the formation of reasonably stable Feshbach molecules during the first decade of the twenty-first century. The early developments in the production of Feshbach molecules are reviewed in two articles [49, 50]. The created molecules can be detected by radio-frequency spectroscopy. A radio-frequency photon can dissociate the molecules into atoms, which can then be detected via absorption imaging.

It was observed that the molecules formed from two-component fermionic atoms are extraordinarily stable compared to those formed from bosonic atoms. This stability stems from Pauli's exclusion principle or fermionic symmetry [51, 52]. Fermionic symmetry allows s-wave resonance only between fermions of two different spin states. When two fermionic atoms with different spin components come close to each other, a third fermion cannot come very close to the pair, because its spin component must be the same as the spin component of one of the atoms in the pair. The lowest partial wave allowed for low-energy collisions between two fermions of the same spin component is $\ell = 1$ (a p-wave). But because of the centrifugal barrier, the third fermion cannot enter the vicinity of an atom pair undergoing an s-wave resonant collision. So, the third atom is effectively repelled by the pair, reducing the chance of dissociation of the molecule due to three-body collisions. Conversely, for bosonic atoms, this is not the case, and their molecules are susceptible to three-body collisions. In 2003, three groups reported the formation of BECs of Feshbach molecules formed from two-component fermions [53–55]. The following year, a crossover from the Bardeen–Cooper–Schrieffer (BCS) state of fermionic atoms to a BEC of Feshbach molecules was experimentally observed [56]. These observations of macroscopic coherent effects in atom–molecule conversion have opened new opportunities to explore quantum many-body physics [57, 58] with ultracold atoms.

Following the progress in magnetoassociation, researchers have turned their attention to optimizing or influencing the Raman photoassociative formation of cold molecules through an MFR. In the previous chapter, we discussed magneto-optical Feshbach resonance, where both optical and magnetic Feshbach resonances affect the resonant collision. Magneto-optical association is based on a similar idea, but uses an additional laser that couples the excited molecular bound state with another molecular bound state at the electronic ground-state potential. This method of forming ground-state molecules works best in the case of two heteronuclear atoms rather than homonuclear atoms because the bound–bound FC factor between the excited and ground molecular wave functions of a heteronuclear molecule is usually much larger than that of a homonuclear molecule. At large separations, both ground and excited heteronuclear molecular bound states are supported by a van der Waals potential, while a homonuclear excited molecular bound state is asymptotically supported by a much stronger resonant dipole–dipole interaction potential.

Magneto-optical association has become an important method for producing ground-state cold polar molecules. Over the years, several experimental groups have reported the efficient production of ultracold polar molecules by magneto-optical association. The production efficiency is enhanced due to the resonant enhancement of the amplitude of the scattering wave function, which in turn leads to an increased free–bound FC factor. Polar molecules are important for exploring the pure rotational spectra of ground-state molecules. Furthermore, polar molecules can interact via electric dipole–dipole interaction, which may be utilized for generating quantum entanglement between the rotational levels of a molecule. Another alternative way is to first prepare an ensemble of Feshbach molecules and then convert these molecules into deeply bound molecules in the electronic ground-state potential by STIRAP via an excited molecular state. In this way, in 2008, Ni *et al* [59] produced ultracold potassium–rubidium (KRb) polar molecules in the electronic, vibrational, and rotational ground states (absolute ground state) with high phase-space density. Danzl *et al* produced a high-density sample of ultracold Cs_2 molecules in the absolute ground state. They first created weakly bound Feshbach molecules in an optical lattice using a magnetic field sweep and then coherently converted these Feshbach molecules into absolute-ground-state molecules by four-photon sequential Raman processes. In 2024, sodium–cesium polar molecules in the absolute ground state (meaning the electronic, vibrational, and rotational ground states) were produced by Bigagli *et al* [2]. By suppressing three- and two-body collisional losses using microwave shielding, they evaporatively cooled the molecular cloud in a dipole trap into quantum degeneracy and thereby realized a BEC of polar molecules for the first time.

6.3 Efimov trimers

In 1970, Vitaly Efimov [60] predicted the formation of a weakly bound trimer composed of three particles that undergo s-wave resonant binary collisions. The most interesting result was that the trimer can form even in a parameter regime where no two-boson bound state exists, implying that the Efimov trimer results solely from three-body interactions. Efimov presented his work in the context of nuclear physics, predicting the halo state of a carbon nucleus (^{12}C) formed from three α particles and a similar state for ^{3}H formed from three resonantly interacting nucleons (H). Thirty-five years after Efimov's work, researchers realized an Efimov trimer in ultracold ^{133}Cs atoms, which are bosons. They demonstrated the formation of the trimer by tuning the s-wave resonant interactions in the ^{133}Cs BEC.

The basic Efimov physics can be described in terms of the universality of the s-wave resonant two-body interaction. Universality implies that in a resonantly interacting system, the physical properties of the system are primarily determined by the scattering length. In chapter 4, we discussed resonant interactions. We know that near a resonance, the scattering length becomes so large that for all practical purposes, one can ignore the range of the interaction. The mean-field interaction of such a system can be modeled by a zero-range potential that is parametrized solely by the scattering length. Efimov showed that for a system of three particles interacting via two-body resonant forces, one can derive an effective three-body long-range attractive potential that gives rise to three-body binding. To derive the Efimov bound state, one uses the coordinates

$$\mathbf{r}_{ij} = \mathbf{r}_j - \mathbf{r}_i \tag{6.33}$$

$$\boldsymbol{\rho}_k = \mathbf{r}_k - \frac{m_i\mathbf{r}_i + m_j\mathbf{r}_j}{m_i + m_j}, \tag{6.34}$$

where $i, j, k \equiv 1, 2, 3$ refers to the indexing of the three particles, 1, 2, and 3, in cyclic order, with masses m_1, m_2, and m_3, respectively. Since the three particles interact via two-particle contact interaction, for $\mathbf{r}_{ij} \neq 0$, where i, j refers to all three possible pair combinations, the particles move freely. Only for $\mathbf{r}_{ij} \equiv \mathbf{r} \to 0$ do two particles undergo scattering. We have seen in chapter 2 that in the limit $k \to 0$, the two-particle s-wave scattering wave function $r\psi(r)$ takes the form of a straight line $\sim r - a_s$, provided that $kr \ll 1$. Thus, one has the low-energy boundary condition

$$\lim_{r\to 0} \frac{1}{r\psi}\frac{\partial}{\partial r} r\psi(r) - \frac{1}{a_s}, \tag{6.35}$$

which will be utilized in deriving the Efimov bound state. Following Efimov, one can introduce a rescaled set of coordinates

$$\mathbf{r}'_{ij} = \left(\frac{\mu_{ij}}{m_0}\right)^{1/2} \mathbf{r}_{ij} \tag{6.36}$$

$$\boldsymbol{\rho}'_k = \left(\frac{\mu_k}{m_0}\right)^{1/2} \boldsymbol{\rho}_k, \tag{6.37}$$

where μ_{ij} is the reduced mass of the particle pair i, j. Here, μ_k is the reduced mass of the particle pair i, j and particle k, while m_0 is a scaled mass introduced such that the Hamiltonian of the three free particles $H_0 = -\hbar^2\sum_{n=1,2,3}(1/m_n)\nabla_n^2$ takes the form $H_0 = H_{\text{com}} + H_{r\rho}$, where

$$H_{r\rho} = -\frac{\hbar^2}{2m_0}\left[\nabla_r^2 + \nabla_\rho^2\right] \tag{6.38}$$

$$H_{\text{com}} = -\frac{\hbar^2}{2M}\nabla_{R_0}^2 . \tag{6.39}$$

Here, $\mathbf{r} \equiv \mathbf{r}_{ij}$ and $\boldsymbol{\rho} = \boldsymbol{\rho}_k$ for any combination of i, j, k in cyclic order; $M = m_1 + m_2 + m_3$ is the total mass, and $R_0 = \sum_n m_n r_n/M$ is the COM position vector. Since $[H_{r\rho}, H_{\text{com}}] = 0$ and the interaction depends only on the relative coordinates, we can work with $H_{r\rho}$ only.

We now consider the Efimov binding of three identical bosons with masses $m_1 = m_2 = m_3 = m$. It then turns out the scaling mass $m_0 = m/2$. Thus, we have

$$\boldsymbol{\rho}'_k = \frac{2}{\sqrt{3}}\left[\mathbf{r}_k - \frac{\mathbf{r}_i + \mathbf{r}_j}{2}\right] \tag{6.40}$$

$$\mathbf{r}'_{ij} = \mathbf{r}_{ij}. \tag{6.41}$$

Since the three-boson wave function must be symmetric despite the exchange of any two particles, a general time-independent stationary wave function of three resonantly interacting bosons can be written as

$$\Psi(\mathbf{r}_1, \mathbf{r}_2, \mathbf{r}_3) = \psi\left(\mathbf{r}'_{12}, \boldsymbol{\rho}'_3\right) + \psi\left(\mathbf{r}'_{23}, \boldsymbol{\rho}'_1\right) + \psi\left(\mathbf{r}'_{31}, \boldsymbol{\rho}'_2\right), \tag{6.42}$$

where the wave function $\psi(\mathbf{r}, \boldsymbol{\rho})$ of the system in the present context for $r \neq 0$ can be obtained by solving the time-independent Schrödinger equation

$$H_{r\rho}\psi(\mathbf{r}, \boldsymbol{\rho}) = E\psi(\mathbf{r}, \boldsymbol{\rho}). \tag{6.43}$$

Since we are interested in s-wave interactions only, the wave function depends on the radial variables $r = |\mathbf{r}|$ and $\rho = |\boldsymbol{\rho}|$. So, we have

$$-\frac{\hbar^2}{m}\left[\frac{1}{r}\frac{\partial^2}{\partial r^2}r + \frac{1}{\rho}\frac{\partial^2}{\partial \rho^2}\rho\right]\psi(r, \rho) = E\psi(r, \rho). \tag{6.44}$$

Let us now focus on only one set of coordinates, $\mathbf{r} \equiv \mathbf{r}'_{12}$ and $\boldsymbol{\rho} \equiv \boldsymbol{\rho}_3$. Recall that the above equation (6.44) is valid for $r \neq 0$. The wave function at $r = 0$ can be calculated by imposing the boundary condition on the wave function for $r \to 0$. The boundary condition is

$$\lim_{r\to 0}\frac{\partial}{\partial r}r\Psi = -\frac{1}{a_s}r\Psi. \tag{6.45}$$

In order to implement the above condition in the evaluation of the wave function, we first note how the three sets of coordinates are related. From equations (6.40) and (6.41), it is easy to write the following relations:

$$\mathbf{r}'_{23} = -\frac{1}{2}\mathbf{r}'_{12} + \frac{\sqrt{3}}{2}\boldsymbol{\rho}'_3 = \frac{1}{2}\mathbf{r} + \frac{\sqrt{3}}{2}\boldsymbol{\rho} \tag{6.46}$$

$$\boldsymbol{\rho}'_1 = -\frac{\sqrt{3}}{2}\mathbf{r}'_{12} - \frac{1}{2}\boldsymbol{\rho}'_3 = -\frac{\sqrt{3}}{2}\mathbf{r} - \frac{1}{2}\boldsymbol{\rho} \tag{6.47}$$

$$\mathbf{r}'_{31} = -\frac{1}{2}\mathbf{r}'_{12} - \frac{\sqrt{3}}{2}\boldsymbol{\rho}'_3 = \frac{1}{2}\mathbf{r} - \frac{\sqrt{3}}{2}\boldsymbol{\rho} \tag{6.48}$$

$$\boldsymbol{\rho}'_1 = \frac{\sqrt{3}}{2}\mathbf{r}'_{12} - \frac{1}{2}\boldsymbol{\rho}'_3\frac{\sqrt{3}}{2}\mathbf{r} - \frac{1}{2}\boldsymbol{\rho}. \tag{6.49}$$

With these relations, the boundary condition (6.45) translates into

$$\begin{aligned}&\left[\frac{\partial}{\partial r}r\psi(\mathbf{r},\,\boldsymbol{\rho})\right]_{r\to 0}+\psi\left(\frac{\sqrt{3}}{2}\boldsymbol{\rho},\,-\frac{1}{2}\boldsymbol{\rho}\right)+\psi\left(\frac{-\sqrt{3}}{2}\boldsymbol{\rho},\,-\frac{1}{2}\boldsymbol{\rho}\right)\\&=\left[-\frac{r}{a}\left\{\psi(\mathbf{r},\,\boldsymbol{\rho})+\psi\left(\frac{\sqrt{3}}{2}\boldsymbol{\rho},\,-\frac{1}{2}\boldsymbol{\rho}\right)+\psi\left(\frac{-\sqrt{3}}{2}\boldsymbol{\rho},\,-\frac{1}{2}\boldsymbol{\rho}\right)\right\}\right]_{r\to 0}.\end{aligned} \tag{6.50}$$

Next, one transforms the coordinates $(\mathbf{r},\,\boldsymbol{\rho})$ into hyperspherical coordinates $(\mathbf{R},\,\alpha)$:

$$r = R\sin\alpha \tag{6.51}$$

$$\rho = R\cos\alpha, \tag{6.52}$$

where

$$R^2 = r^2 + \rho^2 = \frac{2}{3}\left(r_{12}^2 + r_{23}^2 + r_{31}^2\right). \tag{6.53}$$

R is called the hyperradius and $\alpha = \tan^{-1}(r/\rho)$ is called the hyperangle. Similarly, one can express the other two pairs of Cartesian coordinates in terms of hyperspherical coordinates as given by

$$r_{23}' = R\sin\beta \tag{6.54}$$

$$\rho_{23}' = R\cos\beta \tag{6.55}$$

$$r_{31}' = R\sin\gamma \tag{6.56}$$

$$\rho_{31}' = R\cos\gamma, \tag{6.57}$$

where β and γ are the two hyperangles, by making the transformation

$$\psi(\mathbf{r}_{ij},\,\boldsymbol{\rho}_k) = \frac{\phi_k(\mathbf{r}_{ij},\,\boldsymbol{\rho}_k)}{r_{ij}\rho_k}. \tag{6.58}$$

In the limit $\mathbf{r}\to 0$, $\phi_k(\mathbf{r}_{ij},\,\boldsymbol{\rho}_k)$ must be finite. On the other hand, $(\boldsymbol{\rho}_k)\to 0$, $\psi(\mathbf{r}_{ij},\,\boldsymbol{\rho}_k)$ must remain finite, implying that in this limit, $\phi_k(\mathbf{r}_{ij},\,\boldsymbol{\rho}_k)\to 0$. This means α must be bounded by zero and $\pi/2$. One can write equation (6.44) in the form

$$-\frac{\hbar^2}{2m_0}\left[\frac{\partial^2}{\partial r^2}+\frac{\partial^2}{\partial \rho^2}\right]\phi = E\phi, \tag{6.59}$$

which, in hyperspherical coordinates, takes the form

$$\left[-\frac{\partial^2}{\partial R^2}-\frac{1}{R}\frac{\partial}{\partial R}-\frac{1}{R^2}\frac{\partial^2}{\partial \alpha^2}\right]\phi(R,\,\alpha) = k^2\phi(R,\,\alpha), \tag{6.60}$$

where $k^2 = 2m_0E/\hbar^2$. Thus, the limit $r \to 0$ implies $\alpha \to 0$, $\rho \to R$, $\beta \to \pi/3$ and $\gamma \to \pi/3$. With these limits, the boundary condition (6.50) reduces to the form

$$\left[\frac{\partial}{\partial\alpha}\phi(R,\alpha)\right]_{\alpha\to 0} + \frac{8}{\sqrt{3}}\phi(R,\pi/3) = -\frac{R}{a_s}\phi(R,0). \tag{6.61}$$

Clearly, equation (6.60) is separable into R and α coordinates. Let

$$\phi(R,\alpha) = \chi(R)\xi(\alpha). \tag{6.62}$$

Substituting this into equation (6.60), one obtains

$$-\frac{d^2\xi(\alpha)}{d\alpha^2} = s_n^2\chi(\alpha) \tag{6.63}$$

and

$$\left[-\frac{\partial^2}{\partial R^2} - \frac{1}{R}\frac{\partial}{\partial R} + \frac{s_n^2}{R^2}\right]\chi(R) = k^2\chi(R), \tag{6.64}$$

where s_n^2 is the nth eigenvalue of equation (6.63). The boundary condition at $\alpha = \pi/2$ is $\xi(\pi/2) = 0$, which dictates that the solution of equation (6.63) must be of the form

$$\xi(\alpha) = \sin[s_n(\pi/2 - \alpha)]. \tag{6.65}$$

The value of s_n can be obtained by solving equation (6.61). In the limit $a_s \to \pm\infty$ and on substitution of $\phi(R,\alpha) = \chi(R)\xi(\alpha)$, this equation reduces to

$$-s_n\cos(\pi s_n/2) + \frac{8}{\sqrt{3}}\sin(\pi s_n/6) = 0. \tag{6.66}$$

Equation (6.64) can be recast in the form

$$\left[-\frac{d^2}{dR^2} + V_n(R)\right]\tilde{\chi}(R) = k^2\tilde{\chi}(R), \tag{6.67}$$

where $\tilde{\chi}(R) = \sqrt{R}\chi(R)$ and

$$V_n(R) = \frac{s_n^2 - 1/4}{R^2} \tag{6.68}$$

is called the hyperradial potential. One solution of equation (6.66) is purely imaginary, while all other solutions are real. This particular imaginary solution is $s_0 \simeq \pm 1.00624i$. For this particular solution, $V_0(R)$ becomes attractive, as given by

$$V_0(R) = -\frac{|s_0|^2 + 1/4}{R^2} \tag{6.69}$$

which is responsible for the Efimov binding of the three bosons. This attractive force results from bosonic exchange symmetry. It is the attraction between a pair of bosons mediated by the exchange of a third boson.

Efimov bound states can be calculated numerically for finite scattering lengths and finite-ranged potentials. Efimov showed that the number of bound levels has a logarithmic dependence on the scattering length. Suppose that the Nth three-body bound state with binding energy $E_T^{(N)}$ appears for the scattering length a_N. The ratio of $|a_{N+1}|$ to $|a_N|$ is a constant, namely $e^{\pi/|s_0|} \simeq 22.7$; in other words, the $N+1$th Efimov state appears at a scattering length that is 22.7 times the scattering length at which the Nth state is formed. The binding energies follow the rule

$$\frac{E_T^{(N+1)}}{E_T^{(N)}} = e^{-2\pi/|s_0|} \simeq 1/515.03, \tag{6.70}$$

implying that as $N \to \infty$, $E_T^{(N)} \to 0$ and $a_s \to \pm\infty$.

The variation of the binding energy of a few Efimov states as a function of the inverse of the scattering length is qualitatively illustrated in figure 6.8. In the case of two resonantly interacting particles, a dimer state occurs only when the scattering length is large and positive, as discussed in chapter 4. However, an Efimov state can appear for both positive and negative scattering lengths. Figure 6.8 shows that the first Efimov trimer crosses the three-particle continuum with a negative scattering length where a two-particle bound state cannot exist. Figure 6.9 shows the observation of an Efimov trimer in ultracold collisions of cesium atoms by Kraemer *et al* [61]. They measured three-body recombination losses in an optically trapped sample of cesium atoms as a function of the scattering length, which was

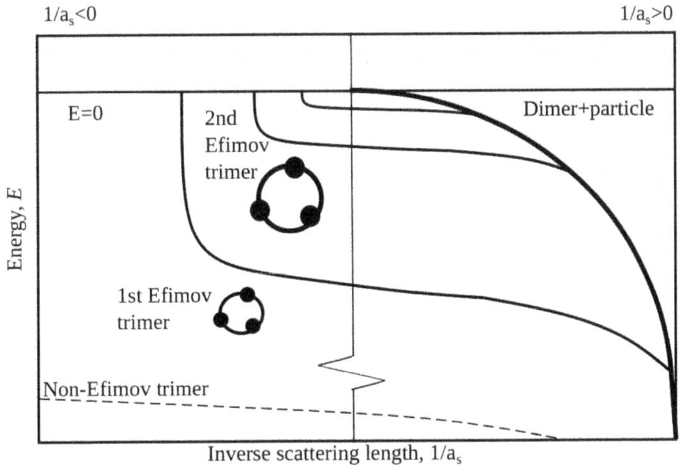

Figure 6.8. Efimov effect: the emergence of a series of Efimov trimer states due to ultracold collisions with resonant two-body interactions. The plot shows the binding energy as a function of the inverse of the s-wave scattering length $1/a$. The shaded portion corresponds to the continuum for scattering between a dimer and one particle ($1/a > 0$).

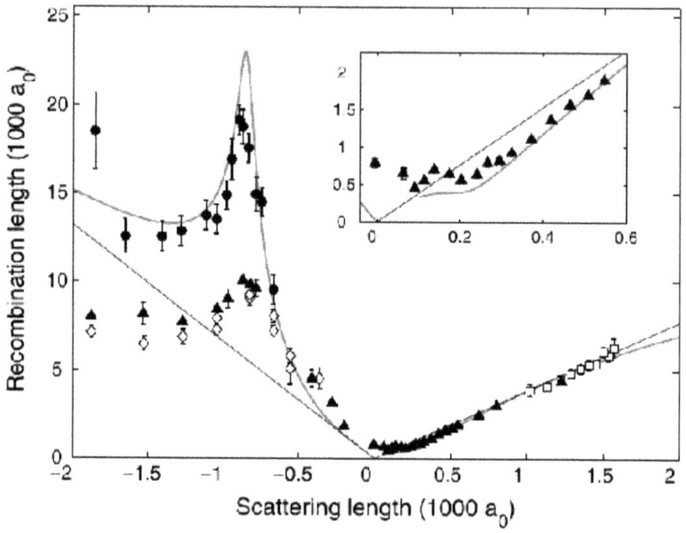

Figure 6.9. Observation of Efimov resonance in ultracold cesium atoms. The plot shows the recombination length as a function of the scattering length for initial temperatures of about 10 nK (dots), 200 nK (filled triangles), and 250 nK (open diamonds). The solid line is a fit to the low-temperature data based on effective field theory. The inset shows a magnified view of the region of positive scattering lengths up to 600 a_0 (a_0 is the Bohr radius). Reproduced from [61], with permission from Springer Nature.

tuned by an MFR. They confirmed the realization of Efimov resonances by observing a giant loss rate in a parameter regime where two-body loss is forbidden, and also by verifying the scaling laws mentioned above.

Exercise

- What is photoassociation? Discuss why cooling of atoms in sub-Kelvin regime is necessary for photoassociation.
- Derive an expression for inelastic scattering cross section in Photoassociation.
- Discuss the method of cold molecule formation by Raman Photoassociation.
- What is magneto-association? Give a brief outline of cold molecule formation by magneto-association.
- What is an Efimov trimer? Derive the effective three-body attractive potential that supports an Efimov trimer.

References and further reading

[1] Huang K 1987 *Statistical Mechanics* 2nd edn (New York: Wiley)

[2] Bigagli N, Yuan W, Zhang S, Bulatovic B, Karman T, Stevenson I and Will S 2024 Observation of Bose-Einstein condensation of dipolar molecules *Nature* **631** 289–93

[3] Micheli A, Brennen G K and Zoller P 2006 A toolbox for lattice-spin models with polar molecules *Nat. Phys.* **2** 341–7

[4] DeMille D 2002 Quantum computation with trapped polar molecules *Phys. Rev. Lett.* **88** 067901

[5] Lett P, Phillips W and Julienne P 1995 Photoassociative spectroscopy of laser cooled atoms *Ann. Rev. Phys. Chem.* **46**

[6] Weiner J, Bagnato V S, Zilio S and Julienne P S 1999 Experiments and theory in cold and ultracold collisions *Rev. Mod. Phys.* **71** 1–85

[7] Jones K M, Tiesinga E, Lett P D and Julienne P S 2006 Ultracold photoassociation spectroscopy: long-range molecules and atomic scattering *Rev. Mod. Phys.* **78** 483–535

[8] Thorsheim H R, Weiner J and Julienne P S 1987 Laser-induced photoassociation of ultracold sodium atoms *Phys. Rev. Lett.* **58** 2420–3

[9] Scheingraber H and Vidal C R 1977 Discrete and continuous Franck Condon factors of the mg2 system and their j dependence *J. Chem. Phys.* **66**(3694-704) 04

[10] Inoue G, Ku J K and Setser D W 1982 Photoassociative laser induced fluorescence of xecl *J. Chem. Phys.* **76**(733-4) 01

[11] Jones R B, Schloss J H and Eden J G 1993 Excitation spectra for the photoassociation of Kr-F and Xe-I collision pairs in the ultraviolet (208-258 nm) *J. Chem. Phys.* **98** 4317–34

[12] Tellinghuisen J 1985 Potentials for weakly bound states in i2 from diffuse spectra and predissociation data *J. Chem. Phys.* **82** 4012–6

[13] Stwalley W C, Uang Y-H and Pichler G 1978 Pure long-range molecules *Phys. Rev. Lett.* **41** 1164–7

[14] Fioretti A, Comparat D, Crubellier A, Dulieu O, Masnou-Seeuws F and Pillet P 1998 Formation of cold cs_2 molecules through photoassociation *Phys. Rev. Lett.* **80** 4402–5

[15] Nikolov A N, Eyler E E, Wang X T, Li J, Wang H, Stwalley W C and Gould P L 1999 Observation of ultracold ground-state potassium molecules *Phys. Rev. Lett.* **82** 703–6

[16] Takekoshi T, Patterson B M and Knize R J 1999 Observation of cold ground-state cesium molecules produced in a magneto-optical trap *Phys. Rev.* A **59** R5–7

[17] Wynar R, Freeland R S, Han D J, Ryu C and Heinzen D J 2000 Molecules in a Bose-Einstein condensate *Science* **287** 1016–9

[18] Bohn J L and Julienne P S 1996 Semianalytic treatment of two-color photoassociation spectroscopy and control of cold atoms *Phys. Rev.* A **54** R4637–40

[19] Sakurai J J and Napolitano J 2017 *Modern Quantum Mechanics* 2nd edn (Cambridge University Press)

[20] Napolitano R, Weiner J, Williams C J and Julienne P S 2004 Line shapes of high resolution photoassociation spectra of optically cooled atoms *Phys. Rev. Lett.* **73** 1352

[21] Kuklinski J R, Gaubatz U, Hioe F T and Bergmann K 1989 Adiabatic population transfer in a three-level system driven by delayed laser pulses *Phys. Rev. A* **40** 6741

[22] Vitanov N V, Fleischhauer M, Shore B W and Bergmann K 2001 Coherent manipulation of atoms molecules by sequential laser pulses *Adv. Atom. Mol. Opt. Phys* **46** 55–190

[23] Vitanov N V, Rangelov A A, Shore B W and Bergmann K 2017 Stimulated Raman adiabatic passage in physics, chemistry, and beyond *Rev. Mod. Phys.* **89** 015006

[24] Sardar D, Naskar S, Pal A, Berriche H and Deb B 2016 Formation of a molecular ion by photoassociative Raman processes *J. Phys. B: At. Mol. Opt. Phys* **49** 245202

[25] Naskar S, Saha S, Dey T N and Deb B 2017 Electromagnetically induced transparency in two-colour ultracold photoassociation *J. Phys. B: At. Mol. Opt. Phys* **50** 125003

[26] Bohn J L and Julienne P S 1999 Semianalytic theory of laser-assisted resonant cold collisions *Phys. Rev.* A **60** 414–25

[27] Jones K M, Julienne P S, Lett P D, Phillips W D, Tiesinga E and Williams C J 1996 Measurement of the atomic Na (3p) lifetime and of retardation in theinteraction between two atoms bound in a molecule *Europhys. Lett.* **35** 85
[28] Stwalley W C 1978 Long-range molecules *Contemp. Phys.* **19** 65–80
[29] Movre M and Pichler G 1977 Resonance interaction and self-broadening of alkali resonance lines. I. Adiabatic potential curves *J. Phys. B: At. Mol. Phys.* **10** 2631
[30] Magnier S, Millié Ph, Dulieu O and Masnou-Seeuws F 1993 Potential curves for the ground and excited states of the na2 molecule up to the (3 s+ 5 p) dissociation limit: results of two different effective potential calculations *J. Chem. Phys.* **98** 7113–25
[31] Abraham E R I, McAlexander W I, Gerton J M, Hulet R G, Côté R and Dalgarno A 1996 Singlet *s* -wave scattering lengths of ^{6}Li and ^{7}Li *Phys. Rev.* A **53** R3713–5
[32] Wang D *et al* 2004 Photoassociative production and trapping of ultracold KRb molecules *Phys. Rev. Lett.* **93** 243005
[33] Mancini M W, Telles G D, Caires A R L, Bagnato V S and Marcassa L G 2004 Observation of ultracold ground-state heteronuclear molecules *Phys. Rev. Lett.* **92** 133203
[34] Haimberger C, Kleinert J, Bhattacharya M and Bigelow N P 2004 Formation and detection of ultracold ground-state polar molecules *Phys. Rev. A* **70** 021402
[35] Deiglmayr J *et al* 2008 Formation of ultracold polar molecules in the rovibrational ground state *Phys. Rev. Lett.* **101** 133004
[36] Sage J, Sainis S, Bergeman T and DeMille D 2005 Optical production of ultracold polar molecules *Phys. Rev. Lett.* **94** 203001
[37] Kiss T, Herzog U and Leonhardt U 1995 Compensation of losses in photodetection and in quantum-state measurements *Phys. Rev. A* **52** 2433
[38] Trachy M L, Veshapidze G, Shah M H, Jang H U and DePaola B D 2007 Photoassociation in cold atoms via ladder excitation *Phys. Rev. Lett.* **99** 043003
[39] Rakshit A and Deb B 2011 Formation of cold molecular ions by radiative processes in cold ion-atom collisions *Phys. Rev.* A **83** 022703
[40] Ladjimi H, Sardar D, Farjallah M, Alharzali N, Naskar S, Mlika R, Berriche H and Deb B 2018 Spectroscopic properties of the molecular ions BeX+ (X = Na, K, Rb): forming cold molecular ions from an ion--atom mixture by stimulated Raman adiabatic process *Mol. Phys.* **116** 1812–26
[41] Deb B and You L 2003 Model study on the photoassociation of a pair of trapped atoms into an ultralong-range molecule *Phys. Rev. A* **68** 033408
[42] Busch T, Englert B-G, Rzazewski K and Wilkens M 1998 *Foundations of Physics* (Springer) pp 549–59
[43] Zelevinsky T, Boyd M M, Ludlow A D, Ido T, Ye J, Ciurylo R, Naidon P and Julienne P S 2006 Narrow line photoassociation in an optical lattice *Phys. Rev. Lett.* **96** 203201
[44] Naidon P and Julienne P S 2006 Optical Feshbach resonances of alkaline-earth-metal atoms in a one-or two-dimensional optical lattice *Phys. Rev. A* **74** 062713
[45] Deb B and Kurizki G 1999 Formation of giant quasibound cold diatoms by strong atom-cavity coupling *Phys. Rev. Lett.* **83** 714
[46] Arpita P, Saha S and Deb B 2016 Manipulating nanoscale atom--atom interactions with cavity QED *J. Phys. B Atom. Mol. Opt. Phys* **49** 245502
[47] Deb B, Rashkit A, Hazra J and Chakraborty D 2013 Quantum effects at low-energy atom–molecule interface *Pramana* **80** 3–19

[48] Timmermans E, Tommasini P, Hussein M and Kerman A 1999 Feshbach resonances in atomic Bose-Einstein condensates *Phys. Rep.* **315** 199–230

[49] Köhler T, Góral K and Julienne P S 2006 Production of cold molecules via magnetically tunable Feshbach resonances *Rev. Mod. Phys.* **78** 1311–61

[50] Chin C, Grimm R, Julienne P and Tiesinga E 2010 Feshbach resonances in ultracold gases *Rev. Mod. Phys.* **82** 1225–86

[51] Petrov D S, Salomon C and Shlyapnikov G V 2004 Weakly bound dimers of fermionic atoms *Phys. Rev. Lett.* **93** 090404

[52] Petrov D S, Salomon C and Shlyapnikov G V 2005 Diatomic molecules in ultracold fermi gases-novel composite bosons *J. Phys. B: At. Mol. Opt. Phys.* **38** S645

[53] Jochim S, Bartenstein M, Altmeyer A, Hendl G, Riedl S, Chin C, Hecker Denschlag J and Grimm R 2003 Bose-Einstein condensation of molecules *Science* **302** 2101–3

[54] ZwierleinM W, Stan C A, Schunck C H, Raupach S M F, Gupta S, Hadzibabic Z and Ketterle W 2003 Observation of Bose–Einstein condensation of molecules *Phys. Rev. Lett.* **91** 250401

[55] Greiner M, Regal C A and Jin D S 2003 Emergence of a molecular Bose-Einstein condensate from a fermi gas *Nature* **426** 537–40

[56] Bourdel T, Khaykovich L, Cubizolles J, Zhang J, Chevy F, Teichmann M, Tarruell L, Kokkelmans S J J M F and Salomon C 2004 Experimental study of the BEC-BCS crossover region in lithium 6 *Phys. Rev. Lett.* **93** 050401

[57] Giorgini S, Pitaevskii L P and Stringari S 2008 Theory of ultracold atomic fermi gases *Rev. Mod. Phys.* **80** 1215–74

[58] Bloch I, Dalibard J and Zwerger W 2008 Many-body physics with ultracold gases *Rev. Mod. Phys.* **80** 885–964

[59] Ni K-K, Ospelkaus S, De Miranda M H G, Pe'Er A, Neyenhuis B, Zirbel J J, Kotochigova S, Julienne P S, Jin D S and Ye J 2008 A high phase-space-density gas of polar molecules *Science* **322** 231–5

[60] Efimov V 1970 Energy levels arising from resonant two-body forces in a three-body system *Phys. Lett.* B **33** 563–4

[61] Kraemer T *et al* 2006 Evidence for efimov quantum states in an ultracold gas of caesium atoms *Nature* **440** 315–8

Chapter 7

Collisions in low dimensions

Quantum collision theory in three dimensions (3D) has been developed since the early days of quantum mechanics, that is, since the early 20th century. Most of the advancements in three-dimensional low-energy collision physics happened in the 1930s and 1940s, in parallel with the developments in atomic and nuclear physics. Collisions in lower dimensions, that is, in one dimension (1D) and two dimensions (2D), began to attract research interest in the latter half of the 20th century. Early studies of the theory of scattering in 1D appeared in the 1960s [1–3]. From an experimental point of view, interest in one-dimensional scattering arose from the analysis of tunneling phenomena in a superlattice [4]. Scattering in 2D began to attract attention following experiments involving helium atoms adsorbed on graphite and spin-polarized hydrogen atoms on a helium film [5, 6]. Research interest in bound states in 2D started with theoretical investigations of the problem of hydrogen in 2D [7–10] and the problem of Coulomb scattering in 2D [11]. These early developments in low-dimensional physics established that the two- and many-body properties of interacting systems in 1D or 2D are qualitatively different from those in 3D.

With progress in the cooling and trapping of atoms in highly anisotropic traps over the last three decades, new perspectives on atomic collisions in low dimensions have arisen [12]. An ensemble of harmonically trapped ultracold atoms becomes kinematically two-dimensional (along the x- and y-directions) when the temperature and the chemical potential of the ensemble are much lower than the harmonic trapping frequency in the tightly confining z-direction. Similarly, an ensemble of trapped atoms becomes one-dimensional along the z-axis if the temperature and the chemical potential are sufficiently lower than the trapping frequency in the transverse (x and y) directions.

The effects of tight anisotropic trapping on cold atomic collisions are an interesting research topic. One of the interesting effects is confinement-induced resonance in quasi-1D [13]. Moritz *et al* [14] experimentally demonstrated that cold

doi:10.1088/978-0-7503-3959-9ch7 © IOP Publishing Ltd 2025. All rights, including for text and data mining (TDM), artificial intelligence (AI) training, and similar technologies, are reserved.

atoms trapped in quasi one-dimensional traps can form confinement-induced dimers. Petrov and Shlyapnikov [15] analyzed quasi two-dimensional and three-dimensional regimes of atom–atom scattering in a trap that was tightly confined in the axial (z) direction. To understand these emerging phenomena of interacting particles in low dimensions from first principles, it is imperative to begin with the fundamentals of collision physics in low dimensions.

7.1 Scattering in two dimensions

7.1.1 Theory

Let $\boldsymbol{\rho}_1$ and $\boldsymbol{\rho}_2$ be the two-dimensional radial position vectors of particle 1 and particle 2, respectively. The relative radial vector is $\boldsymbol{\rho} = \boldsymbol{\rho}_1 - \boldsymbol{\rho}_2$. The two-dimensional polar coordinates (ρ, ϕ) are related to the Cartesian relative coordinates by $x = \rho\cos\phi$ and $y = \rho\sin\phi$, and the square of the two-dimensional Laplacian is given by

$$\boldsymbol{\nabla}^2 = \frac{\partial^2}{\partial x^2} + \frac{\partial^2}{\partial y^2} = \frac{1}{\rho}\frac{\partial}{\partial \rho}\left[\rho\frac{\partial}{\partial \rho}\right] + \frac{1}{\rho^2}\frac{\partial^2}{\partial \phi^2}. \tag{7.1}$$

The time-independent Schrödinger equation for the wave function of relative motion is

$$-\frac{\hbar^2}{2\mu}\boldsymbol{\nabla}^2\Psi + V(\boldsymbol{\rho})\Psi = E\Psi, \tag{7.2}$$

where $E = \hbar^2k^2/(2\mu)$ is the energy, k and μ are the two-dimensional relative wave number and reduced mass, respectively, and $V(\boldsymbol{\rho})$ is the potential. For a central potential, $V(\boldsymbol{\rho}) \equiv V(\rho)$.

The theory of quantum scattering in 2D can be formulated using the two-dimensional counterparts of the partial-wave expansion and the **t**-matrix approach, as discussed in chapter 2. For a free particle and a central potential, the wave function $\Psi = \psi(\rho)\Phi(\phi)$ is separable into the radial and angular wave functions $\psi(\rho)$ and $\Phi(\phi)$, respectively. Thus, one can write

$$\frac{d^2\Phi}{d\phi^2} + m^2\Phi = 0 \tag{7.3}$$

$$-\frac{1}{\rho}\frac{d}{d\rho}\left[\rho\frac{d\psi}{d\rho}\right] + \left[\frac{m^2}{\rho^2} + U(\rho)\right] = k^2\psi, \tag{7.4}$$

where $U(\rho) = 2\mu V(\rho)/\hbar^2$. Two linearly independent solutions of equation (7.3) are $\frac{1}{\sqrt{2\pi}}e^{im\phi}$ and $\frac{1}{\sqrt{2\pi}}e^{-im\phi}$. In order for the solution to be single-valued, m must be a positive or negative integer. A general solution can be written as

$$\Phi_m(\phi) = \frac{1}{\sqrt{2\pi}}[Ae^{im\phi} + Be^{-im\phi}], \tag{7.5}$$

where A and B are two constants. If we demand that the solution be symmetric about the x-axis, then $A = B = \sqrt{2}$; therefore,

$$\Phi_m(\phi) = \frac{1}{\sqrt{\pi}}\cos(m\phi). \tag{7.6}$$

In order to obtain an asymptotic form of the scattering wave function, one may start with the two-dimensional counterpart of the Lippmann–Schwinger integral equation

$$\Psi(\boldsymbol{\rho}) = e^{ikx} + \int\int d^2\boldsymbol{\rho}' G_0^+(\boldsymbol{\rho}, \boldsymbol{\rho}') V(\rho')\psi(\boldsymbol{\rho}'), \tag{7.7}$$

where we have assumed that the wave is incident along the x-axis, and the two-dimensional free-particle Green's function can be written as

$$G_0^+(\boldsymbol{\rho}, \boldsymbol{\rho}') = \frac{2\mu}{\hbar^2}\int\int d^2\mathbf{k}' \frac{\exp[i\mathbf{k}'\cdot(\boldsymbol{\rho} - \boldsymbol{\rho}')]}{k^2 - k'^2}. \tag{7.8}$$

Now, following the same procedure as that given in appendix B, one can obtain an explicit expression for the Green's function, as given by [16]:

$$G_0^+(\boldsymbol{\rho}, \boldsymbol{\rho}') = -\frac{i}{4}\frac{2\mu}{\hbar^2} H_0^{(1)}(k|\boldsymbol{\rho} - \boldsymbol{\rho}'|), \tag{7.9}$$

where $H_0^{(1)}$ is the Hankel function of the first kind. Using the asymptotic form of this function [17], we have

$$G_0^+(\boldsymbol{\rho}, \boldsymbol{\rho}') \xrightarrow[\rho\to\infty]{} -\frac{2\mu}{\hbar^2}\frac{1}{4}\sqrt{\frac{2}{\pi k\rho}}e^{i\pi/4}e^{ik\rho}e^{-i\mathbf{k}'\cdot\boldsymbol{\rho}'}. \tag{7.10}$$

Substituting this equation into equation (7.7) yields

$$\Psi(\boldsymbol{\rho}) \xrightarrow[\rho\to\infty]{} e^{ikx} + f_k(\phi)\frac{e^{ik\rho}}{\sqrt{\rho}}, \tag{7.11}$$

where

$$f_k(\phi) = -\frac{2\mu}{\hbar^2}\sqrt{\frac{2\pi i}{k}}\int\int e^{-i\mathbf{k}'\cdot\boldsymbol{\rho}'} V(\rho')\Psi(\boldsymbol{\rho}')d^2\boldsymbol{\rho}'. \tag{7.12}$$

Here, $|\mathbf{k}'| = |\mathbf{k}| = k$, $\hat{k} = \hat{\rho}$ and $\hat{k}'\cdot\hat{\rho} = \cos\phi$. Note that the form $f_k(\phi)\frac{e^{ik\rho}}{\sqrt{\rho}}$ of the scattered part of the two-dimensional wave function is different from its three-dimensional counterpart, which is of the form $f_k(\phi)\frac{e^{ikr}}{r}$. The reason for this is that the definition of the scattering amplitude or cross section in 2D differs from that in 3D. The two-dimensional scattering cross sections have the dimensions of length, unlike those in 3D; however, we still call them cross sections in analogy with those in 3D. Let us recall the definition of the scattering cross section: it is the number of scattered particles per unit time per unit incident flux. In 3D, the unit of flux is the number of particles per unit time per unit area; in 2D, it is the number of particles per unit time per unit length, and in 1D, it is the number of particles per unit time.

As in the case of 3D, the two-dimensional plane wave e^{ikx} ($x = \rho\cos\phi$) can be expanded in terms of two-dimensional partial waves $\Phi_m(\phi)$, and the expansion coefficients $c_m(\rho)$ can be obtained analytically. These coefficients can be expressed in terms of Bessel functions $J_m(k\rho)$. The result is

$$e^{ikx} = e^{ik\rho\cos\phi} = J_0(k\rho) + 2\sum_{m=1}^{m=\infty}(i)^m\cos(m\phi)J_m(k\rho). \tag{7.13}$$

For a short-range potential, $U \to 0$ in the asymptotic limit $\rho \to \infty$. The two linearly independent solutions of equation (7.4) in the potential-free region (free space or asymptotic regime) are the familiar Bessel and Neumann functions $J_m(\rho)$ and $N_m(\rho)$, respectively. Next, we recall the asymptotic forms of these functions [17]:

$$J_m(k\rho) \underset{\rho\to\infty}{\longrightarrow} \frac{2}{\sqrt{\pi k\rho}}\cos\left[k\rho - \frac{m\pi}{2} - \frac{\pi}{4}\right] \tag{7.14}$$

$$N_m(k\rho) \underset{\rho\to\infty}{\longrightarrow} \frac{2}{\sqrt{\pi k\rho}}\sin\left[k\rho - \frac{m\pi}{2} - \frac{\pi}{4}\right]. \tag{7.15}$$

Similarly, one can write

$$\Psi(\rho, \phi) \underset{\rho\to\infty}{\longrightarrow} \frac{1}{\sqrt{k\rho}}\sum_{m=0}^{\infty}A_m\cos\left[k\rho - \frac{m\pi}{2} - \frac{\pi}{4} + \delta_m\right], \tag{7.16}$$

where A_m are the unknown coefficients to be determined. We now substitute equations (7.13), (7.14), (7.15), and (7.10) into equation (7.7) and equate the resulting equation with equation (7.16). By equating the coefficients of $e^{ik\rho}$ and $e^{-ik\rho}$ on both sides of the resultant equation, one obtains

$$\begin{aligned} A_0 &= \sqrt{\frac{2}{\pi}}e^{i\delta_0} \\ A_m &= 2\sqrt{\frac{2}{\pi}}i^m e^{i\delta_m}\ m \neq 0. \end{aligned} \tag{7.17}$$

Expanding $f_k(\phi)$ in partial waves in the form

$$f_k(\phi) = f_0(k, \phi) + 2\sum_{m=1}^{\infty}f_m(k, \phi), \tag{7.18}$$

we obtain

$$f_m(k, \phi) = \sqrt{\frac{2i}{\pi k}}\cos(m\phi)e^{i\delta_m}\sin\delta_m. \tag{7.19}$$

The above expression for the partial-wave scattering amplitude was derived by Lapidus [18]. Adhikari [19] defined this amplitude without the prefactor of $\sqrt{i}$, as the scattered wave in the asymptotic form of the wave function already contained this prefactor. Thus, the mth partial-wave scattering amplitude is given by $(2-\delta_{m,0})f_m(k,\phi)$, where $\delta_{m,0}$ is a Kronecker delta. Therefore, the partial-wave differential scattering cross section is

$$\sigma_m(k,\phi) = |(2-\delta_{m,0})f_m(k,\phi)|^2. \tag{7.20}$$

The total cross section σ_t is given by

$$\sigma_t = \int_0^{2\pi} d\phi \sum_m \sigma_m(k,\phi). \tag{7.21}$$

Thus, we have

$$\sigma_t = \int_0^{2\pi} d\phi |\sqrt{i/k} f_k(\phi)|^2 = \sum_m \int_0^{2\pi} d\phi (2-\delta_{m,0})\sigma_m(k,\phi), \tag{7.22}$$

which results in

$$\sigma_t = \frac{4}{k}\left[\sin^2\delta_0 + 2\sum_{m=1}^{\infty}\sin^2\delta_m\right]. \tag{7.23}$$

The partial-wave phase shifts δ_m can be calculated by either solving the two-dimensional Lippmann–Schwinger integral equations for the wave functions and the **t**-matrix [20] perturbatively or solving the two-dimensional partial-wave Schrödinger equation (7.4) numerically, subject to scattering boundary conditions. The first-order Born approximation of the scattering amplitude $f_k(\phi)$, as given by equation (7.12), replaces the scattering wave function $\Psi(\rho)$ with the incident plane wave $e^{i\mathbf{k}\cdot\boldsymbol{\rho}}$. As a result, the first-order Born approximation amplitude can be written as

$$f_{k\mathrm{B}}(\phi) = -\frac{2\mu}{\hbar^2}\sqrt{\frac{i}{2k}}\int\int e^{-i(\mathbf{k}-\mathbf{k}')\cdot\boldsymbol{\rho}'} V(\rho')d^2\boldsymbol{\rho}'. \tag{7.24}$$

The plane wave $e^{-i(\mathbf{k}-\mathbf{k}')\cdot\boldsymbol{\rho}'}$ can be expanded in partial waves, as in equation (7.13). For an isotropic potential $V(\rho)$, the angular integration on the rhs of the above equation is finite only for $m=0$, yielding

$$f_{k\mathrm{B}}(\phi) = -\frac{2\mu}{\hbar^2}\sqrt{\frac{i}{2k}}\int_0^{2\pi} d\phi' \int \rho' d\rho' J_0(q\rho')V(\rho'), \tag{7.25}$$

where $q = |\mathbf{k}-\mathbf{k}'| = 2k\sin(\phi/2)$. Thus, one can write

$$f_{k\mathrm{B}}(\phi) = -\frac{2\mu}{\hbar^2}2\pi e^{i\pi/4}\sqrt{\frac{1}{2k}}\int \rho' d\rho' J_0(2k\sin(\phi/2)\rho')V(\rho'). \tag{7.26}$$

7.1.2 Two-dimensional zero-range potential

We have seen that in 3D, a zero-range model potential has a significant role in describing the mean-field interactions of an ensemble of dilute and weakly interacting ultracold atoms. Similarly, one can introduce a model zero-range potential in 2D. A two-dimensional zero-range or delta potential can be written as $V = \xi\delta^2(\boldsymbol{\rho}) = \frac{\xi}{\rho}\delta(\rho)\delta(\phi)$, where ξ is a parameter that has the dimensions of energy $\times$ area. From the Lippmann–Schwinger equation, we know that the scattering amplitude for a zero-range potential is determined by the wave function at $\rho = 0$, that is, by $\Psi(0)$.

In order to define a mean-field interaction at low energy in 2D as the Fourier transform of a model potential into momentum space, or equivalently as the first-order Born approximation **t**-matrix element of the potential, one such model potential is a zero-range or a delta potential in 2D. Now, substituting $V(\rho') = \frac{1}{\rho'}\delta(\rho')\delta(\phi)$ into equation (7.26), one obtains

$$f(\phi) = -\frac{2\mu}{\hbar^2}2\pi\sqrt{\frac{1}{2k}}e^{i\pi/4}\xi. \tag{7.27}$$

Therefore, the parameter ξ is related to the scattering amplitude.

Deriving an effective contact or zero-range potential at low energy for single-channel scattering with zero relative angular momentum in 3D, 2D, and 1D depends on the expansion of $\cot[\delta(k)]$, where $\delta(k)$ is the phase shift, in terms of the wave number k, and retaining the leading-order term. The reason for this is that the scattering amplitude in all dimensions is related to the function

$$F(k) = \frac{1}{\cot\delta - i}. \tag{7.28}$$

In 3D, the scattering amplitude is given by $f_k = F(k)/k$; in 1D, it is given by $f_k = F(k)$, and in 2D, it is $f_k = F(k)/\sqrt{k}$. If one can approximate f_k as a constant (energy-independent) in the limit $k \to 0$, then it is possible to define an effective zero-range or delta function potential whose first-order Born approximation term is $F(k)$. Thus, the expansion of $F(k)$ is important for understanding the low-energy scattering properties in 3D, 2D, and 1D. While in 3D, it is straightforward to derive such a low-energy effective zero-range potential for a broad class of interaction potentials, including atom–atom potentials, this is complicated in 2D due to the logarithmic dependence of the function $F(k)$ on the parameters (range and depth) of the potential. This can be ascertained by solving the two-dimensional scattering problem for a hard-core potential with a hard-core radius d using the method described in section 7.1. The detailed solution of this problem can be found in the book by Morse and Feshbach [16]. For this problem, $\cot\delta$ is given by

$$\cot\delta = \frac{2}{\pi}[\ln(kd/2) + C + (kd/2)^2 + O(k^4)], \tag{7.29}$$

where $C = 0.5772...$ is the Euler constant. So, in the limit $k \to 0$, the scattering amplitude

$$f_k \sim \frac{\pi}{2}[\ln(kd/2)]^{-1}. \tag{7.30}$$

For a finite-range potential with range r_0, one can obtain a similar logarithmic dependence [21] by making the replacement

$$d \to a_{2\mathrm{D}} = r_0 \exp[-1/r_0 D_0], \tag{7.31}$$

where D_0 is the logarithmic derivative of the wave function at $r = r_0$. Here, the parameter a_{2D} can be treated as the two-dimensional scattering length [22, 23]. The complexity of solving the two-dimensional scattering problem with a zero-range potential that may support a bound state has been discussed in the literature [24].

7.1.3 Coulomb scattering in 2D

Coulomb scattering at low energy in 2D is qualitatively different from its three-dimensional counterpart [25]. While low-energy classical and quantum scattering cross sections are identical for a Coulomb potential C/r in 3D, the two-dimensional quantum scattering cross section for collision energy $E < \mu|C|^2/(2\hbar^2)$ deviates significantly from the classical one. We have seen in chapter 1 that the Coulomb scattering cross section diverges in the limit $k \to 0$, but in 2D, it remains finite. The two-dimensional Coulomb potential for a point charge can be obtained by solving the Poisson equation in 2D. This results in a potential in the form $\ln[\rho]$. However, we can think of a physical situation where the charged particles are confined to a plane but experience the usual three-dimensional Coulomb potential.

Following the same procedure as in the three-dimensional Coulomb scattering problem, one can use parabolic coordinates $\xi = \rho - x$ and $\eta = \rho + x$. We assume that the wave is incident along the x-axis. The two-dimensional Laplacian is given by

$$\nabla^2 = \frac{4}{\xi + \eta}\left[\sqrt{\xi}\frac{\partial}{\partial \xi}\left(\sqrt{\xi}\frac{\partial}{\partial \xi}\right) + \sqrt{\eta}\frac{\partial}{\partial \eta}\left(\sqrt{\eta}\frac{\partial}{\partial \eta}\right)\right]. \tag{7.32}$$

Substituting the ansatz $\Psi(\xi, \eta) = e^{ikx}F(\xi)$ into the time-independent Schrödinger equation with potential $V(\rho) = C/\rho = C/(\xi + \eta)$, one obtains [25]

$$\xi\frac{d^2F}{d\xi^2} + \left(\frac{1}{2} - ik\xi\right)\frac{dF}{d\xi} - \nu kF = 0, \tag{7.33}$$

where $\nu = \mu C/(\hbar^2 k)$. This is an equation for the confluent hypergeometric function given by

$$F = \Phi\left(-i\nu, \frac{1}{2}; ik\xi\right). \tag{7.34}$$

Thus, the solution is given by

$$\Psi = \mathcal{N} e^{ikx}\Phi\left(-i\nu, \frac{1}{2}; ik\xi\right), \tag{7.35}$$

where $\mathcal{N}$ is a normalization constant chosen in such a way that as $\xi \to \infty$, Ψ reduces to

$$\Psi \sim e^{ikx+i\nu\log(k\xi)} + \frac{1}{\sqrt{k\xi}} e^{ik\rho - i\nu\log(k\xi) - i\pi/4} \frac{\Gamma(1/2 + i\nu)}{\Gamma(-i\nu)}. \tag{7.36}$$

Writing $\xi = \rho - x = \rho(1 - \cos\phi) = 2\rho\sin^2(\phi/2)$ and $\frac{1}{\Gamma(-i\nu)} = -i\nu/\Gamma(1 - i\nu)$, one can rewrite the above equation in the following form:

$$\Psi \sim e^{ikx+i\nu\log(k\xi)} + f_k(\phi)e^{ik\rho - i\nu\log(k\xi)}/\sqrt{\rho}, \tag{7.37}$$

from which one finds $\mathcal{N} = \frac{\Gamma(1/2 + i\nu)}{\sqrt{\pi}} e^{-\pi\nu/2}$. The scattering amplitude f_k is given by

$$f_k = -\frac{\nu}{\sqrt{2k}\sin(\phi/2)} e^{i\pi/4} \frac{\Gamma(1/2 + i\nu)}{\Gamma(1 - i\nu)}. \tag{7.38}$$

Thus, the differential scattering cross section is $\frac{d\sigma}{d\phi} = |f_k(\phi)^2|$. Since $|\Gamma(1/2 + i\nu)|^2 = \pi/\cosh\pi\nu$ and $|\Gamma(1 - i\nu)|^2 = |\nu|\pi/\sinh\pi\nu$, we have

$$\frac{d\sigma}{d\phi} = \frac{\nu}{2k\sin^2(\phi/2)}\tanh\pi\nu. \tag{7.39}$$

Substituting $\nu = \mu C/(\hbar^2 k)$ and $E = \hbar^2 k^2/(2\mu)$, we find that

$$\frac{d\sigma}{d\phi} = \frac{C}{4E\sin^2(\phi/2)}\tanh[\pi\mu C/(\hbar^2 k)]. \tag{7.40}$$

In the classical limit, $\hbar \to 0$, $\nu \to \pm\infty$, $\tanh\pi\nu \to \pm 1$, where '$+(-)$' corresponds to $C > 0 (C < 0)$, the differential cross section reduces to

$$\frac{d\sigma}{d\phi} = \frac{|C|}{4E\sin^2(\phi/2)}, \tag{7.41}$$

which coincides with the classical expression. Since for finite ν, $|\tanh\pi\nu| < 1$, the value of the quantum mechanical cross section in 2D is lower than the classical value. On the other hand, for $k \to \infty$ or equivalently $\nu \to 0$, it reduces to

$$\frac{d\sigma}{d\phi} = \frac{\pi\mu^2 C^2}{2\hbar^3 k^3 \sin^2(\phi/2)}, \tag{7.42}$$

which deviates from the corresponding three-dimensional expression, which is proportional to $|C|$. Note that the angular distribution of the differential scattering cross sections is the same in both 2D and 3D. The two-dimensional scattering wave function, as given by equation (7.35), is finite at $\rho = 0$ for an attractive Coulomb potential, in contrast to the corresponding three-dimensional case, for which it diverges at $\rho = 0$ [25].

Now, let us derive the scattering amplitude f_{kB} under the first-order Born approximation. Substituting $V = C/r$ in equation (7.26), we obtain

$$\begin{aligned} f_{kB}(\phi) &= -\frac{2\mu}{\hbar^2}2\pi C e^{i\pi/4}\sqrt{\frac{1}{2k}}\int d\rho' J_0(2k\rho' \sin[\phi/2]) \\ &= -\frac{2\mu}{\hbar^2}\frac{2\pi C e^{i\pi/4}}{2k\,\sin(\phi/2)}, \end{aligned} \tag{7.43}$$

where we have used $\int_0^\infty dx J_0(x) = 1$. The above expression for f_{kB} coincides with that of equation (7.38) in the limit $\nu \to 0$, that is, in the high-energy limit.

7.2 Scattering in one dimension

The Hamiltonian for a one-dimensional potential problem is $H = H_0 + V(x)$, where $H_0 = -\frac{\hbar^2}{2\mu}\frac{d^2}{dx^2}$ is the kinetic energy operator and $V(x)$ is the potential. The time-independent Schrödinger equation in 1D is

$$\left[-\frac{d^2}{dx^2} + U(x)\right]\psi(k, x) = k^2\psi(k, x), \tag{7.44}$$

where $U(x)$ is related to $V(x)$ by $U(x) = 2\mu V(x)/\hbar^2$. Assuming that $V(x)$ is centrally symmetric and short-ranged, we have $V(-x) = V(x)$ and $V(x) = 0$ for $|x| > x_0$, where x_0 stands for the range of the potential. Here, $k \geqslant 0$, since we are interested in scattering solutions. The normalization condition for the wave functions is given by

$$\int_{-\infty}^{\infty} \psi^*(k, x)\psi(k', x)dx = \delta(k - k'). \tag{7.45}$$

The solutions $\psi^\pm(k, x)$ with outgoing (+) and incoming (−) scattering boundary conditions can be expressed in integral form as

$$\psi^\pm(k, x) = \phi(k, x) + \int_{-\infty}^{\infty} dx' G_0^\pm(x, x')U(x')\psi^\pm(k, x'), \tag{7.46}$$

where $\phi(k, x) = (1/\sqrt{2\pi})e^{ikx}$ is the one-dimensional plane wave. Here, $G_0^\pm(x, x;) = \lim_{\eta\to 0}[(2\mu/\hbar^2)(E - H_0 \pm i\eta)]^{-1}$ is the Green's function corresponding to H_0, which can be obtained by solving the equation

$$\left[\frac{d^2}{dx^2} + k^2\right]G_0^\pm(x, x') = \delta(x - x'). \tag{7.47}$$

The solution of the above equation is given by

$$G_0^\pm(x, x') = \mp\frac{i}{2k}e^{\pm ik|x-x'|}. \tag{7.48}$$

In one-dimensional scattering, there are only two possible directions of scattering: forward (the positive x-direction) and backward (the negative x-direction). We designate these two directions using the symbol $\varepsilon = \pm 1$, where $+1$ corresponds to the positive x-direction and -1 to the negative x-direction. In order to calculate the scattering amplitude, we require the asymptotic form of $G_0^{\pm}(x, x')$ for $x \to \pm\infty$. For this purpose, we write $|x - x'| = \{(x - x')^2\}^{1/2} \simeq |x| - \varepsilon x'$. Substituting this asymptotic form of $G_0^{\pm}(x, x)$ into equation (7.46), we can write the asymptotic form of $\psi^{\pm}(k, x)$ in the following form:

$$\psi^{\pm}(k, x) \xrightarrow[|x|\to\infty]{} \frac{1}{\sqrt{2\pi}}\left[e^{ikx} + \frac{i}{k} f_k^{\pm}(\varepsilon) e^{\pm ik|x|}\right], \tag{7.49}$$

where

$$f_k^{\pm}(\varepsilon) = \mp\frac{1}{2}\int_{-\infty}^{\infty} \exp[\mp i\varepsilon kx'] U(x') \psi^{\pm}(k, x') dx'. \tag{7.50}$$

Note that $f_k^{\pm}(\varepsilon)$, which has the dimension of the inverse of length (L^{-1}), is not the scattering amplitude. In analogy with the partial-wave analysis of three-dimensional scattering, as discussed in chapter 2, the physical scattering amplitude in the direction ε is given by

$$f_{1\mathrm{D}}(k, \varepsilon) = \frac{i}{k} f_k^{+}(\varepsilon), \tag{7.51}$$

which can be determined from the asymptotic form of the outgoing wave function $\psi^{+}(k, x)(\varepsilon)$, as given in equation (7.49). The differential cross section σ_ε is given by

$$\sigma_\varepsilon = |f_{1D}(k, \varepsilon)|^2, \tag{7.52}$$

while the total cross section is given by

$$\sigma_t = \sum_{\varepsilon=\pm 1} \sigma_\varepsilon. \tag{7.53}$$

Note that the scattering amplitude and cross section in 1D are dimensionless quantities, since the flux in 1D is measured in units of the inverse of time (T^{-1}). The transmission and reflection coefficients T and R, respectively, are related to these discrete scattering amplitudes by

$$T = \left| 1 + if_k^{+}(+1)/k \right|^2 \tag{7.54}$$

$$R = \left| if_k^{+}(-1)/k \right|^2. \tag{7.55}$$

Again, the Lippmann–Schwinger integral equations for the one-dimensional wave function and the on-shell **t** element have the same mathematical forms as in the case of 3D and 2D, but the integration has to be performed over one-dimensional space.

7.2.1 One-dimensional zero-range potential

Here, we consider a one-dimensional contact or delta potential

$$V(x) = V_0\delta(x), \tag{7.56}$$

where V_0 has the dimensions of energy times length. Accordingly, $U_0 = 2\mu V_0/\hbar^2$ has the dimension of the inverse of length. Equation (7.50) for the delta potential provides the scattering amplitude

$$f_k^{\pm}(\varepsilon) = \mp\frac{1}{2}U_0\psi^{\pm}(k, 0), \tag{7.57}$$

implying that the amplitude is determined by the boundary condition on the wave function at $x = 0$. For a one-dimensional delta potential, $f_k^{\pm}(\varepsilon)$ is independent of ε. Substituting equations (7.56) and 7.48 into equation (7.46), we obtain

$$\psi^{\pm}(k, 0) = \phi(k, 0) \mp \frac{iU_0}{2k}\psi^{\pm}(k, 0), \tag{7.58}$$

which yields

$$\psi^{\pm}(k, 0) = \frac{2k}{(2k \pm iU_0)}\phi(k, 0). \tag{7.59}$$

Substituting $\phi(k, 0) = \frac{1}{\sqrt{2pi}}$ and equation (7.59) into equation (7.57), we have

$$f_k^{\pm} = \mp\frac{kU_0}{(2k \pm iU_0)}. \tag{7.60}$$

Thus, the scattering amplitude as given by equation (7.51) can be written as

$$f_{1\mathrm{D}}(k) = -\frac{iU_0}{2k + iU_0} = \frac{1}{1 + ika_{1\mathrm{D}}}, \tag{7.61}$$

where

$$a_{1\mathrm{D}} = -\frac{2}{U_0} \tag{7.62}$$

is introduced by analogy with the standard expression for low-energy scattering amplitude in 3D based on the effective range expansion of $k\cot[\delta_0(k)]$ as given by

$$f(k \to 0) = \lim k \to 0\frac{1}{k\cot\delta_0 - i} \simeq -\frac{1}{1 + ika_s}. \tag{7.63}$$

Thus, in terms of $a_{1\mathrm{D}}$, one can write the one-dimensional contact potential

$$V(x) = -\frac{\hbar^2}{\mu a_{1\mathrm{D}}}\delta(x). \tag{7.64}$$

The transmission and reflection coefficients are determined by $f_k^+(\pm\varepsilon)$. Substituting the above equation into equations (7.54) and (7.55), we obtain the transmission and reflection coefficients:

$$T = \left| 1 + if_k^+/k \right|^2 = \frac{4k^2}{4k^2 + U_0^2} \tag{7.65}$$

$$R = \left| if_k^+/k \right|^2 = \frac{U_0^2}{4k^2 + U_0^2}. \tag{7.66}$$

The two expressions above show that, in the limit $k \to 0$, $R \to 1$ and $T \to 0$. This is a remarkable result of one-dimensional scattering—it means that for a system of particles interacting with a delta potential in 1D, there is total reflection of the particles at low energy, regardless of whether the particles interact attractively ($U_0 < 0$) or repulsively ($U_0 > 0$).

We next discuss the applications of the low-dimensional scattering methods to trapped cold atoms. In reality, no system is strictly two- or one-dimensional; only under certain limiting conditions can a system be approximated as a two-dimensional or one-dimensional one. In the case of trapped atomic systems, an atom in a highly anisotropic trap at a sufficiently low temperature can be treated as a two-dimensional or one-dimensional system, depending on the ratio between the transverse and axial trapping frequencies in comparison to the minimum excitation energy gap (in kelvin) to the temperature.

7.3 Cold atoms in low dimensions

In 1998, Olshanii [13] theoretically considered low-energy atomic scattering in trap-induced confined geometry leading to effects that can be attributed to low dimensionality. Specifically, he studied one-dimensional scattering between bosonic atoms in a cigar-shaped trap or atom waveguide with strong confinement in the transverse directions. By approximating the interaction potential with a one-dimensional regularized delta potential, he showed that, in the low-energy limit, the scattering processes lead to total reflection, as we noted in the previous section. A system of particles interacting in 1D showing total reflection is called a Tonks–Girardeau gas [26] of impenetrable bosons; such bosons behave like fermions as far as quantum statistics is concerned.

Let us consider the motion of two interacting atoms in a harmonic oscillator. We assume that the atoms interact with a regularized delta pseudopotential

$$V_{\text{pseu}}(\mathbf{r}) = g\delta(\mathbf{r})\frac{\partial}{\partial r}\mathbf{r}. \tag{7.67}$$

where $g = \frac{2\pi\hbar^2 a_s}{\mu}$, $\mu = m/2$ is the reduced mass, and $\mathbf{r} = \mathbf{r}_1 - \mathbf{r}_2$ is the relative position vector between atoms 1 and 2. In a harmonic trap, the center of mass and the relative motions between the two atoms become separable. As a result, only the relative motion is affected by the isotropic interaction potential of equation (7.67). For an axially symmetric harmonic trap, the Schrödinger equation for relative motion is

$$\left[-\frac{\hbar^2}{2\mu}\left(\frac{d^2}{d\rho^2}+\frac{d}{\rho d\rho}-\frac{|m|^2}{\rho^2}+\frac{d^2}{dz^2}\right)+\frac{\mu}{2}\left(\omega_\rho^2\rho^2+\omega_z^2z^2\right)\right.$$
$$\left.+V_{\text{pseu}}(\mathbf{r})\right]\psi(\rho, z)=E\psi(\rho, z) \tag{7.68}$$

where $\rho^2=(x_1-x_2)^2+(y_1-y_2)^2$ and $z=z_1-z_2$; ω_ρ and ω_z are the radial and axial trapping frequencies, respectively. The width of the probability function corresponding to the ground state of the axial harmonic oscillator is $a_z=\sqrt{\frac{\hbar}{m\omega_z}}$, and that corresponding to the radial harmonic oscillator is $a_z=\sqrt{\frac{\hbar}{m\omega_\rho}}$. Therefore, the aspect ratio between the widths is $\alpha=a_z/a_\rho=\sqrt{\omega_\rho/\omega_z}$. For $\omega_\rho\gg\omega_z$, the trap looks like a cigar. In the absence of interaction, the energy eigenvalues of the isotropic two-dimensional harmonic oscillator

$$E_{n,m}=(n+1)\omega_\rho \tag{7.69}$$

where $n=2n_\rho+|m|$ is the two-dimensional principal quantum number and n_ρ and m are the two-dimensional radial and magnetic quantum numbers, respectively. If the atoms are cooled to the radial harmonic oscillator ground state, temperature $T\ll\hbar\omega_\rho/k_B$ (k_B being the Boltzmann constant), and the mean interaction energy

$$V_{\text{mean}}=\frac{4\pi\hbar^2a_s}{m}\int d^3\mathbf{r}_i|\psi(\mathbf{r}_i)|^4 \tag{7.70}$$

is much smaller than $\hbar\omega_\rho$, then the motion of the system is confined only in the axial direction, provided the average kinetic energy in the z-direction, $\hbar^2k_z^2/2\mu$, is much smaller than the energy gap between the ground state of the two-dimensional harmonic oscillator ($n=0$) and the excited state $n=2$ ($n_\rho=1$, $m=0$). These conditions lead to an effective one-dimensional system where the motion of the atoms in the z-direction can be approximated as that in free space; that is, the effects of the trap in this direction can be ignored. The wave function is then given by the following ansatz:

$$\Psi(z, \rho)=\psi(z)\phi_{n=0,\ m=0}(\rho), \tag{7.71}$$

which implicitly assumes that the interaction potential affects the motion only in the z-direction. Here, $\phi_{n=0,m=0}(\rho)$ is the ground state of the two-dimensional transverse harmonic oscillator, and $\psi(z)$ is the one-dimensional scattering wave function that satisfies the asymptotic boundary condition as given in the previous section. The scattering amplitude is primarily determined by the limit

$$g\delta(\mathbf{r})\frac{\partial}{\partial r}\mathbf{r}\Psi(z, \rho)|_{\mathbf{r}\to 0}, \tag{7.72}$$

which may be calculated term by term using the perturbation method. The higher-order terms involve virtual excitations to the states $\phi_{n\neq 0,m=0}$. In this way, one can derive an effective one-dimensional scattering length $a_{1\text{D}}$ by equating the calculated scattering amplitude with the standard expression for the low-energy one-dimensional scattering amplitude given by (7.61). The result is [13]

$$a_{1\mathrm{D}} = -\frac{a_\rho^2}{2a_s}\left[1 - C\frac{a_s}{a_\rho}\right], \tag{7.73}$$

where $C = 1.4603...$ is a constant. Substituting this into equation (7.64), one can write the effective one-dimensional potential

$$V_{1\mathrm{D}}(z) = \frac{\hbar^2}{\mu}\frac{2a_s}{a_\rho^2\left[1 - C\frac{a_s}{a_\rho}\right]}\delta(z), \tag{7.74}$$

which shows that the effective strength of the potential diverges if $a_s = a_\rho / C$. Therefore, the effective one-dimensional coupling for an atom–atom interaction in a cylindrical, cigar-shaped trap with tight confinement in the transverse directions goes to $\pm\infty$ if one tunes the s-wave scattering length a_s close to the width of the ground state of the transverse harmonic oscillations. As a result, the system shows confinement-induced resonance (CIR) when a_s is tuned across Ca_ρ.

Haller *et al* [27, 28] observed CIR in a quantum gas of Cs atoms in a highly anisotropic quasi one-dimensional configuration by tuning an MFR. The mechanism of CIR is schematically shown in figure 7.1. They detected CIR through

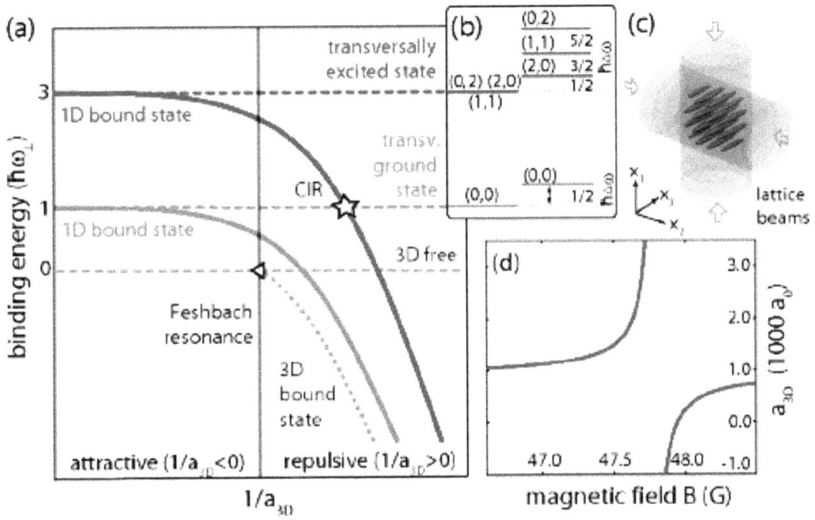

Figure 7.1. Physical processes that lead to confinement-induced resonance (CIR) in 1D. The plot shows the binding energy (in units of $\hbar\omega_\perp \equiv \hbar\omega_\rho$) (a) near a resonance as a function of the inverse of $a_s \equiv a_{3\mathrm{D}}$, $1/a_{3\mathrm{D}}$. (b) The shift and splitting of the two-dimensional transverse harmonic oscillator bound states for the indicated quantum numbers in Cartesian coordinates. (c) Experimental configuration where two lasers generate an optical lattice with an array of about 3000 horizontally oriented one-dimensional tubes. (d) MFR with a resonant magnetic field of $B = 47.78$ G. Figure reprinted with permission from [29], Copyright (2010) by the American Physical Society.

measurements of the collective modes of oscillation of the trapped gas [27] and also the trap loss spectra [28] as a function of the magnetic field or a_s tuned across the CIR for different trap parameters.

Zürn *et al* [30] experimentally demonstrated the fermionization of two strongly interacting atoms in 1D using fermionic ^{6}Li atoms. They showed that two interacting distinguishable fermionic atoms in a quasi one-dimensional trap with tight transverse confinement behave like two indistinguishable noninteracting fermions when the effective one-dimensional interaction strength diverges. They confirmed this effect by measuring the mean particle number and the tunneling time constants for different magnetic field strengths around a CIR for both two distinguishable fermions and by comparing the results with those for two non-interacting identical fermions, as shown in figure 7.2.

The early theoretical studies [12, 32] of the quasi two-dimensional regime of trapped ultracold atomic gases showed that the effective two-dimensional interaction is sensitive to the trapping frequency in the axial direction. It was shown that if the trapping frequency in the axial direction largely exceeds the temperature of the trapped gas, then the scattering phenomena exhibit characteristics of typical two-dimensional motion [12]. The Berezinskii–Kosterlitz–Thouless (BKT) crossover in

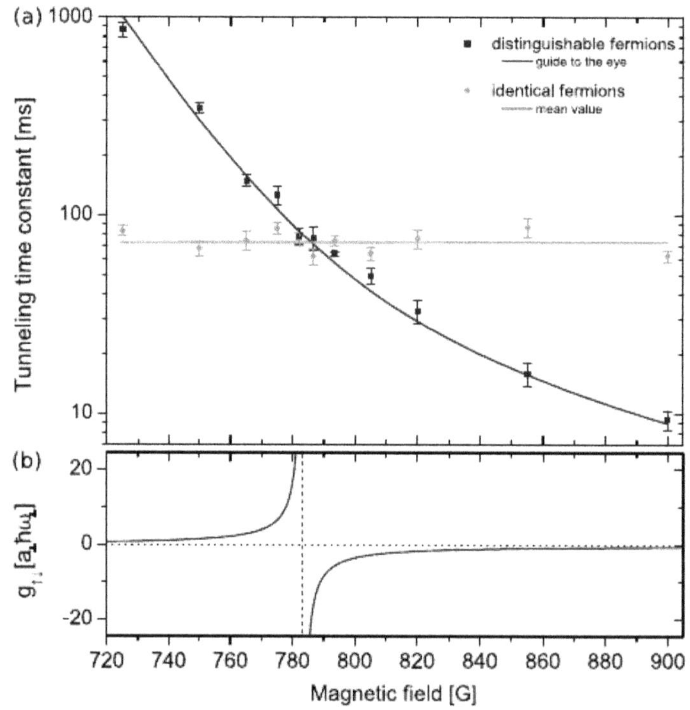

Figure 7.2. (a) The tunneling time constant of two repulsively interacting distinguishable fermions (blue) for different one-dimensional interaction strengths (b) is compared with that of two noninteracting indistinguishable fermions as the magnetic field is tuned across a CIR. Figure reprinted with permission from [31], Copyright (2012) by the American Physical Society.

the two-dimensional regime of trapped bosonic atoms was experimentally demonstrated in 2006 [33]. The BKT phase arises in two-dimensional quantum fluids due to vortex–antivortex pairing below a certain critical temperature; as the temperature increases, the system enters a new phase characterized by free vortices and a quasi-condensate.

As mentioned above, the dimensional crossover from 3D to 2D or 1D for trapped ultracold atoms is achieved by changing the aspect ratio between the axial and radial sizes of a cylindrical harmonic trap. The purpose is to explore low-energy many-body quantum physics at low dimensions with ultracold atoms. In such quantum simulation studies, it is generally considered that atoms interact with a zero-range or contact three-dimensional potential. For contact interactions, the dimensional crossover depends on the interplay between the three-dimensional scattering length and the radial and axial harmonic oscillator length scales. One important question is what the minimum aspect ratio should be to access the two-dimensional or one-dimensional regime for a given a_s. Furthermore, for resonantly or strongly interacting atoms due to a tunable Feshbach resonance, the effects of the finite range of interaction cannot always be ignored [34–37]. In this case, the finite range of interaction introduces another length scale into the consideration of low-dimensional effects on the bound states of cold atoms in tightly confined traps [38, 39]. It has been theoretically shown that if the range is much smaller than the axial size of the trap, then one can reach the one-dimensional regime for $\eta > 10000$, where η is the ratio of the radial frequency to the axial frequency of a harmonic trap [38] for a wide range of a_s. However, for a large range, the one-dimensional regime can be achieved for smaller values of η.

Exercise

- Discuss two dimensional partial-wave expansion. Derive an expression for scattering amplitude due to central potential in two dimension in polar coordinates as an integral over the potential.
- Derive an expression for a two-dimensional contact or delta potential.
- Discuss Coulomb scattering in two dimension.
- Derive transmission and reflection coefficients in terms of the scattering amplitude in one dimension.
- Derive an expression for one dimensional contact potential in terms of one dimensional scattering length.

References

[1] Kahn A H 1961 Phase-shift method for one-dimensional scattering *Am. J. Phys.* **29** 77–80

[2] Eberly J H 1965 Quantum scattering theory in one dimension *Am. J. Phys.* **33** 771–3

[3] Lapidus I R 1969 Phase shifts for scattering by a one-dimensional delta-function potential *Am. J. Phys.* **37** 930–1

[4] Tsu R and Esaki L 1973 Tunneling in a finite superlattice *Appl. Phys. Lett.* **22** 562–4

[5] Bennemann K H and Ketterson J B 1978 (Chichester: John Wiley) Physics of liquid and solid helium. Part II

[6] Dash J G 1978 Helium films from two to three dimensions *Phys. Rep.* **38** 177–226
[7] Alliluev S P 1958 On the relation between 'accidental' degeneracy and 'hidden' symmetry of a system *Sov. J. Exp. Theor. Phys.* **6** 156 http://www.jetp.ras.ru/cgi-bin/dn/e_006_01_0156.pdf
[8] Zaslow B and Zandler M E 1967 Two-dimensional analog to the hydrogen atom *Am. Am. J. Phys.* **35** 1118–9
[9] Cisneros A and McIntosh H V 1969 Symmetry of the two-dimensional hydrogen atom *J. Math. Phys.* **10** 277–86
[10] Wen-Kuang Huang J and Kozycki A 1979 Hydrogen atom in two dimensions *Am. J. Phys.* **47** 1005–6
[11] Barton G 1983 Rutherford scattering in two dimensions *Am. J. Phys.* **51** 420–2
[12] Petrov D S and Shlyapnikov G V 2001 Interatomic collisions in a tightly confined bose gas *Phys. Rev.* A **64** 012706
[13] Olshanii M 1998 Atomic scattering in the presence of an external confinement and a gas of impenetrable bosons *Phys. Rev. Lett.* **81** 938–41
[14] Moritz H, Stöferle T, Günter K, Köhl M and Esslinger T 2005 Confinement Induced Molecules in a 1D Fermi Gas *Phys. Rev. Lett.* **94** 210401
[15] Petrov D S and Shlyapnikov G V 2001 Interatomic collisions in a tightly confined bose gas *Phys. Rev.* A **64** 012706
[16] Morse P M and Feshbach H 1953 *Methods of Theoretical Physics* (McGraw-Hill)
[17] Gradshteyn I S and Ryzhik I M 2014 *Table of Integrals, Series, and Products* 8th edn (New York, NY: Academic)
[18] Lapidus I R 1982 Quantum-mechanical scattering in two dimensions *Am. J. Phys.* **50** 45–7
[19] Adhikari S K 1986 Quantum scattering in two dimensions *Am. J. Phys.* **54** 362–7
[20] Adhikari S K 1986 Quantum scattering in two dimensions *Am. J. Phys.* **54** 362–7
[21] Adhikari S K, Gibson W G and Lim T K 1986 Effective-range theory in two dimensions *J. Chem. Phys.* **85** 5580–3
[22] Verhaar B J, van den Eijnde J P H W, Voermans M A J and Schaffrath M M J 1984 Scattering length and effective range in two dimensions: application to adsorbed hydrogen atoms *J. Phys. A: Math. Gen* **17** 595
[23] Verhaar B J, de Goey L P H and Vredenbregt E J D 1985 Scattering length and effective range for charged-particle scattering in a plane and in higher dimensions *Phys. Rev.* A **32** 1430
[24] Fernando Perez J and Coutinho F A B 1991 Schrödinger equation in two dimensions for a zero-range potential and a uniform magnetic field: An exactly solvable model *Am. J. Phys* **59** 52–4
[25] Barton G 1983 Rutherford scattering in two dimensions *Am. J. Phys.* **51** 420–2
[26] Tonks L 1936 The complete equation of state of one, two and three-dimensional gases of hard elastic spheres *Phys. Rev.* **50** 955
[27] Haller E, Gustavsson M, Mark M J, Danzl J G, Hart R, Pupillo G and Nägerl H-C 2009 Realization of an excited, strongly correlated quantum gas phase *Science* **325** 1224–7
[28] Haller E, Mark M J, Hart R, Danzl J G, Reichsöllner L, Melezhik V, Schmelcher P and Nägerl H-C 2010 Confinement-induced resonances in low-dimensional quantum systems *Phys. Rev. Lett.* **104** 153203
[29] Höhmann R, Kuhl U, Stöckmann H-J, Kaplan L and Heller E J 2010 Freak Waves in the Linear Regime: A Microwave Study *Physical Review Letters* **104** 093901

[30] Zürn G, Serwane F, Lompe T, Wenz A N, Ries M G, Bohn J E and Jochim S 2012 Fermionization of two distinguishable fermions *Phys. Rev. Lett.* **108** 075303
[31] Yu Liping and Zunger Alex 2012 Identification of Potential Photovoltaic Absorbers Based on First-Principles Spectroscopic Screening of Materials *Physical Review Letters* **108** 068701
[32] Petrov D S, Holzmann M and Shlyapnikov G V 2000 Bose-Einstein Condensation in Quasi-2D Trapped Gases *Phys. Rev. Lett.* **84** 2551
[33] Hadzibabic Z, Krüger P, Cheneau M, Battelier B and Dalibard J 2006 Berezinskii–Kosterlitz–Thouless crossover in a trapped atomic gas *Nature* **441** 1118–21
[34] Hazlett E L, Zhang Y, Stites R W and O'Hara K M 2012 Realization of a resonant Fermi gas with a large effective range *Phys. Rev. Lett.* **108** 045304
[35] Dyke P, Pollack S E and Hulet R G 2013 Finite-range corrections near a Feshbach resonance and their role in the Efimov effect *Phys. Rev. A* **88** 023625
[36] Roy S, Landini M, Trenkwalder A, Semeghini G, Spagnolli G, Simoni A, Fattori M, Inguscio M and Modugno G 2013 Test of the universality of the three-body Efimov parameter at narrow Feshbach resonances *Phys. Rev. Lett.* **111** 053202
[37] Blackley C L, Julienne P S and Hutson J M 2014 Effective-range approximations for resonant scattering of cold atoms *Phys. Rev. A* **89** 042701
[38] Partha G and Deb B 2016 A model study on a pair of trapped particles interacting with an arbitrary effective range *Phys. Scripta* **91** 085401
[39] Partha G, Rashkit A and Deb B 2015 A model study on atom--atom interactions with large scattering length in quasi-two dimensional traps *Ind. J. Phys* **89** 773–81

IOP Publishing

Low Energy Collision Physics
Applications to cold atoms
Bimalendu Deb

Chapter 8

Numerical methods

In this chapter, we present numerical methods for calculating scattering and bound states. Numerically computing scattering and bound states implies numerically solving the time-independent radial Schrödinger equation subject to two-sided boundary conditions at $r \to 0$ and $r \to \infty$. In order to write a program, one must first devise a strategy and select a suitable algorithm. The strategies for computing scattering and bound states are different. A general strategy is as follows:

1. Divide the space from $r = r_{\text{min}}$ to $r = r_{\text{max}}$ into a large number of discrete points, where r_{min} is a point in the regime $r \to 0$ and r_{max} is a point in the asymptotic regime $r \to \infty$ where the potential is zero or negligibly small. Let $r_0 = r_{\text{min}}$ and $r_{\text{max}} = r_N$. The nth point is denoted by r_n, which means that the entire space is divided into $N + 1$ discrete points. The spacing between two adjacent points is $h_n = r_{n+1} - r_n$. The spacing between all adjacent points may be equal or unequal. For equal spacing, the spacing is denoted by h.
2. Let the wave function $\psi(r)$ at the point $r = r_n$ be denoted by ψ_n. Thus, the wave function is also made discrete, having $N + 1$ points.
3. To calculate the scattering wave function, one must begin calculating the wave function point by point, starting from the boundary conditions at $r = r_0$. The code or subroutine that does this is called a forward propagator.
4. Since the Schrödinger equation is a second-order differential equation in space variables, one requires at least two boundary conditions at $r \to 0$ to start the forward propagator. These two boundary conditions may be (i) given values of $\psi(r_o)$ and $\psi(r_1)$ or (ii) given values of $\psi(r_o)$ and $\psi(r_o)'$. These given values are determined analytically.
5. The derivative of a function $f(x)$ is defined by $df(x)/dx = \lim\limits_{h \to 0}[f(x + h) - f(x)]/h$. Therefore, the discrete version of the derivative at point r_n is $f_n' = (f_{n+1} - f_n)/h$, provided h is sufficiently small. For $h > 0$, one can equivalently define another version of the discrete derivative, $f_n' = (f_n - f_{n-1})/h$. The fact that a derivative exists at $r = r_n$ implies that both

 © IOP Publishing Ltd 2025. All rights, including for text and data mining (TDM), artificial intelligence (AI) training, and similar technologies, are reserved.

derivatives should converge in the limit $h \to 0$. However, the convergence depends on the nature of the function and the step size h—if the function is smooth, then convergence is good for any arbitrarily small h. On the other hand, if the function is highly oscillatory, then the convergence is poor. It is therefore desirable to use an average of these two derivatives, that is $f'_n = [f_{n+1} - f_{n-1}]/(2h)$, and check the convergence of this form of the derivative against the variation in h. However, a more accurate derivative can be calculated using Newton's interpolation formula, as we discuss below.

6. A discrete version of the second-order derivative must be fixed. Since the Schrödinger equation provides the exact second-order derivative of the wave function, its discrete version at $r = r_n$ is equated with that at $r = r_n$ provided by the Schrödinger equation. Using Numerov's recursion formula, which we derive below, one can calculate the wave function at $r = r_n$ from the values of the wave function calculated at points $r_{n-1}, r_{n-2}, \ldots$.
7. The wave function calculated by the forward propagator must be normalized so that it matches the scattering boundary conditions at $r = r_{\max}$; thereby, the scattering matrices are deduced.
8. To calculate the bound states, the space must be made discrete, as in the case of computing the scattering wave function, but the number of discrete points may be much lower than that used in the case of the scattering wave function. Though the boundary conditions at $r = r_0$ are the same as in the scattering wave function, the boundary conditions for the bound states at $r = r_{\max}$ are $\psi(r_{\max}) \sim e^{-k_n r_{\max}} \simeq 0$, where $k_n = \sqrt{2\mu|E_n|}/hbar$, with E_n being the energy of the nth bound state and n being the number of nodes in the wave function.
9. If one wishes to calculate bound states by the propagator method as discussed above, then the forward propagator alone cannot do the job, because at large r, the propagator code has be propagated through the classically forbidden region ($E < V(r)$), where the code will find an exponentially diverging solution, which is wrong. Thus, the computation of a bound state requires a different strategy, which we will discuss below.

We first discuss the Numerov method [1]. Its improved version, described by Cooley [2], is also known as the Numerov–Cooley method or algorithm. We will describe how to apply this method to the calculation of both single- and multichannel scattering wave functions. We will also discuss the calculation of Green's functions and bound states using this method. A detailed discussion of the Numerov method can be found in [3]. There are other improved versions of the Numerov–Cooley algorithm [4–6], which are reviewed in [7].

8.1 Numerov–Cooley algorithm

Here, we follow the method of derivation given by Smith [3].

To derive the Numerov–Cooley algorithm, let us first consider Newton's formula for the interpolation of a function $\psi(r)$. The main idea is that, over a given interval,

$\psi(r)$ can be replaced by a suitable polynomial. Let $\phi(r)$ be a polynomial which can be written as

$$\begin{aligned}\phi(r) = a_0 + a_1(r - r_n) + a_2(r - r_n)(r - r_{n-1}) + a_3(r - r_n)(r - r_{n-1})(r - r_{n-2}) + \cdots \\ + a_{n-1}(r - r_n)(r - r_{n-1}) \cdots (r - r_2),\end{aligned} \tag{8.1}$$

where r_i are the points at which $\psi(r_i)$ are tabulated. The coefficients a_j are determined as follows:

$$\phi(r_i) = \psi_i,\ i = 1, 2, 3, \ldots, n;$$

hence

$$\begin{aligned}\phi(r_n) &= a_0 = \psi_n; \\ \phi(r_{n-1}) &= \psi_n + a_1(r_{n-1} - r_n) = \psi_{n-1}\end{aligned}$$

or

$$a_1 = \frac{\psi_{n-1} - \psi_n}{r_{n-1} - r_n} \equiv \frac{\Delta_1\psi_n}{h}, \tag{8.2}$$

where the notation $\Delta_1\psi$ denotes the first differences and h is the step size. Let the step size be equal throughout the range; it is defined as $r_i - r_{n-1} = h$. $\Delta_1\psi$ is called the first difference of the function $\psi(r)$. At $r = r_{n-2}$, equation (8.1) provides

$$\phi(r_{n-2}) = \psi_n + \frac{\Delta_1\psi_n}{h}(-2h) + a_2(-2h)(-h) = \psi_{n-2}, \tag{8.3}$$

yielding

$$\begin{aligned}a_2 &= \frac{\psi_{n-2} - \psi_n + 2\Delta_1\psi_n}{2h^2} \\ &= \frac{\psi_{n-2} - 2\psi_{n-1} + \psi_n}{2h^2} = \frac{\Delta_2\psi_n}{2h^2},\end{aligned} \tag{8.4}$$

where $\Delta_2\psi$ is the second difference of ψ. In this way, one can write

$$a_n = \frac{1}{n!h^n}\Delta_n\psi_n. \tag{8.5}$$

Substituting the values of a_j into equation (8.1) yields

$$\begin{aligned}\phi(r) = \psi_n + \frac{\Delta_1\psi_n}{h}(r - r_n) + \frac{\Delta_2\psi_n}{2h^2}(r - r_n)(r - r_{n-1}) + \cdots \\ + \frac{\Delta_{n-1}\psi_n}{(n-1)!h^{n-1}}(r - r_n)(r - r_{n-1}) \cdots (r - r_2),\end{aligned} \tag{8.6}$$

which is Newton's formula for backward interpolation. By making the change of variable

$$u \equiv \frac{r - r_n}{h} \text{ or } r = r_n + hu, \tag{8.7}$$

we obtain

$$\begin{aligned} \frac{r - r_{n-1}}{h} &= \frac{r - (r_n - h)}{h} = u + 1, \\ \frac{r - r_{n-2}}{h} &= \frac{r - (r_n - 2h)}{h} = u + 2. \end{aligned} \tag{8.8}$$

Substituting these into equation (8.6), one can write

$$\begin{aligned} \phi(u) = \psi_n + \Delta_1\psi_n u + \frac{\Delta_2\psi_n}{2}u(u+1) + \frac{\Delta_3\psi_n}{2}u(u+1)(u+2) + \cdots \\ + \frac{\Delta_{n-1}\psi_n}{(n-1)!}u(u+1)(u+2)\cdots(u+n-2). \end{aligned} \tag{8.9}$$

Let $\psi(r)$ satisfy the second-order ordinary differential equation of the form

$$\frac{d^2\psi}{dr^2} = \psi'' = f(r, \psi), \tag{8.10}$$

subject to two-sided boundary conditions. We approximate $\psi''(r)$ using the polynomial of equation (8.9), as follows:

$$\psi'' = \psi_n'' + u\Delta_1\psi_n'' + \frac{u(u+1)}{2}\Delta_2\psi_n'' + \frac{u(u+1)(u+2)}{6}\Delta_3\psi_n'' + \dots . \tag{8.11}$$

Since r_n and h are fixed, $dr = hdu$. Integrating equation (8.11) with respect to r, one obtains

$$\int \psi'' dr = h \int du \left[\psi_n'' + u\Delta_1\psi_n'' + \frac{u(u+1)}{2}\Delta_2\psi_n'' + \cdots \right], \tag{8.12}$$

which yields

$$\psi' = C_1 + h\left[u\psi_n'' + \frac{u^2}{2}\Delta_1\psi_n'' + \frac{1}{2}\left(\frac{u^2}{2} + \frac{u^3}{3}\right)\Delta_2\psi_n'' + \cdots \right], \tag{8.13}$$

where C_1 is an integration constant. For $r = r_n$, $u = 0$, and $\psi' \to \psi' = C_1$,

$$\psi(r)' = \psi_n' + h\left[u\psi_n'' + \frac{u^2}{2}\Delta_1\psi_n'' + \left(\frac{u^2}{4} + \frac{u^3}{6}\right)\Delta_2\psi_n'' + \cdots \right]. \tag{8.14}$$

Upon integrating this equation a second time with respect to r, we obtain

$$\psi = hu\psi_n' + h^2\left[\frac{u^2}{2}\psi_n'' + \frac{u^3}{6}\Delta_1\psi_n'' + \left(\frac{u^3}{12} + \frac{u^4}{24}\right)\Delta_2\psi_n'' + \cdots \right] + C_2. \tag{8.15}$$

The second integration constant C_2 can be evaluated by noting that at $r = r_n$, $u = 0$; thus, $\psi \to \psi_n = C_2$. Therefore, we have

$$\begin{aligned}\psi(u) \;=\; & \psi_n + hu\psi'_n + h^2\left[\frac{u^2}{2}\psi''_n + \frac{u^3}{6}\Delta_1\psi''_n + \left(\frac{u^3}{12} + \frac{u^4}{24}\right)\Delta_2\psi''_n\right.\\ & + \frac{1}{6}\left(\frac{u^3}{3} + \frac{u^4}{4} + \frac{u^5}{20}\right)\Delta_3\psi''_n \\ & \left. + \frac{1}{24}\left(u^3 + \frac{11u^4}{12} + \frac{3u^5}{10} + \frac{u^6}{30}\right)\Delta_4\psi''_n + \cdots\right].\end{aligned} \tag{8.16}$$

Setting $u = -1, -2$, one obtains the values ψ_{n-1} and ψ_{n-2}, respectively, as given by

$$\psi_{n-1} = \psi_n - h\psi'_n + h^2\left[\frac{1}{2}\psi''_n - \frac{1}{6}\Delta_1\psi''_n - \frac{1}{24}\Delta_2\psi''_n - \frac{1}{45}\Delta_3\psi''_n - \frac{7}{480}\Delta_4\psi''_n + \cdots\right] \tag{8.17}$$

$$\psi_{n-2} = \psi_n - 2h\psi'_n + h^2\left[2\psi''_n - \frac{4}{3}\Delta_1\psi''_n - \frac{2}{45}\Delta_3\psi''_n - \frac{1}{30}\Delta_4\psi''_n + \cdots\right]. \tag{8.18}$$

Upon multiplying equation (8.17) by two, subtracting the result from equation (8.18), and truncating the higher-order terms, we obtain

$$\psi_n - 2\psi_{n-1} + \psi_{n-2} \approx h^2\left[\psi''_n - \Delta_1\psi''_n - \frac{1}{12}\Delta_2\psi''_n\right], \tag{8.19}$$

which can also be written as

$$\psi_n - 2\psi_{n-1} + \psi_{n-2} \approx h^2\left[\psi''_{n-1} - \frac{1}{12}\Delta_2\psi''_n\right]. \tag{8.20}$$

Now, let us consider second-order ordinary differential equations of the general form

$$\psi''(r) = T(r)\psi(r) + G(r), \tag{8.21}$$

whose discrete version can be readily written as

$$\psi'' = T(r_n)\psi_n + G(r_n) \equiv T_n\psi_n + G_n, \tag{8.22}$$

where T_n and G_n are the two known functions $T(r)$ and $G(r)$, respectively, at the nth point. Upon substituting this equation into equation (8.20), we obtain

$$\psi_n - 2\psi_{n-1} + \psi_{n-2} \approx h^2\left[T_{n-1}\psi_{n-1} + G_{n-1} - \frac{1}{12}\Delta_2(T_n\psi_n + G_n)\right]. \tag{8.23}$$

After some minor algebra, one finally obtains the Numerov formula:

$$\left[1-\frac{1}{12}h^2T_{n+1}\right]\psi_{n+1}=2\left[1-\frac{1}{12}h^2T_n\right]\psi_n-\left[1-\frac{1}{12}h^2T_{n-1}\right]\psi_{n-1} \\ +h^2\left[T_n\psi_n+G_n+\frac{1}{12}\Delta_2G_{n+1}\right]. \tag{8.24}$$

8.2 Scattering-state calculations

Let us first consider solving the single-channel partial-wave time-independent Schrödinger equation with the spherically symmetric potential $V(r)$:

$$\psi''=\left[\frac{\ell(\ell+1)}{r^2}+U(r)-k^2\right]\psi, \tag{8.25}$$

where $U(r)=2\mu V(r)/\hbar^2$. Comparing this equation with equation (8.21), we see that

$$T(r)=\frac{\ell(\ell+1)}{r^2}+U(r)-k^2 \tag{8.26}$$

and $G(r)=0$. Thus, the Numerov formula in equation (8.24) reduces to

$$\psi_{n+1}=F_{n+1}^{-1}[2F_n\psi_n-F_{n-1}\psi_{n-1}+h^2T_n\psi_n], \tag{8.27}$$

where $F_n=1-(1/12)h^2T_n$. Therefore, given values of ψ_{n-1} and ψ_n, one can calculate the value of ψ_{n+1}. To propagate the code in the forward direction, one must specify the values of ψ_0 and ψ_1. Assuming that the potential $V(r)$ is less singular than $1/r^2$ as $r\to0$, one can obtain two linearly independent analytical solutions in the vicinity of $r=0$. The regular solution is $\psi(r)\sim r^{\ell+1}$. Since the initial normalization of the wave functions is arbitrary, one can set any arbitrary value ψ_0 (that is, the value of $\psi(r)$ at $r=r_{\min}$) but fix the value ψ_1 $(\psi(r=r_{\min}+h))$ at $\psi_1=[(r_{\min}+h)/r_{\min}]^{\ell+1}\,\psi_0$. In this way, one can calculate the wave function point by point up to $r=r_{\max}$, which is a point in the asymptotic regime. An asymptotic separation can be chosen by requiring that $|U(r)|\ll k^2$ for $\ell=0$ and $|U(r)|\ll\ell(\ell+1)/r^2$ for $\ell\neq0$, so that in that region, the wave function can be expressed as a superposition of the two linearly independent spherical Bessel functions $j_\ell(kr)$ and $n_\ell(kr)$ as given by

$$\psi(r)=A_\ell j_\ell(r)-B_\ell n_\ell(r). \tag{8.28}$$

Now, recall the asymptotic behavior ($r\to\infty$) of the spherical Bessel functions discussed in chapter 2. One can readily identify the ratio B_ℓ/A_ℓ as the matrix element of the **K**-matrix or **R**-matrix given by

$$\tan\delta_\ell=\frac{B_\ell}{A_\ell}. \tag{8.29}$$

To evaluate A_ℓ and B_ℓ, one must equate the numerically calculated asymptotic wave functions at two space points with the analytical forms given above. Another way to find the phase shift is to equate the logarithmic derivatives of the numerical and analytical wave functions. To find the s-wave phase shift, one must match the numerically calculated wave function with the asymptotic analytical solution at a large separation that may extend to several thousands of a_0. To calculate $a_s = -\lim_{k\to 0} \tan \delta_0/k$, one must check the constancy of $-\tan \delta_0/k$ as a function of k in the low-k regime; by extrapolating the function to $k = 0$, one can determine the scattering length. For $\ell \neq 0$, the asymptotic boundary conditions can be imposed at a much shorter separation.

The Numerov formula can be generalized to calculate a multichannel scattering wave function by considering the wave function and all the Hamiltonian terms in matrix form, as discussed in chapter 3. If all channels are open, then it is straightforward to calculate the multichannel **K** matrix as introduced in chapter 3 by propagating the multichannel Numerov code to the asymptotic regime and matching the numerical multichannel wave function with the analytical form of the asymptotic multichannel wave function. Each column vector of the matrix wave function represents a multicomponent wave function—each component corresponds to a channel state. The multicomponent wave functions form a complete set of orthonormal functions and are thus linearly independent. However, while solving them numerically by propagating the code through the classically forbidden region, the different wave vectors may lose linear independence. This can be remedied by making the vectors orthogonal from time to time using the Gram–Schmidt orthogonalization procedure.

If some of the channels are closed, one can again propagate the code from $r_{\rm min}$ in matrix form. Let there be N_o open channels and N_c closed channels, with $N = N_o + N_c$ being the total number of channels. As discussed in section 3.2.2, in the asymptotic regime, one must eliminate the closed-channel components by performing a transformation on the numerically calculated wave matrix, leaving a physical matrix with dimensions of $N \times N_o$. The scattering is described by the $N_o \times N_o$ submatrix of this physical matrix, and so the multichannel scattering boundary conditions are imposed on this submatrix. Practically, propagating the code to a large separation in the presence of closed channels is problematic because closed-channel components of the wave matrix may diverge. One remedy for this problem is to apply the transformation as mentioned in section 3.2.2, eliminating the closed-channel part not at an asymptotic separation but at a relatively shorter separation r_t, where the magnitude of the potential of the lowest-threshold closed channel is quite large compared to the collision energy. The implicit assumption in this procedure is that the closed-channel components reach their asymptotic forms at a shorter separation than the open-channel ones. In this way, one is then left with the task of propagating all open-channel $N_o \times N_o$ wave matrices from r_t to the asymptotic regime. Upon making the transformation at $r = r_t$, one is left with an $N \times N_o$ reduced matrix. Its $N_c \times N_o$ submatrix then represents the closed-channel wave matrix whose elements should decay exponentially for $r \geqslant r_t$.

8.3 Green's function calculation

Once a partial-wave regular solution of a single-channel potential is calculated using the Numerov method, as discussed above, an irregular solution can be found by numerically solving the Wronskian equation. Thus, a partial-wave Green's function can be constructed using regular and irregular solutions, as discussed in chapter 2. A Wronskian equation is a first-order differential equation and can be solved numerically utilizing the numerical regular solution and its derivative. Let us recall the Wronskian equation

$$\psi^{(1)}_{k\ell}(r)\frac{d}{dr}\psi^{(2)}_{k\ell} - \psi^{(2)}_{k\ell}(r)\frac{d}{dr}\psi^{(1)}_{k\ell} = C, \tag{8.30}$$

where C is a constant that depends on the normalization of the functions. The superscript '1' denotes the regular solution, and '2' denotes the irregular solution. Since the irregular solution is defined by its boundary condition at $r \to \infty$, it is convenient to solve the above equation numerically, point by point, starting from the asymptotically large r. At the node of the regular solution, the irregular function is given by

$$\psi^{(2)}_{k\ell}(r) = -C\left[\frac{d}{dr}\psi^{(1)}_{k\ell}\right]^{-1}. \tag{8.31}$$

This means that once C is fixed, the irregular solution at the node of the regular solution is uniquely determined.

One may observe that while solving for the single-channel regular wave function using the Numerov method, as described above, one may easily calculate the derivative of the wave function in the intermediate steps. The Numerov method calculates the wave function at the $n + 1$th grid point, given the wave functions at the previous two points. Thus, the derivative at the nth point can be recorded as $\psi' = (\psi_{n+1} - \psi_{n-1})/2h$. Using these data for the derivative of the regular wave function, one can readily calculate the irregular wave function by solving the Wronskian equation point by point, starting from the boundary condition at the asymptotically large r or from any nodal point of the regular wave function.

8.4 Bound state calculation by the Numerov method

The bound states of a single-channel potential can be calculated using the 'renormalized Numerov' method [6]. Let us consider the radial Schrödinger equation, written as

$$[d^2/dr^2 + Q(r)]\phi(r) = 0, \tag{8.32}$$

where

$$Q(r) = (2\mu/\hbar)^2[E - V(r)] + l(l+1)/r^2. \tag{8.33}$$

The potential $V(r)$ is assumed to be capable of supporting bound states that satisfy the boundary conditions

$$\phi(r) \to 0, \text{ at} r \to 0, \quad r \to \infty. \tag{8.34}$$

We consider that the eigenvalues are arranged in ascending order. The function $\phi_n(r)$ corresponding to the nth eigenvalue, E_n, has n nodes.

The method used to calculate bound states is as follows. The r coordinate is divided into $N + 1$ equally spaced grid points from r_0 to r_N. The endpoints r_0 and r_N are located in the classically forbidden region. A trial energy E is chosen. The solution is obtained in two stages: first, the Numerov code is propagated inward, starting from r_N; then, an outward propagation is carried out, starting from r_0. Both propagated solutions meet at a common matching point $r_M (r_0 < r_M < r_N)$, located within the classically allowed region. Since the outward solution approaches r_M from the left, it is designated ϕ_l(r), while the inward solution, which approaches from the right, is designated $\phi_r(r)$. The points r_0 and r_N are located in classically forbidden regions on either side of the potential well and are far enough from classical turning points that the calculated eigenvalues are insensitive to the initial values of the wave function at these points.

The inward solution is first obtained up to the grid point r_M where the relation

$$\phi_r(r_M) \geqslant 0 \tag{8.35}$$

first occurs. The outward propagation from r_0 to r_M is then carried out. The number of nodes in the outward solution is counted. If the number of nodes is greater than n, the trial energy is too high; if less than n, it is too low. If the count is equal to n, then a correction to E is determined by the difference between the slopes of the two curves at r_M, and the process is repeated until the two curves meet with the same derivative. The correction is defined as $D(E)$. If $D(E)$ is negative, the trial energy is below the desired eigenvalue; if it is positive, then the trial energy is above the desired value. If the trial energy is low, it is denoted by E_L, and if it is high, it is denoted by E_H. A correction $\Delta E$is then added to the trial energy, E, to give an improved estimate of the eigenvalue. This procedure is iterated and usually converges rapidly to the eigenvalue E_n nearest the value of the initial guess.

A 'bisection procedure' can be used initially to isolate a single eigenvalue with a specified node. Let us assume two energies E_L and E_H, such that the desired eigenvalue lies between them: $E_L < E_n < E_H$. We set the energy parameter in the Schrödinger equation equal to $E = 0.5(E_L + E_H)$, carry out the inward and outward propagations, and take the node number, n_E. If $n_E = n$, then we calculate $D(E)$; if $D(E)$ is negative, we set $E_L = E$; if $D(E)$ is positive, we set $E_H = E$. This procedure converges linearly and is iterated until E is close enough to E_n.

Another method used to calculate bound states is the discrete variable representation (DVR).

8.5 Discrete variable representation method

Another method of solving time-independent Schrödinger equation to evaluate bound states is the discrete value representation (DVR) [8, 9]. This method is based on the expansion of wave function in a complete set of orthonormal basis functions and a quadrature rule in a discrete set of space points. The choice of basis function depends on the nature of the potential. Usually, for most of the problems, Gaussian quadrature basis is chosen. The problem of solving for the bound states then reduces to diagonalization of a matrix and the negative eigenvalues correspond to the bound state energies. DVR method is more suited for calculating bound states unlike Numerov–Cooley method which can be used for solving both continuum and bound states. However, DVR method can yield all the bound states in one step of diagonalization while Numerov–Cooley solves for bound states one by one. Here we do not discuss DVR method any further. For further details about this method, we refer to the articles [10].

8.6 Quantum defect theoretic method

Quantum defect theory (QDT) was formulated by Seaton [11] to explain the spectra of atomic Rydberg states. Its multichannel version, namely, multichannel quantum defect theory (MQDT) has found applications in atomic and molecular collisions [12–15]. MQDT method is particularly suited for calculating resonances at near-zero energies and bound states close to the threshold of a potential.

MQDT starts by propagating outward the multichannel wave function or its log-derivative in a matrix form from a short separation, and the propagation is stopped at an intermediate and usually classically allowed separation known as matching point r_m. Then, this numerically obtained solution at r_m is matched with the analytical solutions of the long-range form of the potential matrix. Implicit assumptions or conditions in this method are that: (i) for separations $r > r_m$, the off-diagonal elements of the potential matrix (which are basically inter-channel couplings) become negligible, rendering the potential matrix essentially into a diagonal form; and (ii) the diagonal elements of the potential matrix for $r > r_m$ should be expressible, at least in leading order, in analytical forms that should also admit analytical solutions. The latter condition is the most stringent one, restricting the application of MQDT methods to certain specific classes of long-range potentials, such as Coulomb ($1/r$), dipolar ($1/r^2$), van der Waals ($1/r^6$), charge-neutral ($1/r^4$), etc type potentials. However, for long-range potentials which do not admit analytical solutions, one can always evaluate a numerical solution of such potentials by propagating wave function backward from an asymptotic separation and calculate another solution which is linearly independent to the former one by solving the corresponding Wronskian equation [16].

In the context of cold atomic collisions, a simple MQDT formalism was developed by Burke *et al.* [14] using Milne phase-amplitude solutions. The analytical solutions of $-C_6/r^6$ [17] and $-C_4/r^4$ potentials have been used to develop MQDT methods [18, 19] to treat ultracold atomic collisions.

Exercise

- Derive Numerov-Cooley formula for calculating scattering wave functions.
- Discuss how Numerov-Cooeley algorithm can be used to calculate a bound state.

References

[1] Numerov B 1933 *Publ. Observatoire Central Astrophys. Russ.* **2** 188
[2] Cooley J W 1961 An improved eigenvalue corrector formula for solving the Schrödinger equation for central fields *Math. Comput.* **15** 363–74
[3] Smith K 1972 *The Calculation of Atomic Collision Processes* (New York: Wiley-Interscience)
[4] Gordon R G 1969 New method for construction wavefunctions for bound states and scattering *J. Chem. Phys.* **51** 14–25
[5] Johnson B R 1973 The multichannel log-derivative method for scattering calculations *J. Comput. Phys.* **13** 445–9
[6] Johnson B R 1977 New numerical methods applied to solving the one-dimensional eigenvalue problem *J. Comput. Phys.* **67** 4086–93
[7] Thomas L D *et al* 1981 Comparison of numerical methods for solving the second-order differential equations of molecular scattering theory *J. Comput. Phys.* **41** 407–26
[8] Lill J V, Parker G A and Light J C 1982 Discrete variable representations and sudden models in quantum scattering theory *Chemical Physics Letters* **89** 483–9
[9] Bacić Z and Light J C 1986 *J. Chem. Phys.* **85** 4594–604
[10] Balakrishnan N, Kalyanaraman C and Sathyamurthy N 1997 Time-dependent quantum mechanical approach to reactive scattering and related processes *Physics reports* **280** 79–144
[11] Seaton M J 1966 *Proc. Phys. Soc.* **88** 801
[12] Mies F H and Julienne P S 1984 A multichannel quantum defect analysis of two-state couplings in diatomic molecules *J. Chem. Phys.* **80** 2526–2536
[13] Julienne P S and Mies F H 1989 Collisions of ultracold trapped atoms *Journal of the Optical Society of America B* **6** 2257–2269
[14] Burke J P, Greene C H, Bohn J L and John L 1998 Multichannel cold collisions: Simple dependences on energy and magnetic field *Phys. Rev. Lett.* **81** 3355
[15] Gao B, Tiesinga E, Williams C J and Julienne P S 2005 Multichannel quantum-defect theory for slow atomic collisions *Phys. Rev. A* **72** 042719
[16] Sardar D, Rakshit A, Naskar S and Deb B 2023 arXiv:2004.09091
[17] Gao B 1998 Solutions of the schrödinger equation for an attractive $1/r^6$ potential *Phys. Rev. A* **58** 1728–1734
[18] Gao B 1998 *Phys. Rev. A.* **58** 4222
[19] Gao B 2010 Universal properties in ultracold ion-atom interactions *Phys. Rev. Lett.* **104** 213201

Appendix A

Dirac delta functions and their properties

Dirac delta functions and their properties find wide applications throughout the entire theory of scattering. They are introduced in physics to handle the integrals involving continuum wave functions such as plane waves, spherical waves, etc. A delta function $\delta(x - x')$ in 1D is a symmetric function of x and zero everywhere except $x \to x'$ such that

$$\int_a^b \delta(x - x')dx = \int_{-\infty}^{\infty} \delta(x - x')dx = 1 \tag{A.1}$$

where $a < (x - x') < b$ with a and b are two finite real numbers. The 2D and 3D delta functions in Cartesian coordinates are the products of 1D delta functions as given by

$$\delta^{(2)}(\rho - \rho') = \delta(x - x')\delta(y - y') \tag{A.2}$$

$$\delta^{(3)}(\mathbf{r} - \mathbf{r}') = \delta(x - x')\delta(y - y')\delta(z - z') \tag{A.3}$$

Delta functions are closely related to the normalization of plane waves. Consider a 3D plane wave $\psi_{\mathbf{k}}(\mathbf{x}) = e^{i\mathbf{k}\cdot\mathbf{r}}$ where $\mathbf{p} = \hbar\mathbf{k}$ is the momentum. Here $\hbar = h/2\pi$ and h are the Planck constant. The probability amplitude that the quantum states represented by this plane wave will occupy an element of phase-space volume $d^3\mathbf{p}d^3\mathbf{r}$ around a phase-space point $(\mathbf{r}', \mathbf{p})$ is given by

$$\frac{1}{h^3}e^{\mathbf{k}\cdot(\mathbf{r}-\mathbf{r}')}d^3\mathbf{p}d^3\mathbf{r} \tag{A.4}$$

The total occupancy of a single-particle single-quantum state over the entire phase space must be equal to unity. Thus we have

$$\int\int\int d^3\mathbf{r}\left[\frac{1}{(2\pi)^3}\int\int\int d^3\mathbf{k}e^{i\mathbf{k}\cdot(\mathbf{r}-\mathbf{r}')}\right] = 1 \tag{A.5}$$

doi:10.1088/978-0-7503-3959-9ch9

© IOP Publishing Ltd 2025. All rights, including for text and data mining (TDM), artificial intelligence (AI) training, and similar technologies, are reserved.

which implies that

$$\frac{1}{(2\pi)^3}\int\int\int d^3\mathbf{k}e^{i\mathbf{k}\cdot(\mathbf{r}-\mathbf{r}')} = \delta^{(3)}(\mathbf{r} - \mathbf{r}'). \tag{A.6}$$

So, a delta function in 1D can be defined by

$$\delta(x - x') = \frac{1}{(2\pi)}\int_{-\infty}^{\infty}\frac{1}{(2\pi)}dke^{ik(x-x')} \tag{A.7}$$

Since a delta function is symmetric, one can write $\delta(x - x') = \delta(x' - x)$. So, $\delta(c\{x - x'\}) = \delta(c\{x' - x\})$ where c is a constant, implying that $\delta(c\{x - x'\})$ will depend only on $|c|$. In general, delta function $\delta(f(x))$ of a function $f(x)$ of x can be written in terms of $\delta(x)$ as given by

$$\delta(f(x)) = \frac{1}{|D(f(x))|}\delta(x) \tag{A.8}$$

where

$$D(f(x)) = \frac{d}{dx}f(x) \tag{A.9}$$

Delta function as defined by equation (A.7) can be expressed in alternative forms One can write this equation in the form

$$\begin{aligned}\delta(x - x') &= \lim_{q\to\infty}\int_{-q}^{q}\frac{1}{(2\pi)}dke^{ik(x-x')}\\ ' &= \frac{1}{\pi}\lim_{q\to\infty}\frac{\sin[q(x - x')]}{x - x'}\end{aligned} \tag{A.10}$$

Note that here the parameter q approaches infinity from the positive side. One can write equation (A.7) in another form

$$\begin{aligned}\delta(x - x') &= \frac{1}{(2\pi)}\left[\int_{-\infty}^{0} dke^{ik(x-x')} + \int_{0}^{\infty} dke^{ik(x-x')}\right]\\ &= \frac{1}{(2\pi)}\lim_{\epsilon\to 0}\left[\int_{0}^{\infty} dke^{-ik(x-x'-i\epsilon)} + \int_{0}^{\infty} dke^{ik(x-x'+i\epsilon)}\right]\\ &= \frac{1}{(\pi)}\lim_{\epsilon\to 0}\frac{\epsilon}{(x - x')^2 + \epsilon^2}\end{aligned} \tag{A.11}$$

Note that here the parameter approaches zero from positive side.

By expanding the plane waves in terms of spherical waves as discussed in chapter 2, one can express the 3D delta function as defined by equation (A.6) in spherical polar coordinates (r, θ, ϕ). The result is

$$\delta^{(3)}(\mathbf{r} - \mathbf{r}') = \frac{1}{r^2}\delta(r - r')\frac{1}{\sin\theta}\delta(\theta - \theta')\delta(\phi - \phi') \tag{A.12}$$

The angular part of the 3D delta function can be expressed in terms of spherical harmonics in the form

$$\frac{1}{\sin\theta}\delta(\theta - \theta')\delta(\phi - \phi') = \sum_{\ell=0}^{\infty}\sum_{m_\ell=-\ell}^{\ell} Y_{\ell m_\ell}^{*}(\hat{r}')\, Y_{\ell m_\ell}(\hat{r}) \tag{A.13}$$

and the radial part is given by

$$\frac{1}{r^2}\delta(r - r') = \frac{2}{\pi}\int k^2 dk j_\ell(r')j_\ell(r) \tag{A.14}$$

Similarly, 2D delta function can be expressed in the cylindrical coordinates in the following form

$$\delta^{(2)}(\rho - \rho') = \frac{1}{\rho}\delta(\rho - \rho')\delta(\phi - \phi') = \frac{1}{2\pi\rho}\delta(\rho - \rho')\sum_m e^{im(\phi-\phi')} \tag{A.15}$$

Apart from the three forms of the delta functions given by equations (A.7), (A.10) and (A.11), there exist some other functions which reduce to delta function in some conditions when the parameter ϵ approaches zero from the positive side. These are

$$\frac{1}{2\epsilon}\exp[-|x|\epsilon] \tag{A.16}$$

$$\frac{1}{\epsilon\sqrt{\pi}}e^{-x^2\epsilon^2} \tag{A.17}$$

$$\frac{\epsilon}{\pi}\frac{\sin^2[x/\epsilon]}{x^2} \tag{A.18}$$

Appendix B

Green's function of a free particle

Here, we derive the Green's function $G_0^+(\mathbf{r}, \mathbf{r}')$ of a free particle in 3D. Its expression is given in equation (2.7). A convenient way to derive it is to use the complete momentum basis states of a free particle and then carry out complex integration over the momentum variable, as we discuss below. Let us recall the definition of this Green's function:

$$G_0^{\pm}(\mathbf{r}, \mathbf{r}') = \langle \mathbf{r} \left| \frac{1}{E - \hat{H} \pm i\varepsilon} \right| \mathbf{r}' \rangle. \tag{B.1}$$

Note that here, $\varepsilon \to 0^+$, and the Hamiltonian $\hat{H} = -\hbar^2/(2\mu)\nabla^2$. Now, using the identity operator $\mathbf{1} = \int d^3\mathbf{k} |\mathbf{k}\rangle\langle\mathbf{k}|$, one can express equation (B.1) in the following form:

$$G_0^{\pm}(\mathbf{r}, \mathbf{r}') = \frac{2\mu}{\hbar^2} \int d^3\mathbf{k}' \frac{\phi_{\mathbf{k}'}(\mathbf{r})\phi^*_{\mathbf{k}'}(\mathbf{r}')}{k^2 - k'^2 \pm i\varepsilon'}, \tag{B.2}$$

where $\varepsilon' = 2\mu\varepsilon/\hbar^2$. Now, replacing the wave functions $\phi_{\mathbf{k}}(\mathbf{r})$ with plane waves, as given by equation (2.6), we have

$$\begin{aligned} G_0^{\pm}(\mathbf{r}, \mathbf{r}') &= \frac{2\mu}{(2\pi)^3\hbar^2} \int d^3\mathbf{k}' \frac{\exp\left[i\mathbf{k}' \cdot (\mathbf{r} - \mathbf{r}')\right]}{k^2 - k'^2 \pm i\varepsilon'} \\ &= \frac{2\mu}{(2\pi)^3\hbar^2} \int_0^{\infty} k'^2 dk' \int_0^{\pi} \sin\theta d\theta \int_0^{2\pi} d\phi \frac{\exp\left[ik'|\mathbf{r} - \mathbf{r}'|\cos\theta\right]}{k^2 - k'^2 \pm i\varepsilon'}, \end{aligned} \tag{B.3}$$

where θ is the angle between the vector $\mathbf{k}'$ and $\mathbf{r} - \mathbf{r}'$. Integrating over θ and ϕ, we obtain

$$G_0^{\pm}(\mathbf{r}, \mathbf{r}') = \frac{4\mu}{(2\pi)^2\hbar^2} \int_0^{\infty} k' dk' \frac{1}{k^2 - k'^2 \pm i\varepsilon'} \frac{\sin[k'|\mathbf{r} - \mathbf{r}'|]}{|\mathbf{r} - \mathbf{r}'|}. \tag{B.4}$$

Writing $\sin[k'|\mathbf{r} - \mathbf{r}'|] = (1/2i)[\exp(ik'|\mathbf{r} - \mathbf{r}'|) - \exp(-ik'|\mathbf{r} - \mathbf{r}'|)]$, the above expression can be readily expressed in the following form:

doi:10.1088/978-0-7503-3959-9ch10 © IOP Publishing Ltd 2025. All rights, including for text and data mining (TDM), artificial intelligence (AI) training, and similar technologies, are reserved.

$$G_0^{\pm}(\mathbf{r}, \mathbf{r}') = \frac{2\mu}{(2\pi\hbar)^2 i} \int_{-\infty}^{\infty} k' dk' \frac{1}{k^2 - k'^2 \pm i\varepsilon'} \frac{\exp[ik'|\mathbf{r} - \mathbf{r}'|]}{|\mathbf{r} - \mathbf{r}'|}. \tag{B.5}$$

We now write

$$\frac{1}{k^2 - k'^2 \pm \varepsilon'} = -\frac{1}{2k}\left[\frac{1}{k' - (k \pm i\delta)} - \frac{1}{k' + (k \pm i\delta)}\right], \tag{B.6}$$

where δ is a small positive number given by $\sqrt{k^2 \pm i\varepsilon'} \simeq k \pm i\delta$. The contour of integration can be closed in the upper half-plane. The result for G_0^+ is given by

$$G_0^+(\mathbf{r}, \mathbf{r}') = -\frac{2\mu}{4\pi\hbar^2} \frac{\exp[ik|\mathbf{r} - \mathbf{r}'|]}{|\mathbf{r} - \mathbf{r}'|}, \tag{B.7}$$

which is the same as in equation (2.7).

Appendix C

Expansion of the full 3D Green's function in terms of partial-wave Green's functions

Consider a general Green's function $G(\mathbf{r}, \mathbf{r}')$ of the Hamiltonian $\hat{H}$ given by

$$G(\mathbf{r}, \mathbf{r}') = \langle \mathbf{r} \left| \frac{1}{E - \hat{H}} \right| \mathbf{r}' \rangle. \tag{C.1}$$

Using the complete basis set (including the bound states that the Hamiltonian can support) as embodied in the identity operator of equation (2.5), we obtain

$$G^{+}(\mathbf{r}, \mathbf{r}') = \frac{2\mu}{\hbar^2}\left[\sum_n \frac{\psi_n(\mathbf{r})\psi_n(\mathbf{r}')}{k^2 - k_n^2} + \int d^3\mathbf{k}' \frac{\psi_{\mathbf{k}'}(\mathbf{r})\psi^*_{\mathbf{k}'}(\mathbf{r}')}{k^2 - k'^2 + i\varepsilon'}\right], \tag{C.2}$$

where k_n is related to the binding energy E_n of the nth bound state by $E_n = \hbar^2 k_n^2/(2\mu)$. Note that since E_n is negative, k_n is a purely imaginary number. Using the partial-wave expansion of $\psi_{\mathbf{k}'}(\mathbf{r})$ given by equation (2.36), we have

$$\begin{aligned}\psi_{\mathbf{k}'}(\mathbf{r})\psi_{\mathbf{k}'}(\mathbf{r}') = & \frac{1}{(2\pi)^3}(4\pi)^2 \sum_{\ell,\ell'} i^{\ell}(-i)^{\ell'} \sum_{m_\ell, m_{\ell'}} \psi_\ell(k'r)\psi^*_{\ell'}(k'r') \\ & \times Y_{\ell, m_\ell}(\hat{r}) Y^*_{\ell', m_{\ell'}}(\hat{r}') Y^*_{\ell, m_\ell}(\theta_{k'}, \phi_{k'}) Y_{\ell', m_{\ell'}}(\theta_{k'}, \phi_{k'}).\end{aligned} \tag{C.3}$$

Angular integration over $(\theta_{k'}, \phi_{k'})$ yields $\delta_{\ell,\ell'}\delta_{m_\ell,m_{\ell'}}$. Thus, after carrying out angular integration, we have

$$\int d^3\mathbf{k}' \frac{\psi_{\mathbf{k}'}(\mathbf{r})\psi^*_{\mathbf{k}'}(\mathbf{r}')}{k^2 - k'^2 + i\varepsilon'} = \frac{2}{\pi}\sum_\ell \sum_{m_\ell} Q_{k,\ell}(r, r') Y_{\ell, m_\ell}(\hat{r}) Y^*_{\ell, m_\ell}(\hat{r}'), \tag{C.4}$$

where

$$Q_{k,\ell}(r, r') = \int_0^\infty k'^2 dk' \frac{\psi_\ell(k'r)\psi^*_\ell(k'r')}{k^2 - k'^2 + i\varepsilon'}. \tag{C.5}$$

doi:10.1088/978-0-7503-3959-9ch11

© IOP Publishing Ltd 2025. All rights, including for text and data mining (TDM), artificial intelligence (AI) training, and similar technologies, are reserved.

Suppose r and r' are very large. We can then use the asymptotic form of the wave function of $\psi_\ell(k'r)$ given by equation (2.45), obtaining

$$\psi_\ell(k'r)\psi_\ell^*(k'r') = \frac{1}{4k'^2 rr'} \left[e^{-i(k'(r-r'))} - 2\,\mathrm{Re}\,\{S_\ell(k')e^{-i(k'(r+r'))-i\ell\pi}\} + |S_\ell(k)|^2 e^{ik'(r-r')} \right]. \quad \text{(C.6)}$$

Since $|S_\ell(k)|^2 = 1$, substituting this into equation (C.5) yields

$$Q_{k,\ell}(r, r') = \frac{1}{4rr'} \int_{-\infty}^{\infty} dk' \frac{e^{+i(k'(r-r'))} - (-1)^\ell S_\ell(k')e^{+i(k'(r+r'))}}{k^2 - k'^2 + i\varepsilon'}. \quad \text{(C.7)}$$

For $r > r'$, the contour of integration can be closed in the upper half-plane, while for $r < r'$, the contour can be closed in the lower half-plane, as in appendix B. Then, following the same procedure as that used in appendix B, the complex integration can be performed using Cauchy's residue theorem. The S-matrix has simple poles corresponding to the bound states on the imaginary axis in the upper half-plane. Thus, for $r > r'$, the result is

$$\begin{aligned} Q_{k,\ell}(r, r') = & -\frac{1}{4rr'} \left[(-1)^\ell \sum_n \frac{\mathrm{Res}\; S_\ell(k)\big|_{k=ik_n}}{k^2 - k_n^2} \exp[-|k_n|(r + r')] \right] \\ & - \frac{1}{4krr'} \pi i \left[(-1)^\ell S_\ell(k) \exp[ik(r + r)] - \exp[ik(r - r')] \right]. \end{aligned} \quad \text{(C.8)}$$

The first term on the rhs of the above equation cancels with the first term on the rhs of equation (C.1). Similarly, for $r < r'$, the result is

$$Q_{k,\ell}(r, r') = -\frac{1}{4krr'} \pi i \left[(-1)^\ell S_\ell(k) \exp[ik(r + r)] - \exp[-ik(r - r')] \right]. \quad \text{(C.9)}$$

Thus, replacing $S_\ell(k) = e^{2i\delta_\ell}$ and substituting all the above results into equation (C.1), one obtains

$$\begin{aligned} G^+(\mathbf{r}, \mathbf{r}') = & \frac{2\mu}{\hbar^2} \frac{1}{krr'} \sum_\ell e^{i\delta_\ell} \exp\,[ikr_> - \ell\pi/2 + \delta_\ell] \sin\,(kr_< - \ell\pi/2 + \delta_\ell) \\ & \times \sum_{m_\ell} Y_{\ell, m_\ell}(\hat{r}) Y^*_{\ell, m_\ell}(\hat{r}'). \end{aligned} \quad \text{(C.10)}$$

The above expression for $G^+(\mathbf{r}, \mathbf{r}')$ holds good for $kr \gg 1$ and $kr' \gg 1$. One can identify the function $-e^{i\delta_\ell} \exp\,[ikr - \ell\pi/2 + \delta_\ell]/(kr)$ as the asymptotic form of the ℓ-wave irregular wave function and the function $\sin\,(kr - \ell\pi/2 + \delta_\ell)/(kr)$ as the ℓ-wave regular wave function of the Hamiltonian $\hat{H} = \hat{H}_0 + V(r)$. Therefore, the Green's function for any arbitrary values of r and r' can be expressed as

$$G^{+}(\mathbf{r}, \mathbf{r}') = -\frac{2\mu k}{\hbar^2}\sum_{\ell}\psi^{\mathrm{irr}}(r_>)\psi^{\mathrm{reg}}(r_<)\sum_{m_\ell} Y_{\ell, m_\ell}(\hat{r})\, Y^{*}_{\ell, m_\ell}(\hat{r}'). \tag{C.11}$$

For a free particle ($V(r) = 0$, $\delta_\ell = 0$), this reduces to

$$G^{+}(\mathbf{r}, \mathbf{r}') = -\frac{2\mu k}{\hbar^2}\sum_{\ell} h_\ell^{+}(r_>) j_\ell(r_<)\sum_{m_\ell} Y_{\ell, m_\ell}(\hat{r})\, Y^{*}_{\ell, m_\ell}(\hat{r}'). \tag{C.12}$$

www.ingramcontent.com/pod-product-compliance
Ingram Content Group UK Ltd.
Pitfield, Milton Keynes, MK11 3LW, UK
UKHW052039301026
9729IPUK00019B/307

9 780750 339605